Aut

31020

D1465392

BASIC ELECTRICAL AND
ELECTRONIC ENGINEERING

BASIC ELECTRICAL AND ELECTRONIC ENGINEERING

E.C. BELL and R.W. WHITEHEAD

FOURTH EDITION
REVISED BY

W. BOLTON

OXFORD
BLACKWELL SCIENTIFIC PUBLICATIONS
LONDON   EDINBURGH   BOSTON
MELBOURNE   PARIS   BERLIN   VIENNA

© 1981, 1987, 1993 by E.C. Bell & R.W.·
Whitehead
© 1993 by W. Bolton for revisions for 4th edition

Blackwell Scientific Publications
Editorial Offices:
Osney Mead, Oxford OX2 0EL
25 John Street, London WC1N 2BL
23 Ainslie Place, Edinburgh EH3 6AJ
238 Main Street, Cambridge,
  Massachusetts 02142, USA
54 University Street, Carlton
  Victoria 3053, Australia

Other Editorial Offices:
Librairie Arnette SA
2, rue Casimir-Delavigne
75006 Paris
France

Blackwell Wissenschafts-Verlag
Meinekestrasse 4
D-1000 Berlin 15
Germany

Blackwell MZV
Feldgasse 13
A-1238 Wien
Austria

First edition published by Granada
  Publishing 1977
Reprinted 1978, 1979
Second edition 1981
Reprinted 1982
Reprinted by Collins Professional and
  Technical Books 1985
Third edition 1987
Fourth edition published by Blackwell
  Scientific Publications 1993

Set by Best-set Typesetter Ltd., Hong Kong
Printed and bound in Great Britain by
Hartnolls Ltd, Bodmin, Cornwall

DISTRIBUTORS

Marston Book Services Ltd
PO Box 87
Oxford OX2 0DT
(*Orders*: Tel: 0865 791155
       Fax: 0865 791927
       Telex: 837515)

USA
Blackwell Scientific Publications, Inc.
238 Main Street
Cambridge, MA 02142
(*Orders*: Tel: 800 759-6102
          617 876-7000)

Canada
Oxford University Press
70 Wynford Drive
Don Mills
Ontario M3C 1J9
(*Orders*: Tel: 416 441-2941)

Australia
Blackwell Scientific Publications
(Australia) Pty Ltd
54 University Street
Carlton, Victoria 3053
(*Orders*: Tel: 03 347-5552)

British Library
Cataloguing in Publication Data

A catalogue record for this book is
available from the British Library

ISBN 0-632-03493-9

Library of Congress
Cataloging in Publication Data

Bell, E.C. (Ernest Clifford)
    Basic electrical and electronic engineering /
  E.C. Bell and R.W. Whitehead. – 4th ed. /
  revised by W. Bolton.
       p.   cm.
    Rev. ed. of: Basic electrical engineering and
  instrumentation for engineers. 2nd. 1981.
    Includes index.
    ISBN 0-632-03493-9
    1. Electric engineering.   2. Electronics.
  I. Whitehead, R.W. (Roland Ward)
  II. Bolton, W. (William)   III. Bell, E.C.
  (Ernest Clifford). Basic electrical engineering
  and instrumentation for engineers.   IV. Title.
  TK145.B418   1993
  621.3 – dc20   92-26725
                                           CIP

Contents

Preface

The first edition of this highly successful textbook appeared in 1977. With the production of a fourth edition, the opportunity has been taken to change the title, from *Basic Electrical Engineering* to *Basic Electrical and Electronic Engineering*, to indicate the usefulness of the book to both electrical and electronic engineers. Changes include a reorganisation of the early chapters so that there can be more practice on basic concepts, the inclusion of chapters on attenuators and filters, and oscillators, as well as the incorporation of more problems with all chapters.

The book is aimed at first year undergraduates in physics, electrical and electronic engineering at universities and polytechnics. It is also seen as being relevant to those taking technician qualifications, such as BTEC Higher National Certificate and Higher National Diploma. It assumes a basic knowledge of physics/electrical principles and calculus.

W. Bolton

Chapter 1
Basic Principles

1.1 Electrical and electronic circuits

Electrical and electronic circuits consist of networks of passive components and electronic devices. Passive components are such items as resistors, inductors and capacitors. Electronic devices are such items as diodes and transistors. At the atomic level these all rely on the movement of electric charge with the charge being carried by electrons. This section is an introduction to the basic terms and laws of circuits, circuit analysis being considered in more detail in Chapter 2 and later chapters, and the principles involved with passive components; electronic devices are introduced in Chapter 6.

1.1.1 Circuit terms

A simple electrical circuit may consist of a torch bulb connected across a battery. We can explain the results in terms of the battery acting as a source of energy and driving charge through the circuit. The battery is a chemical system which separates positive and negative charge so that the positive terminal of the battery has a positive charge and the negative terminal a negative charge. We can liken the situation to a person lifting objects off the floor and on to a table. Work is done in giving the objects potential energy relative to the floor. This is because the objects are attracted to the floor (masses attract each other) and they are having to be pulled apart. We could talk of the potential energy produced by the separation per 1 kg of mass lifted from the floor. The situation in a battery is comparable with charges being pulled apart. The terminology used is to say that the battery has an *electromotive force* (e.m.f.), with this being defined as the potential energy produced in the battery per coulomb (C) of charge moved between the terminals. The e.m.f. is 1 volt (V) when 1 joule (J) of energy is used to move 1 C. With sources of e.m.f., the adopted convention is to draw an arrow on the circuit diagram pointing from the negative terminal to the positive terminal to indicate the uphill direction to the higher potential energy.

The term *current* is used for the rate of movement of charge through the circuit. The symbol i is used for the current at an instant of time when currents are likely to be changing and I when steady values of current are involved.

$$i = \frac{dq}{dt} \tag{1.1}$$

The current has the unit ampere (A) when the charge is in coulombs and the time in seconds. The current direction is defined as being from the positive terminal of the battery to the negative terminal. This convention was adopted before it was known what was moving through the circuit. Electrons are in fact moving though the circuit, from the negative terminal to the positive terminal and so in the opposite direction to that used for the current direction.

The movement of the charge through the circuit is like the falling of the masses in our earlier analogy from the table on to the floor. Both the masses and the charge steadily loose potential energy. When 1 J of energy is lost in 1 C of charge moving between two points in the circuit then there is said to be a *potential difference*, or voltage, of 1 V between them. Potential difference is the energy dissipated per coulomb of charge moved. Thus the potential difference v at some instant is

$$v = \frac{dw}{dq} \tag{1.2}$$

On circuit diagrams an arrow is often drawn alongside a component to indicate the uphill direction of the potential difference. The arrow points from the end of the component connected to the negative side of the e.m.f. source to the end connected to the positive side (see Fig. 1.1 later in this chapter for examples of such markings).

The rate at which energy is dissipated is called the *power*. Thus the power p is

$$p = \frac{dw}{dt}$$

and so, using Equation (1.2)

$$p = \frac{dw}{dq}\frac{dq}{dt} = vi \tag{1.3}$$

The unit of power is the watt W, with 1 W being 1 J/s. For Equation (1.3), when the power is in watts then the potential difference v is in volts and the current i in amps.

Example   What is the power dissipated in a lamp when the potential difference across it is 6 V and the current through it 0.3 A? What charge is moved through the lamp in 10 s?

Using Equation (1.3) then

$$p = vi = 6 \times 0.3 = 1.8 \, W$$

Since current is the rate of movement of charge, 0.3 C pass through the lamp every 1 s. Thus the charge moved in 10 s is 3 C.

1.1.2 Circuit laws

When there is a junction in an electrical circuit then, since charge does not accumulate at the junction, the rate at which charge enters a junction must be equal to the rate at which it leaves it. This means that the sum of the currents entering a junction must equal the sum of the currents leaving it. This is known as *Kirchoff's current law* and is discussed again in Chapter 2.

If we consider a circuit with an e.m.f. driving a current through a number of series components then we will have a potential difference drop across each component with the result that the e.m.f. is equal to the sum of the potential difference drops round the circuit. This is one way of expressing *Kirchoff's voltage law*, see Chapter 2 for further discussion.

1.2 Resistance

In the 19th century the German scientist Ohm discovered that, at a constant temperature, the current i flowing in a conductor is directly proportional to the potential difference v across it (Fig. 1.1). This constant of proportionality is called the resistance R.

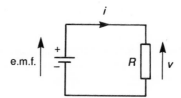

Fig. 1.1 Resistor in circuit.

$$v = Ri \tag{1.4}$$

When the current is in amps, the potential difference in volts, then the resistance is in ohms (Ω).

Since the power dissipation p for a circuit element with a current through it of i and potential difference across it of v is vi (Equation (1.3)), then

$$p = vi = Ri^2 = \frac{v^2}{R} \tag{1.5}$$

Example What is the resistance of an electric heater that dissipates a power of 2.0 kW when connected to a 240 V supply?

Since $p = v^2/R$ (Equation (1.5)), then

$$R = \frac{v^2}{p} = \frac{240^2}{2.0 \times 10^3} = 28.8\,\Omega$$

1.2.1 *Resistors in series and parallel*

For two resistors in series, as in Fig. 1.2, the potential difference V across the two will be the sum of the potential differences across each one. But the potential difference across R_1 is IR_1 and that across R_2 is IR_2. The current is the same through both. Thus

$$V = IR_1 + IR_2$$

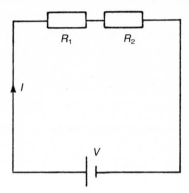

Fig. 1.2 Resistors in series.

If we replaced the two resistors by a single resistor R, then for it to give the same total potential difference V and current I we must have $V = RI$. Thus

$$R = R_1 + R_2 \qquad (1.6)$$

The total resistance of a number of series resistors is the sum of their resistances.

For two resistors in parallel, as in Fig. 1.3, the total current I entering the parallel arrangement must be the sum of the currents through each resistor.

$$I = I_1 + I_2$$

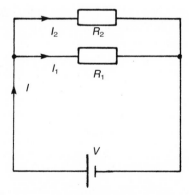

Fig. 1.3 Resistors in parallel.

Thus since $I_1 = V/R_1$ and $I_2 = V/R_2$, and the potential difference V is the same across both,

$$I = \frac{V}{R_1} + \frac{V}{R_2}$$

If we replaced the two resistors by a single resistor R, then for it to give the same potential difference V and current I we must have $I = V/R$ and so

$$\frac{1}{R} = \frac{1}{R_1} + \frac{1}{R_2} \tag{1.7}$$

For resistors in parallel, the sum of the reciprocals of the individual resistances is equal to the reciprocal of the total resistance.

Example   What is the total resistance when $4\,\Omega$ and $2\,\Omega$ resistors are connected in (a) series, (b) parallel?

(a) For the two resistors in series the total resistance is, using Equation (1.6), $4 + 2 = 6\,\Omega$.

(b) For the two resistors in parallel, Equation (1.7) gives

$$\frac{1}{R} = \frac{1}{4} + \frac{1}{2}$$

Hence the total resistance R is $1.3\,\Omega$.

1.2.2 Resistivity and conductivity

For wires of one material it is found that the resistance R is directly proportional to the length l and inversely proportional to the cross-sectional area A.

$$R = \frac{\rho l}{A} \tag{1.8}$$

where ρ is the resistivity of the material. Resistivity has the unit $\Omega\,\mathrm{m}$. Copper, a good conductor, has a resistivity at 20°C of about $1.7 \times 10^{-8}\,\Omega\,\mathrm{m}$.

Sometimes it is more useful to talk of the conductance G of a conductor.

$$G = \frac{1}{R} \tag{1.9}$$

and so

$$i = Gv \tag{1.10}$$

Conductance has the unit of ohm^{-1}. This unit is called the siemen (S). Likewise we can refer to the conductivity σ.

$$\sigma = \frac{1}{\rho} = \frac{l}{RA} = \frac{Gl}{A} \qquad\qquad (1.11)$$

Conductivity has the unit S/m.

Example   What are the resistances of 10 m lengths of copper wire and nichrome wire if they have the same cross-sectional area of 1 mm²? Resistivity of copper at 20°C = $1.76 \times 10^{-8}\,\Omega\,\mathrm{m}$, resistivity of nichrome at 20° = $108 \times 10^{-8}\,\Omega\,\mathrm{m}$.

Using Equation (1.8), for the copper

$$R = \frac{\rho l}{A} = \frac{1.76 \times 10^{-8} \times 10}{1 \times 10^{-6}} = 0.176\,\Omega$$

and for the nichrome

$$R = \frac{\rho l}{A} = \frac{108 \times 10^{-8} \times 10}{1 \times 10^{-6}} = 10.8\,\Omega$$

Copper is the material that is used for electric cables where good conduction is required. Nichrome is used to make heating elements where significant resistance is required.

1.3 Flux

There are many basically similar systems involving flow. Thus we can have the flow of liquid through a pipe (Fig. 1.4(a)), the flow of heat through a bar (Fig. 1.4(b)) or the flow of charge through a circuit (Fig. 1.4(c)). In each system there is a relationship between the flow and the cause of flow. Let the quantity that flows be termed *flux* and the quantity that causes the flow *potential difference* (p.d.). Thus for the liquid flowing along the pipe the mass flowing through per second, i.e. the flux, is proportional to the pressure difference

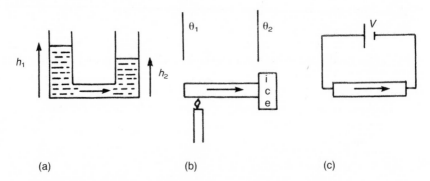

(a)   (b)   (c)

Fig. 1.4 Analogous systems: (a) flow of liquid through a pipe, (b) flow of heat through a bar, (c) flow of charge through a conductor.

between the ends of the pipe, this being proportional to the height difference $(h_1 - h_2)$. For the heat flow through a bar the rate at which heat flows through is proportional to the temperature difference between the ends $(\theta_1 - \theta_2)$. For the charge flowing through the circuit, the flux, i.e. the current, is proportional to the potential difference V (just a statement of Ohm's law).

Flux is proportional to p.d.

When we have, say, a liquid flowing through a pipe, then if we have two parallel pipes with the same pressure difference we have twice the rate of flow. Two parallel pipes is equivalent to having a single pipe with twice the cross-sectional area. Thus the rate of flow is proportional to the cross-sectional area. If we have two pipes of equal cross-sectional area and length, and with the same pressure difference between the ends, then the rate of flow through each is the same. If we now connect these two pipes in series, then we will have the same rate of flow if the pressure difference across each length is unchanged, i.e. the pressure difference across the double length pipe is double that across the single pipe. If we had the same pressure difference across the double length pipe as the single length pipe then we would have half the pressure difference per single length pipe and so half the rate of flow. Thus the rate of flow, for the same pressure difference, is inversely proportional to the length.

In general, for a fixed value of potential difference, the flux is proportional to the cross-sectional area of the pipe or conductor and inversely proportional to its length. The flux is also dependent on the material, e.g. the flow of liquid will depend on whether the liquid was water or syrup, the flow of charge on whether the conductor was copper or nichrome. In general, therefore,

$$\text{flux} = \frac{\text{medium factor} \times \text{area}}{\text{length}} \times \text{p.d.} \tag{1.12}$$

Thus for the liquid flowing through the pipe

$$Q = \frac{(1/\eta)A}{l}(p_2 - p_1) \tag{1.13}$$

where Q is the mass of liquid flowing through per second, η is the viscosity, A the cross-sectional area of the pipe, l its length and $(p_2 - p_1)$ the pressure difference.

For heat flow through a bar

$$Q = \frac{kA}{l}(\theta_1 - \theta_2) \tag{1.14}$$

where Q is the quantity of heat flowing per second, k the thermal conductivity, A the cross-sectional area of the bar, l its length and $(\theta_1 - \theta_2)$ the temperature difference.

For the current in the electrical circuit

$$I = \frac{\sigma A}{l}V \tag{1.15}$$

where I is the current or rate of flow of charge, σ the conductivity, A the cross-sectional area of the conductor and l its length. This can be expressed alternatively as

$$I = \frac{(1/\rho)A}{l}V \tag{1.16}$$

where ρ is the resistivity.

1.3.1 Flux density

The term *flux density* is used for the flux per unit cross-sectional area, i.e.

$$\text{flux density} = \frac{\text{flux}}{\text{area}} \tag{1.17}$$

Thus Equation (1.15) can be rewritten as

$$\frac{I}{A} = \frac{\sigma V}{l}$$

where I/A is the flux density. The term more usually used with electrical circuits is *current density* (J). The term V/l is the potential difference per unit length, i.e. the potential gradient. The potential gradient is called the *electric field strength E*.

$$J = \sigma E \tag{1.18}$$

Example What is the electric field strength in copper wire of cross-sectional area $1\,\text{mm}^2$ and resistivity $1.6 \times 10^{-8}\,\Omega\,\text{m}$ when the current is $5.0\,\text{A}$?

The current density J is

$$J = \frac{I}{A} = \frac{5.0}{1.0 \times 10^{-6}} = 5.0 \times 10^6\,\text{A/m}^2$$

The conductivity $\sigma = 1/\rho$ and so the electric field strength, given by Equation (1.18), is

$$E = \frac{J}{\sigma} = J\rho = 5.0 \times 10^6 \times 1.6 \times 10^{-8} = 0.08\,\text{V/m}$$

1.4 Magnetic flux

When a current flows through a wire then a magnetic field is set up in the space around the wire. The direction of a magnetic field is taken as being the direction in which the north-seeking pole of a suspended magnet, e.g. a

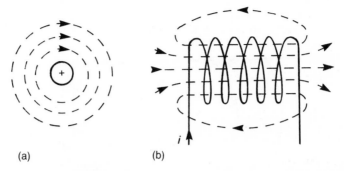

Fig. 1.5 Magnetic field due to (a) a current-carrying wire, (b) a solenoid.

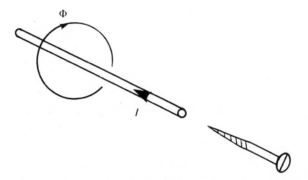

Fig. 1.6 The screw rule.

compass needle, would point when placed in the field. If the field pattern is plotted round the current-carrying wire then it is found to be a series of concentric circles round the wire (Fig. 1.5(a)). A convenient method of representing the relationship between the current direction and that of the magnetic field is to consider the direction a right-handed screw or corkscrew would have to be rotated to move the screw forward in the direction of the current (Fig. 1.6); the screw rotates in the same direction as the field. For a current-carrying solenoid the magnetic field pattern is as shown in Fig. 1.5(b). The direction of the magnetic field along the axis of the solenoid can be deduced by the use of the screw rule for the single wire or by a slight modification of that rule. If a right-handed screw is driven along the axis of the solenoid so that it turns in the direction of the current, it travels in the direction of the field.

A useful way of considering magnetic fields is in terms of magnetic flux. The magnetic field lines can be considered to be paths along which magnetic flux flows. Thus in the case of a single current-carrying wire the magnetic flux flows in concentric circles round the wire, while with a current-carrying solenoid the flux flows in the solenoid along its axis.

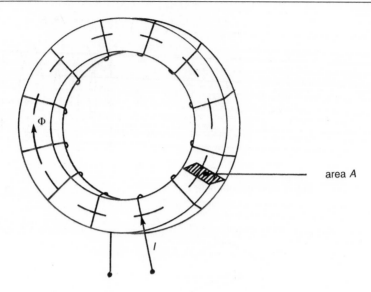

area *A*

Fig. 1.7 A magnetic circuit.

1.4.1   *Magnetomotive force*

Consider a coil of *N* turns wound on a circular core and carrying a current *i* (Fig. 1.7). A magnetic flux Φ is set up in the core. Following our earlier discussion of flux analogies, there will be a potential difference which causes the flux to flow. In this case this is provided by the current through the coil of wire, the greater the current and the greater the number of turns the more flux is produced. Thus the potential difference, or *magnetomotive force* m.m.f. (*F*) as it is called, is defined as being

$$F = IN \tag{1.19}$$

With the current in amps then the unit of m.m.f. is amps (A). However, to indicate that it is the product of the number of turns and a current it is often given the unit ampere-turns.

A simple electrical circuit consists of a source of e.m.f. driving a current through the circuit. Figure 1.5 is comparable in that a source of m.m.f. drives magnetic flux through the magnetic circuit. By analogy with Equation (1.10) we can write for the flux flow in the circuit

$$\Phi = \frac{\mu A}{l} F \tag{1.20}$$

where μ, the factor representing the medium, is called the *permeability*, *A* is the cross-sectional area of the flux path and *l* its length, in this case the mean circumference of the ring. The unit of flux is the weber (defined by electromagnetic induction, see Section 1.4.3). With *A* in m², *l* in m, *F* in A, the unit of permeability is Wb m⁻¹ A⁻¹. A more usual unit that is used is the henry/metre (H/m), since $1\,\text{H} = 1\,\text{Wb/A}$ (see Section 1.5).

The permeability of a vacuum, called the *permeability of free space* with the symbol μ_0, has the value $4\pi \times 10^{-7}\,\text{H/m}$. It is usual to express the permeabilities of other materials relative to this value. Thus

$$\mu = \mu_r \mu_0 \tag{1.21}$$

where μ_r is the relative permeability. The relative permeability of air is effectively 1. The relative permeability, however, is generally not a constant but depends on the flux density (see Section 1.6).

With an electrical circuit we have the concept of resistance as v/i. In a similar way we can define for a magnetic circuit *reluctance S* as F/Φ

$$S = \frac{F}{\Phi} \tag{1.22}$$

The unit of reluctance is A/Wb or H^{-1}. Hence, using Equation (1.20)

$$S = \frac{1}{\mu A} \tag{1.23}$$

The reciprocal of resistance is conductance. For a magnetic circuit the reciprocal of reluctance is *permeance* (Λ), unit H.

$$\Lambda = \frac{1}{S} = \frac{\mu A}{l} \tag{1.24}$$

Example   A toroidal coil (as in Fig. 1.7) has 100 turns and carries a current of 1.0 A. If the flux in the core is $1\,\mu\text{Wb}$ what is (a) the m.m.f. and (b) the circuit reluctance?

(a) The m.m.f. is given by Equation (1.19) as

$$F = IN = 1.0 \times 100 = 100\,\text{A}$$

(b) The reluctance in the circuit is given by Equation (1.22) as

$$S = \frac{F}{\Phi} = \frac{100}{1 \times 10^{-6}} = 1.0 \times 10^8\,\text{A/Wb}$$

1.4.2   Flux density

Equation (1.20) can be written in a different form as

$$\frac{\Phi}{A} = \mu \frac{F}{l}$$

The term Φ/A is the flux density, the symbol for magnetic flux density being B. It has the unit Wb/m^2. This unit is given a special name of the tesla (T), $1\,\text{T} = 1\,\text{Wb/m}^2$. The quantity F/l is the m.m.f. per unit length of flux path and is called the *magnetic field strength H*.

$$H = \frac{F}{l} \tag{1.25}$$

Thus Equation (1.20) can be written as

$$B = \mu H \tag{1.26}$$

Example   A toroidal coil has 200 turns and a magnetic flux path of length 40 mm. What will be the current required to produce a magnetic field strength of 100 A/m?

Since $F = IN$ then the magnetic field strength H, given by Equation (1.25), is

$$H = \frac{F}{l} = \frac{IN}{l}$$

Then the current is

$$I = \frac{Hl}{N} = \frac{100 \times 0.040}{200} = 0.020 \, A$$

1.4.3   Electromagnetic induction

When the magnetic flux linked by a circuit changes then an e.m.f. is induced in that circuit, the e.m.f. being proportional to the rate of change of flux linked. This statement is called *Faraday's law* and the effect called *electromagnetic induction*. For a coil of N turns, each turn linked by flux changing at the rate $d\Phi/dt$, then the e.m.f. e is

$$e = N\frac{d\Phi}{dt} \tag{1.27}$$

The unit used for flux is the weber (Wb), the induced e.m.f. being 1 V when, for a single turn coil, the flux changes at the rate of 1 Wb/s.

The current set up as a result of the induced e.m.f. always produces magnetic flux which opposes the change of magnetic flux responsible for the generation of the induced e.m.f. This is known as *Lenz's law*. For this reason, Equation (1.27) is often written as

$$e = -N\frac{d\Phi}{dt} \tag{1.28}$$

The change in flux linked by a coil can be achieved by having a static coil linked by a field which is varying with time or by having a coil moving relative to the field (Fig. 1.8).

1.4.4   Force on a current-carrying conductor

Consider a conductor of length l carrying a current I and lying perpendicular to a magnetic field of flux density B, as in Fig. 1.9. If as a result of this a force F

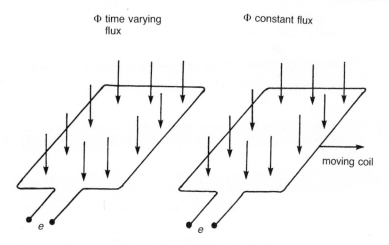

Fig. 1.8 Changing flux linkage.

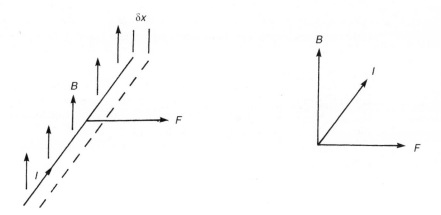

Fig. 1.9 Force on a current-carrying conductor.

acts on the conductor and causes it to move through a small distance δx then the work done is $F\delta x$. The movement of the conductor results in an increase in the flux linked. The area swept out is $l\delta x$ and thus the increase in flux linked is $Bl\delta x$. Suppose this movement takes a small time δt. The rate of change of flux is $Bl\delta x/\delta t$. An e.m.f. will be induced in the conductor. Faraday's law states the induced e.m.f. is the rate of change of flux and thus is $Bl\delta x/\delta t$. The induced e.m.f. is the source of the energy responsible for the work done. The change in electrical energy due to this is thus $eI\delta t$ and so is $BlI\delta x$. Thus

$$F\delta x = BIL\,\delta x$$

$$F = BIL \tag{1.29}$$

The force is in newtons (N) when the flux density is in webers, the current in amps and the length in metres.

A useful way of remembering the directions of the field, current and force is *Fleming's left-hand rule*. If the first finger represents the direction of the magnetic field, the second finger the current direction, and they are held at right angles to each other and the thumb, then the direction of the thumb represents the direction of the force.

Example A conductor carrying a current of 1.5 A has a length of 100 mm in a magnetic field which has a flux density of 0.5 T at right angles to it. What is the force on the conductor?

Using Equation (1.29),

$$F = BIL = 0.5 \times 1.5 \times 0.100 = 0.075\,N$$

1.5 Inductance

The circuit element known as an *inductor* is essentially just a coil of wire. When a current flows through the coil then magnetic flux is set up. This flux links the turns of the coil. Thus when the current through the coil changes then the magnetic flux linked by the coil changes. There is thus an e.m.f. induced in the coil. This induced e.m.f. produces a current which will be in such a direction as to oppose the change responsible for its production. Thus if the current through the coil is growing then the induced current will oppose the growth of the current. If the current through the coil is decreasing then the induced current will oppose the decrease in current.

Consider a coil of N turns carrying a current i. The flux produced in the coil will be, using Equation (1.22), F/S and since F is iN (Equation (1.20)) and S is $l/\mu A$ (Equation (1.23)),

$$\Phi = \frac{F}{S} = \frac{iN}{l/\mu A} = \frac{iN\mu A}{l} \tag{1.30}$$

The induced e.m.f., when this flux changes at the rate $d\Phi/dt$, is given by Faraday's law as

$$e = N\frac{d\Phi}{dt} = N\frac{d}{dt}(iN\mu A/l)$$

For any given coil N, A and l are constant, and assuming that μ is constant, then

$$e = N^2\frac{\mu A}{l}\frac{di}{dt} \tag{1.31}$$

The term $N^2 (\mu A/l)$ or $N^2\Lambda$ is called the *self inductance* of the coil and is given the symbol L. Thus Equation (1.31) can be written as

$$e = L\frac{di}{dt} \tag{1.32}$$

When the e.m.f. is in volts, the current in amps and the time in seconds, then the inductance is in henries (H). The self inductance is thus

$$L = \frac{N^2 \mu A}{l} \tag{1.33}$$

or by substituting for Φ from Equation (1.30)

$$L = \frac{N\Phi}{i} \tag{1.34}$$

Inductance can thus be considered to be the flux linkage per amp.

Example   What is the mean e.m.f. induced in a coil of inductance 0.1 H when the current is increased from 1.0 A to 1.5 A in 0.01 s?

Using Equation (1.32)

$$e = L\frac{di}{dt} = 0.1 \times \frac{0.5}{0.01} = 5.0\,\text{V}$$

Example   What is the flux produced in a coil of 500 turns and constant inductance 10 mH by a current of 2.0 A?

Using Equation (1.34) then

$$\Phi = \frac{Li}{N} = \frac{0.010 \times 2.0}{500} = 4.0 \times 10^{-5}\,\text{Wb}$$

1.5.1   Mutual inductance

When the current through a coil changes then the magnetic flux produced by that coil changes. If there is another coil in the vicinity then the magnetic flux can link turns of that coil and so induce an e.m.f. in that coil. The two coils are said to be *magnetically coupled* and the effect is called *mutual inductance*. The mutual inductance M is defined as

$$M = \frac{\text{change of flux linkages with secondary coil}}{\text{change in current in the primary coil}} \tag{1.35}$$

Thus if the secondary coil has N_2 turns, each of which is linked by a flux change of Φ from the primary coil produced as a result of a current change i_1 then

$$M = \frac{N_2 \Phi}{i_1} \tag{1.36}$$

In the above it has been assumed that the flux is proportional to the current, i.e. the coil has a constant reluctance as a result of μ being a constant.

The e.m.f. induced in the secondary coil e_2 is given by Faraday's law as $N_2\,d\Phi/dt$. Hence, using Equation (1.36)

$$e_2 = N_2\frac{\mathrm{d}}{\mathrm{d}t}(Mi_1/N_2) = M\frac{\mathrm{d}i_1}{\mathrm{d}t} \tag{1.37}$$

The principle involved in the *transformer* is mutual inductance. The transformer consists of two coils wound on the same iron core. When an alternating current is applied to the primary coil then an alternating flux is produced in the core. This results in e.m.f.s being induced in both the primary and secondary coils. Since the same flux links both coils, the e.m.f. induced per turn of each coil will be the same. Thus the ratio of the induced e.m.f.s in the two coils is the ratio of their numbers of turns. In most practical transformers, the voltage applied to the primary coil is equal to the e.m.f. induced in the primary coil. Consequently, the ratio of the input voltage in the primary coil to the output voltage from the secondary coil is the ratio of the number of primary turns to the number of secondary turns. A transformer can thus be used to step-up or step-down an alternating voltage. See Chapter 12 for further discussion of transformers.

Example   A pair of coils has a mutual inductance of $100\,\mu\mathrm{H}$. What will be the e.m.f. induced in the secondary coil when the current in the primary coil changes at the rate of $1000\,\mathrm{A/s}$?

Using Equation (1.37) then

$$e_2 = M\frac{\mathrm{d}i_1}{\mathrm{d}t} = 100 \times 10^{-6} \times 1000 = 0.1\,\mathrm{V}$$

1.6   Magnetic circuits

Almost all practical applications of electromagnetic systems involve iron (ferrous) cores of some form or other, as it is much easier to establish magnetic flux in an iron cored coil than it is in an air cored coil. Consider two identical toroidal coils, one wound on a wooden former (any non-ferrous material can be considered as having the magnetic properties of air) and the other on a cast steel former. If the current is increased in steps and the flux density in the core measured at each step, then a plot of B against H can be obtained.

From the $B–H$ or *magnetization* curves shown in Fig. 1.10, it can be seen that for the same magnetic field strength H_1 the flux density produced in the cast steel core is about 600 times that produced in the air core. It is also seen from Fig. 1.10 that the ratio B/H, i.e. the permeability, is not constant in the case of cast steel. This is true for all ferrous cored coils, where at large values of flux density the core is said to *saturate* and a large increase in magnetic field strength will produce only a small change in flux density.

The permeability is usually written as

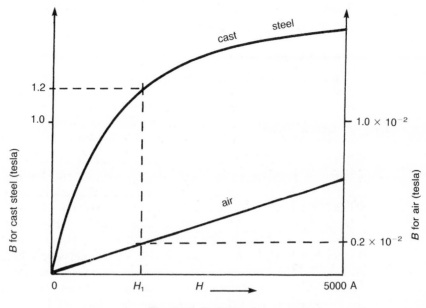

Fig. 1.10 *B–H* graph.

$$\mu = \mu_r \mu_0 \qquad (1.38)$$

where the permeability μ is the ratio of B to H at a point on the magnetization curve, μ_0 is a constant called the permeability of free space and has the value $4\pi \times 10^{-7}\,\text{H/m}$, and μ_r is the relative permeability which expresses how the permeability of a material compares with that of a vacuum, or effectively air, and depends on the point on the magnetization curve considered. For the magnetic field strength H_1 considered in the graph, since the flux density at this value for cast steel is $1.2\,\text{T}$ and for air $0.2 \times 10^{-2}\,\text{T}$ then the relative permeability of the steel is $1.2/0.2 \times 10^{-2} = 600$. The relative permeability for cast steel varies from about 400 to 900. Silicon iron has relative permeabilities varying from about 500 to 7000, mumetal from about 140 000 to 350 000.

Example A coil of 500 turns and resistance $3\,\Omega$ is connected to a d.c. supply of $4.0\,\text{V}$. The coil is wound on a steel ring of uniform cross-sectional area $500\,\text{mm}^2$ and mean diameter $150\,\text{mm}$. What is the flux in the ring? The steel can be assumed to have a relative permeability of 600.

The current through the coil is $V/R = 4.0/3 = 1.3\,\text{A}$. The magnetic field strength H is

$$H = \frac{NI}{l} = \frac{500 \times 1.3}{\pi \times 0.150} = 1379\,\text{A/m}$$

The flux density B is given by

$$B = \mu H = \mu_0 \mu_r H = 600 \times 4\pi \times 10^{-7} \times 1379 = 1.04\,\text{T}$$

The flux Φ is thus

$$\Phi = BA = 1.04 \times 500 \times 10^{-6} = 5.2 \times 10^{-4}\,\text{Wb}$$

1.6.1 Laws of magnetic circuits

When there is a junction in a magnetic circuit then, since flux does not accumulate at the junction, the total magnetic flux entering the junction is equal to the flux leaving the junction. This is the magnetic circuit equivalent of Kirchoff's current law for electrical circuits (see Section 1.1.2).

If we consider a magnetic circuit with a m.m.f. driving flux through a number of series components then we will have a m.m.f. drop across each component, which is equal to the product of the magnetic field strength and the length of the flux path for the component, with the result that the m.m.f. is equal to the sum of the m.m.f. drops round the circuit. This is the magnetic circuit equivalent of *Kirchoff's voltage law* (see Section 1.1.2).

A consequence of the laws of the magnetic circuit and electrical circuit being similar is that similar relationships occur for reluctances in series, and reluctances in parallel, as resistances in series or resistances in parallel. Thus for two reluctances S_1 and S_2 in series, the total reluctance is given by

$$S = S_1 + S_2 \tag{1.39}$$

and for them in parallel

$$\frac{1}{S} = \frac{1}{S_1} + \frac{1}{S_2} \tag{1.40}$$

Example   A magnetic circuit is of the form shown in Fig. 1.11. What is the flux in the air gap when there is a current of 1.5 A through the 500 turn coil. The core has a relative permeability of 400.

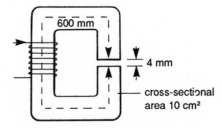

600 mm

4 mm

cross-sectional
area 10 cm²

Fig. 1.11 Example.

The reluctance of the core is

$$S = \frac{l}{\mu A} = \frac{0.600}{400 \times 4\pi \times 10^{-7} \times 10 \times 10^{-4}} = 1.19 \times 10^6 \, \text{A/Wb}$$

The reluctance of the air gap is

$$S = \frac{l}{\mu A} = \frac{0.004}{4\pi \times 10^{-7} \times 10 \times 10^{-4}} = 3.18 \times 10^6 \, \text{A/Wb}$$

Because the two reluctances are in series, the total reluctance is 4.37×10^6 A/Wb. The flux through the air gap is the same as that through the rest of the circuit, if it is assumed that there is no flux leakage at the air gap. Thus

$$\Phi = \frac{F}{S} = \frac{1.5 \times 500}{4.37 \times 10^6} = 1.72 \times 10^{-4} \, \text{Wb}$$

Example   Consider the two C cores shown in Fig. 1.12. The magnetic flux path length in each core is 500 mm and the cross-sectional area is 10 cm². A coil of 1000 turns is wound on the cores and they are clamped together with non-magnetic spacers to form an air gap of g mm in each limb. If the flux density in the core has not to exceed 1.2 T, calculate the current in the coil and the inductance of the coil for a range of spacers 0.5, 1, 2, 5 mm.

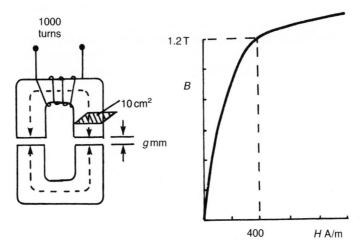

Fig. 1.12 Example.

From the $B-H$ curve of the core material the magnetic field strength H_c to produce a flux density of 1.2 T in the core is 400 A/m. Therefore the m.m.f. required to produce the flux in the core is

$$F_c = Hl = 400 \times (50 + 50) \times 10^{-2} = 400 \, \text{A}$$

The magnetic field strength required to produce a flux density of 1.2 T in the air gap is

$$H_g = \frac{B}{\mu_0} = \frac{1.2}{4\pi \times 10^{-7}} = 9.55 \times 10^5 \, \text{A/m}$$

Therefore the m.m.f. required to produce the flux in the air gap, for a gap width of 0.5 mm, is

$$F_g = Hl = 9.55 \times 10^5 \times 2 \times 0.5 \times 10^{-3} = 955 \, \text{A}$$

The total m.m.f. required to produce the flux in the core and the air gap is 400 + 955 = 1355 A. The current in the coil is thus

$$I = \frac{F}{N} = \frac{1355}{1000} = 1.355 \, \text{A}$$

The inductance of the coil may be obtained from Equation (1.34).

$$L = \frac{N\Phi}{I} = \frac{NBA}{I} = \frac{1000 \times 1.2 \times 10 \times 10^{-4}}{1.355} = 0.89 \, \text{H}$$

The complete set of answers for the different air gaps is:

  g 0.5 mm, I 1.4 A, L 0.89 H
  g 1.0 mm, I 2.3 A, L 0.52 H
  g 2.0 mm, I 4.2 A, L 0.29 H
  g 5.0 mm, I 9.9 A, L 0.12 H

It can be seen that as the air gap is increased, the current required to produce a given flux density is increased, whereas the inductance is decreased. Why then are air gaps introduced in iron-cored inductors? They are introduced to provide a measure of linearity, i.e. to enable the inductance to remain constant over a range of operating currents.

Example Figure 1.13 shows the form of a transformer core. The two outer limbs are the same size and material and each have a flux path length of 200 mm and a cross-sectional area of 4 cm². The inner limb has a flux path length of 80 mm and a cross-sectional area of 8 cm². It is wound with a coil of 400 turns. What is the current required if the flux in the central arm is to be 1.5 mWb. The material used for the core can be assumed to have a constant permeability of 500.

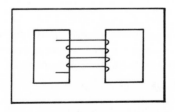

Fig. 1.13 Example.

The two outer limbs will have the same reluctance, this being

$$S_o = \frac{l}{\mu A} = \frac{0.200}{500 \times 4\pi \times 10^{-7} \times 4 \times 10^{-2}} = 7958 \, A/Wb$$

The two outer limbs are in parallel and thus their total reluctance S_p is

$$\frac{1}{S_p} = \frac{1}{S_o} + \frac{1}{S_o} = \frac{2}{S_o}$$

Hence $S_p = 3979 \, A/Wb$. The central limb has a reluctance of

$$S_c = \frac{l}{\mu A} = \frac{0.080}{500 \times 4\pi \times 10^{-7} \times 8 \times 10^{-2}} = 1592 \, A/Wb$$

The total reluctance of the circuit is thus $3979 + 1592 = 5571 \, A/Wb$. The m.m.f. required to give a flux of $1.5 \times 10^{-4} \, Wb$ is thus

$$F = S\Phi = 1592 \times 1.5 \times 10^{-3} = 2.39 \, A \qquad \bullet$$

Thus, since $F = NI$, the required current is

$$I = \frac{F}{N} = \frac{2.39}{400} = 5.98 \, mA$$

1.6.2 *Hysteresis*

Consider a toroidal coil wound on an iron core which is completely demagnetized. Let the current be taken through a cyclic change from zero to $+I$ to zero to $-I$ to zero to $+I$. If the flux density in the core is measured as the current and hence the magnetic field strength H is changed, then a curve of B versus H can be plotted (Fig. 1.14). This curve is termed the *hysteresis loop* for the core material.

For the part of the graph AP when the magnetic field strength H is increased, the flux density in the core increases. When the magnetic field strength is reduced to zero the flux density in the core does not drop to zero but to point C on the graph. This value of flux density is called the *remanent flux density* or *remanence*. To reduce the flux density to zero the magnetic field strength has to be made negative, to the value indicated by point D on the graph. This magnetic field strength value is called the *coercive force*. Further increase in the magnetic field strength will produce a flux acting in the opposite direction to that at P.

Consider some point on the hysteresis loop when the magnetic field strength is H and the flux density B. If now the magnetic field strength is increased by a small amount in a time δt then the flux density increases by a small amount δB. The e.m.f. induced in the coil from this change is given by Faraday's law as

$$e = N\frac{d\Phi}{dt} = \frac{NA \, \delta B}{\delta t}$$

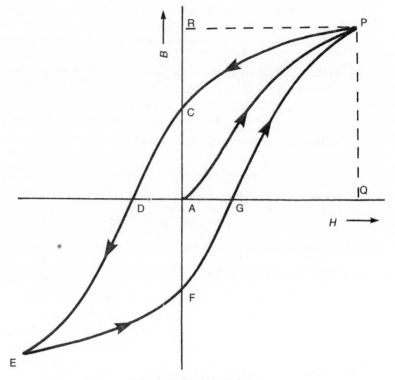

Fig. 1.14 Hysteresis loop.

The applied voltage needed to neutralize this e.m.f. must therefore have this value. Therefore the power that has to be supplied to make this change is

$$p = iv = \frac{iNA\,\delta B}{\delta t}$$

Thus in a time δt the energy that has to be supplied is

$$w = iNA\,\delta B$$

Since the magnetic field strength at which this change in flux density occurs is H, with $H = Ni/l$, then

$$w = HlA\,\delta B$$

But lA is the volume of the core. Thus

$$\text{energy/unit volume} = H\,\delta B$$

The total energy supplied when the flux density is B and the magnetic field strength H is thus

$$\text{energy/unit volume} = \int H\,dB \qquad\qquad (1.41)$$

This is the area between the graph line and the flux density axis. Referring to Fig. 1.14, as the magnetic field strength is increased from zero to Q and the flux density changes from F to R, then the energy supplied to the coil is given by the area FGPRF. The energy returned when the magnetic field strength is reduced to zero is given by the area PCRP. Thus there is a net loss in energy equal to the area FGPCAF. A similar argument may be used as the magnetic field strength is increased in the negative direction. Thus the total loss in energy for one complete cycle of magnetization is the area of the hysteresis loop.

1.6.3   Soft and hard magnetic materials

The term soft was originally applied to magnetic materials which happened to be mechanically soft. However, the term is now used for those magnetic materials which have a small area enclosed by the hysteresis loop, with a low coercive field and small remanence. Loop 1 in Fig. 1.15 illustrates this. The term hard was originally applied to magnetic materials which happened to be mechanically hard. However, the term is now used for those magnetic materials which have a large area enclosed by the hysteresis loop, with a high coercive field and high remanence. Loop 2 in Fig. 1.15 illustrates this.

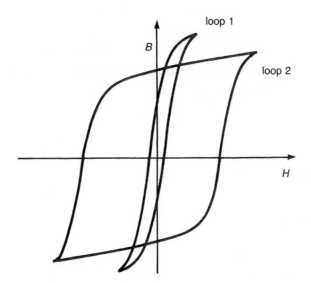

Fig. 1.15 Soft and hard magnetic materials.

There are many applications, such as electrical machines, transformers, electromagnetic lifting magnets, relays, solenoids, etc., which require the properties of soft magnetic materials. Thus, for example, the electromagnetic lifting magnet must lose all, or a very high percentage, of its magnetism when the energizing current is switched off. This means a low remanence. It is also desirable to waste as little energy as possible. Permanent magnets, however, require a high remanence and so need a hard magnetic material.

1.6.4 Eddy current loss

When the flux linking a coil changes with time then an e.m.f. is induced in the coil, the magnitude of which is given by Faraday's law. If the ends of the coil are connected together to form a closed circuit, the induced e.m.f. will cause a current to flow.

Consider now an iron core in which there exists a flux which is varying with time, as illustrated by Fig. 1.16. A typical path PQRS in the core is equivalent to a closed coil. Therefore, because the iron is a conductor, a current will flow. There are of course many such paths and currents in the core. These currents are termed *eddy currents* and, because the core presents a resistance through which the currents flow, there will be an energy loss associated with these currents. This is called the *eddy current loss* and is dissipated as heat in the core.

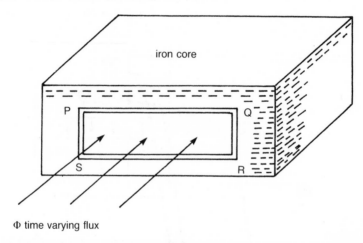

Fig. 1.16 Eddy currents.

To reduce this loss, magnetic circuits which are subjected to varying flux are built up using laminated or sintered cores instead of solid cores. The laminations are thin sheets insulated from each other. Sintered cores are made up of particles insulated from each other. The consequence of this is that eddy currents become confined to a lamination or particle. This reduces the size of the currents produced by the induced e.m.f.s, because the thinner the sheet or the smaller the particle the smaller its cross-sectional area and so the greater its resistance.

The phenomenon of eddy currents is put to good use in several heating applications which go under the name of *induction heating*. The metallic work piece to be heated is placed inside a coil which is supplied with alternating current. This produces an alternating flux and so eddy currents are induced in the work piece and manifest themselves as heat. The depth of penetration of the heat is dependent on the depth of penetration of the eddy currents, which in turn is dependent on the frequency of the alternating current supplied to the coil. Very high frequencies are used to produce surface heating and this type of induction furnace is extensively used to surface harden castings, a process requiring surface heating.

1.7 Capacitance

Consider two parallel metallic plates when a potential difference V is applied to them (Fig. 1.17). The result of applying the potential difference is to make one of the plates positively charged and the other negatively charged. This is because the circuit used to apply the potential difference takes electrons from one plate and moves them round the circuit to the other plate. The bigger the potential difference V the more charge Q that appears on the plates. The relationship between potential difference and charge is expressed as

$$Q = CV \tag{1.42}$$

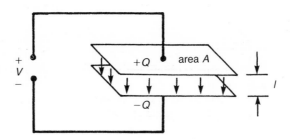

Fig. 1.17 A parallel plate capacitor.

where C is a constant for a particular arrangement of plates and is called the *capacitance*. When the charge is in coulombs and the potential difference in volts, then the capacitance is in farads (F). It should be noted that the farad is a very large unit and thus it is usual to talk of capacitances in terms of microfarads ($1\,\mu\text{F} = 10^{-6}\,\text{F}$) and picofarads ($1\,\text{pF} = 10^{-12}\,\text{F}$).

Example What is the charge on the plates of a $2\,\mu\text{F}$ capacitor when a potential difference of $10\,\text{V}$ is connected across it?

Using Equation (1.42),

$$Q = CV = 2 \times 10^{-6} \times 10 = 20\,\mu\text{C}$$

1.7.1 Capacitance and flux

Surrounding a charge is an electric field, this term being used for a region where another charged body experiences a force. A charged surface can thus be considered to be a source of flux. Thus for the pair of parallel metal plates in Fig. 1.17, we can consider there to be a flux flow between the plates. Since the flux relationship derived earlier in this chapter (Section 1.3) has the flux proportional to the potential difference, then since we have the charge Q proportional to the potential difference (Equation (1.42)), it seems reasonable to consider that the flux is equal to the charge. This statement is called *Gauss's law*. Thus the general flux equation (1.12)

$$\text{Flux} = \frac{\text{medium factor} \times \text{area}}{\text{length}} \times \text{p.d.}$$

can be written for the volume between the parallel charged plates as

$$Q = \frac{\varepsilon A}{l} V \tag{1.43}$$

where A is the plate area, l the distance between the plates and ε the medium factor. This factor is called the *permittivity*. The permittivity of a vacuum is called the permittivity of free space, symbol ε_0, and has the value 8.85×10^{-12} F/m. The permittivity of other materials is usually written in terms of a relative permittivity (or dielectric constant) ε_r, where

$$\varepsilon = \varepsilon_r \varepsilon_0 \tag{1.44}$$

The relative permittivity of waxed paper and mica are about 5 and 6.5 respectively. It is possible to obtain permittivities of the order of 1000 by using high permittivity ceramics.

Since $Q = CV$ (Equation (1.42)), then Equation (1.43) gives the capacitance of the parallel plate capacitor as

$$C = \frac{\varepsilon_r \varepsilon_0 A}{l} \tag{1.45}$$

Equation (1.43) can be rewritten in terms of a flux density Q/A as

$$\frac{Q}{A} = \varepsilon \frac{V}{l}$$

The flux density Q/A is denoted by the symbol D. The quantity V/l is the potential gradient and this is called the electric field strength E. Thus

$$D = \varepsilon E \tag{1.46}$$

Example   Calculate the capacitance of a pair of parallel metal plates, of area $300\,\text{cm}^2$, separated by a distance of $2\,\text{mm}$, when the space between them is filled with (a) air of relative permittivity 1, (b) mica of relative permittivity 6.

(a) Using Equation (1.45),

$$C = \frac{\varepsilon_r \varepsilon_0 A}{l} = \frac{1 \times 8.85 \times 10^{-12} \times 300 \times 10^{-4}}{2 \times 10^{-3}} = 133\,\text{pF}$$

(b) For the mica,

$$C = \frac{\varepsilon_r \varepsilon_0 A}{l} = \frac{6 \times 8.85 \times 10^{-12} \times 300 \times 10^{-4}}{2 \times 10^{-3}} = 797\,\text{pF}$$

1.7.2 Capacitors in series and parallel

Consider three capacitors connected in series, as in Fig. 1.18. The potential difference across the three will be the sum of the potential differences across each, i.e.

$$V = V_1 + V_2 + V_3$$

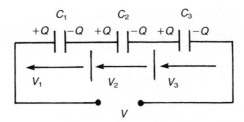

Fig. 1.18 Capacitors in series.

But the charges must be the same on each capacitor. Thus since $Q = C_1 V_1$, $Q = C_2 V_2$ and $Q = C_3 V_3$,

$$V = \frac{Q}{C_1} + \frac{Q}{C_2} + \frac{Q}{C_3}$$

Thus the equivalent capacitor will have a capacitance $C = Q/C$, and so

$$\frac{1}{C} = \frac{1}{C_1} + \frac{1}{C_2} + \frac{1}{C_3} \tag{1.47}$$

For three capacitors in parallel, as in Fig. 1.19, then the total charge Q is the sum of the charges on each capacitor. Thus

$$Q = Q_1 + Q_2 + Q_3$$

Thus, since the potential difference V is the same across each capacitor,

$$Q = C_1 V + C_2 V + C_3 V$$

The equivalent capacitor will have a capacitance $C = Q/V$, thus

$$C = C_1 + C_2 + C_3 \tag{1.48}$$

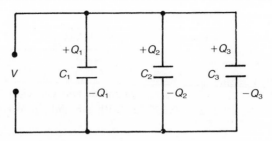

Fig. 1.19 Capacitors in parallel.

Example   What is the capacitance when capacitors of $2\,\mu F$ and $4\,\mu F$ are connected in (a) series, (b) parallel?

(a) For the capacitors in series, Equation (1.47) gives

$$\frac{1}{C} = \frac{1}{C_1} + \frac{1}{C_2} = \frac{1}{2} + \frac{1}{4}$$

Hence $C = 1.3\,\mu F$.

(b) For the capacitors in parallel, Equation (1.48) gives

$$C = C_1 + C_2 = 2 + 4 = 6\,\mu F$$

Example   Calculate the capacitance of a pair of parallel metal plates, of area $300\,\text{cm}^2$, separated by a distance of $2\,\text{mm}$, when the space between them is occupied by (a) a mica sheet of relative permittivity 6, area $300\,\text{cm}^2$ and thickness $1\,\text{mm}$, (b) a mica sheet of relative permittivity 6, area $150\,\text{cm}^2$ and thickness $2\,\text{mm}$.

(a) This capacitor is equivalent to two capacitors of plate area $300\,\text{cm}^2$ and separation $1\,\text{mm}$, one filled with air and the other with mica, connected in series (Fig. 1.20).

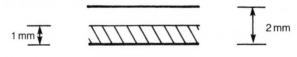

1 mm   2 mm

Fig. 1.20 Example.

$$C_{\text{air}} = \frac{\varepsilon A}{l} = \frac{8.85 \times 10^{-12} \times 300 \times 10^{-4}}{0.001} = 266\,\text{pF}$$

$$C_{\text{mica}} = \frac{6 \times 8.85 \times 10^{-12} \times 300 \times 10^{-4}}{0.001} = 1593\,\text{pF}$$

$$\frac{1}{C} = \frac{1}{C_{air}} + \frac{1}{C_{mica}} = \frac{1}{266} + \frac{1}{1593}$$

Hence $C = 228\,\text{pF}$.

(b) This capacitor is equivalent to two capacitors of plate area $150\,\text{cm}^2$ and separation 2 mm, one filled with air and the other with mica, connected in parallel (Fig. 1.21).

$$C_{air} = \frac{8.85 \times 10^{-12} \times 150 \times 10^{-4}}{0.002} = 66\,\text{pF}$$

$$C_{mica} = \frac{6 \times 8.85 \times 10^{-12} \times 150 \times 10^{-4}}{0.002} = 398\,\text{pF}$$

$$C = C_{air} + C_{mica} = 66 + 398 = 464\,\text{pF}$$

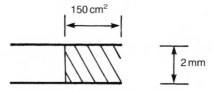

Fig. 1.21 Example.

1.8   Energy stored in inductors and capacitors

When the current through an inductor changes at the rate di/dt then an e.m.f. is induced of $L\,di/dt$. The supply voltage v needed to overcome this and maintain the current i through the inductor is $L\,di/dt$. Thus the power required to maintain the current is

$$p = iv = iL\frac{di}{dt}$$

Thus the energy required to change the current from 0 to I is

$$W = \int_0^I iL\frac{di}{dt}\,dt$$

$$W = \tfrac{1}{2}LI^2 \tag{1.49}$$

When the potential difference across a capacitor is v then the addition of a small increment of charge results in work being done of $v\,dq$. Thus the total energy required to change the charge on the capacitor from 0 to Q is

$$W = \int_0^Q v\,dq = \int_0^Q \frac{q}{C}\,dq$$

$$W = \frac{Q^2}{2C} \quad \text{or} \quad \tfrac{1}{2}CV^2 \tag{1.50}$$

Example   What is the energy stored in a $2\,\mu F$ capacitor when there is a potential difference of $10\,V$ across it?

Using Equation (1.50),
$$W = \tfrac{1}{2}CV^2 = \tfrac{1}{2} \times 2 \times 10^{-6} \times 10^2 = 1.0 \times 10^{-4}\,J$$

1.9   Resistors, inductors and capacitors in circuits

When a potential difference v is applied across a resistor then the current i through it is given by $i = v/R$. When the potential difference is applied across a capacitor then the current i is the rate of change of charge on the plates with time and since $q = Cv$, then i is given by

$$i = C\frac{dv}{dt} \tag{1.51}$$

When the potential difference is applied across an inductor (assumed to have only inductance and no resistance) then since $v = L\,di/dt$ the current i is given by

$$i = \frac{1}{L}\int v\,dt \tag{1.52}$$

It is sometimes useful to have the above equations written in terms of the potential difference produced when a current is applied. For the resistance then $v = iR$. For the capacitor, since $q = Cv$ and q is the integral of $i\,dt$,

$$v = \frac{1}{C}\int i\,dt \tag{1.53}$$

For the inductor

$$v = L\frac{di}{dt} \tag{1.54}$$

Problems

1.   What is the energy supplied to a torch bulb over the period of $1\,h$ if it takes a current of $0.35\,A$ when the potential difference across it is $6.0\,V$?

2.   A $100\,\Omega$ resistor has a power rating of $250\,mW$. What should be the maximum current and voltage used with that resistor?

3. What is the potential difference across a lamp and its resistance if is dissipates 20 W when the current through it is 0.1 A?

4. A 1 kW electric fire is connected to 240 V. What will be the current taken?

5. Three resistors of resistances 2 Ω, 4 Ω and 6 Ω are connected (a) in series, (b) in parallel. What is the total resistance in each case?

6. What is the total power dissipated when a current of 0.2 A passes into a circuit consisting of 5 Ω and 10 Ω resistors connected (a) in series, (b) in parallel?

7. What is the resistance of a 2 m length of manganin wire of diameter 1.0 mm if manganin has a resistivity at 20°C of $42 \times 10^{-8}\,\Omega\,m$?

8. What is the length of nichrome wire, diameter 1 mm, needed to make a resistance of 10 Ω? The resistivity of nichrome at 20°C is $108 \times 10^{-8}\,\Omega\,m$.

9. What is the electric field strength in copper with a conductivity of $5.5 \times 10^7\,S/m$ when the current density is $2.0 \times 10^6\,A/m^2$?

10. What is the electric field strength between two points in an electrical circuit 1 mm apart if the potential difference between them is 10 V?

11. A toroidal coil has 100 turns and a current of 0.5 A. If the length of the magnetic flux path is 100 mm, what is (a) the m.m.f., (b) the magnetic field strength?

12. What current has to be passed through a toroidal coil of 200 turns with a flux path length of 50 mm if the magnetic field strength is to be 120 A/m?

13. A closed square 20 turn coil of side 200 mm and resistance 10 Ω is placed with its plane at right angles to a magnetic field of flux density 0.1 T. Calculate the quantity of charge passing through it when the coil is turned through 180° about an axis in its plane.

14. What is the average e.m.f. generated in a 1000 turn coil when the magnetic flux of 200 μWb passing through it is reversed in 0.1 s?

15. A conductor at right angles to a magnetic field of flux density 1.2 T has a length of 120 mm in the field and carries a current of 2 A. What is the force on the conductor?

16. A conductor at right angles to a magnetic field experiences a force of 0.05 N. If the conductor has a length of 50 mm in the field and carries a current of 250 mA, what is the magnetic flux density?

17. A straight conductor 100 mm long is placed between the faces of a permanent magnet. The flux density of the field produced by the magnet is 1 T. If the mass of the conductor is 10 g, what current must be passed through the conductor so that it just remains stationary? Sketch the field and current directions necessary for this condition to obtain.

18. What is the average e.m.f. induced in a coil of inductance 20 mH when the current through it changes from 20 mA to 50 mA in 1 ms?

19. What is the inductance of a coil of 500 turns if a current through it of 2 A produces a magnetic flux of 20 µWb?

20. Two coils have a mutual inductance of 200 µH. What will be the e.m.f. induced in the secondary coil when the current in the primary coil changes at the rate of 10 000 A/s?

21. A coil of 200 turns is wound on a soft iron core of cross-sectional area 10 cm². A current of 1 A in the coil produces a flux of 2 mW in the core. Calculate the self inductance of the coil. An identical coil is then placed on the core adjacent to the first so that 90% of the flux produced by the first coil links the second coil. Calculate the mutual inductance between the two coils. If the two coils are connected in series, so that the flux they produce is additive, what is the effective inductance of the two coils?

22. A coil of wire of 450 turns is wound on a ring of mean diameter 150 mm and uniform cross-sectional area 600 mm². What will be the flux in the ring if it has a relative permeability of 450 and the current through the coil is 1.6 A?

23. A coil of wire of 400 turns is wound on a wooden ring with a uniform cross-sectional area of 500 mm² and a mean circumference of 600 mm. What is the flux in the ring if the current through the coil is 4 A?

24. A coil of wire of 200 turns is wound on a steel ring with a cross-sectional area of 500 mm² and a mean circumference of 400 mm. What is the current required to produce a flux in the ring of 400 µWb?

25. A coil with 1000 turns is wound on a cast steel ring with a cross-sectional area of 2 cm² and a mean circumference of 200 mm. What is the flux in the core when the current through the coil is 200 mA? For the cast steel:

H in A/m	500	1000	1500
B in T	0.40	1.32	1.41

26. A transformer core, of the form shown in Fig. 1.13, has identical outer limbs with each having half the cross-sectional area of the inner limb. Each outer limb has a flux path length of 200 mm. The inner limb has a flux path length of 100 mm and is wound with a coil of 500 turns. What will be the current required to give a flux density of 0.1 T in the inner limb? The outer and inner limbs are made of the same material and the relative permeability can be assumed to be 600.

27. A mild steel ring has a cross-sectional area of 5 cm² and a mean circumference of 150 mm. It is wound with a coil of 500 turns. What current is needed to produce a flux of 0.5 mWb in the ring when the ring is (a) uncut, (b) cut so that it has a radial gap of 2 mm. For the steel we have:

H in A/m	1200	1400	1500
B in T	0.9	1.0	1.1

28. A rectangular steel core has a cross-sectional area of 4 cm² and a flux path of length 220 mm. What current is required in a coil of 500 turns wound on the

core if the flux in the core is to be 4×10^{-4} Wb? The relative permeability of the core may be taken as 800. A 1 mm air gap is then introduced in the circuit. What is now the current required to produce the same flux?

29. What is the charge on the plates of a $4\,\mu F$ capacitor when a potential difference of 12 V is connected across it?

30. What is the capacitance of a parallel plate capacitor which has plates of area $0.01\,m^2$ and 5 mm apart when the space between them is (a) air with a relative permittivity of 1, (b) polythene with a relative permittivity of 2.3?

31. A $2\,\mu F$ capacitor is to be made of two strips of metal foil, width 100 mm, separated by a paper sheet of thickness 0.01 mm and relative permittivity 3.0. What will be the length of each strip of metal foil?

32. A parallel plate capacitor has plates of area $100\,cm^2$ separated by a dielectric of thickness 1.0 mm and having a relative permittivity of 3.7. A potential difference of 100 V is applied to the plates. What is (a) the capacitance, (b) the electric flux density, (c) the electric field strength between the plates?

33. What is the capacitance when capacitors with capacitances of $2\,\mu F$ and $8\,\mu F$ are connected in (a) series, (b) parallel?

34. A $2\,\mu F$ capacitor is connected in parallel with a $4\,\mu F$ capacitor and this combination is then connected in series with a $1\,\mu F$ capacitor. What is the total capacitance?

35. Two capacitors of $1\,\mu F$ and $4\,\mu F$ are connected in series across a 10 V d.c. supply. What is the potential difference across each capacitor?

36. A capacitor has plates of area $500\,mm^2$ and is filled by two sheets of dielectric, one of thickness 1 mm and relative permittivity 2 and the other of thickness 2 mm and relative permittivity 5. What is the capacitance of the capacitor?

37. A capacitor has plates of area $0.2\,m^2$. The space between the plates if filled with a sheet of paper 2.0 mm thick and relative permittivity 2 and a sheet of glass 3.0 mm thick and relative permittivity 8. What is the capacitance?

38. The capacitance per unit length of a capacitor consisting of two concentric cylinders, the inner with a radius a and the outer with radius b, with the space between the two cylinders occupied by a dielectric is

$$C = \frac{2\pi\varepsilon}{\ln(b/a)}$$

Such an arrangement can be used as a liquid level gauge with the liquid acting as the dielectric between vertical cylinders. For cylinders of height H and with the liquid level at height h, derive an equation for the capacitance in terms of the liquid height h.

39. A capacitor is formed from 10 sheets of metal foil, each 2 cm by 10 cm. The metal sheets are separated by waxed paper 0.1 mm thick and with a relative

permittivity of 2.0. How should the metal foil be connected to give a total capacitance of 39 pF or 3186 pF?

40. The permittivity of the dielectric material between the plates of a parallel plate capacitor varies uniformly from ε_1 to ε_2. Show that the capacitance is

$$C = \frac{A(\varepsilon_2 - \varepsilon_1)}{d \ln(\varepsilon_2/\varepsilon_1)}$$

where d is the distance between the plates and A is the area of the plates.

41. What is the energy stored by a 4 μF capacitor when there is a potential difference of 12 V across it?

42. A 10 μF capacitor is charged by a potential difference of 100 V being applied across it. It is then disconnected from the supply and connected in parallel with a 6 μF uncharged capacitor. What is the energy stored before and after the capacitors are connected in parallel?

43. A 4 μF capacitor is charged by a potential difference of 40 V and then, after being disconnected from the voltage, connected in parallel with an uncharged capacitor of 2 μF. What is (a) the potential difference across each capacitor, (b) the total energy stored in the parallel arrangement?

44. What is the energy stored in a 200 mH inductor when the current through it is 100 mA?

Chapter 2
Electric Circuits

2.1 Kirchoff's laws

Kirchoff's laws state for electrical circuits the basic fundamental laws of the conservation of charge and the conservation of energy (see Section 1.1.2). *Kirchoff's current law* can be stated as: the algebraic sum of the currents at any node in a circuit is zero, the term *node* being used for a point in an electrical circuit at which two or more conductors are joined to form a junction. An alternative way of expressing the law is: the sum of the currents entering a junction equals the sum of the currents leaving it. Thus for Fig. 2.1

$$I_1 + I_2 - I_3 - I_4 - I_5 = 0$$

$$\Sigma I = 0$$

Kirchoff's voltage law can be stated as: the algebraic sum of the voltage drops and e.m.f.s around any closed circuit is zero.

In considering the voltage sources and voltage drops the conventions adopted are that for voltage sources the voltage is positive in going from negative terminal to positive terminal and a voltage drop is positive in going in the opposite direction to the current (see Section 1.1.1).

Example   Find the current in each branch of the network shown in Fig. 2.2.

Let the current in the branches be I_1, I_2 and I_3 as shown. Applying Kirchoff's current law at junction B gives

$$I_3 = I_1 + I_2$$

Applying Kirchoff's voltage law around the closed circuit ABEF gives

$$12 - 1 \times I_1 - 2 \times I_3 = 0$$

Substituting for I_3 then

$$12 = 3I_1 + 2I_2 \qquad (2.1)$$

Applying Kirchoff's voltage law around the closed circuit CBED gives

$$6 - 1 \times I_2 - 2 \times I_3 = 0$$

Substituting for I_3

$$6 = 2I_1 + 3I_2 \qquad (2.2)$$

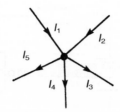

Fig. 2.1 Currents at a node.

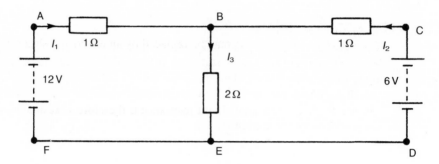

Fig. 2.2 Example.

Solving Equations (2.1) and (2.2) simultaneously gives $I_1 = 4.8\,\text{A}$, $I_2 = -1.2\,\text{A}$, $I_3 = 3.6\,\text{A}$. The negative sign for I_2 means that the assumed direction of I_2 was wrong and the current I_2 is actually flowing into the positive terminal of the 6 V battery.

2.1.1   Series and parallel circuits

For resistors in series the total resistance R is the sum of the resistances of the resistors (see Section 1.2.1),

$$R = R_1 + R_2 + \ldots \tag{2.3}$$

For resistors in parallel, the reciprocal of the total resistance R is equal to the sum of the reciprocal of the resistances of the resistors,

$$\frac{1}{R} = \frac{1}{R_1} + \frac{1}{R_2} + \ldots \tag{2.4}$$

This expression is often more conveniently expressed in terms of the conductances, i.e.

$$G = G_1 + G_2 + \ldots \tag{2.5}$$

Example   Find the current drawn from the battery in Fig. 2.3.

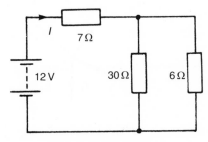

Fig. 2.3 Example.

The $30\,\Omega$ and $6\,\Omega$ resistors may be replaced by an effective resistance R, where

$$\frac{1}{R} = \frac{1}{30} + \frac{1}{6}$$

Hence $R = 5\,\Omega$. The total circuit resistance is therefore $12\,\Omega$ and the current is thus $V/R = 12/12 = 1.0\,\text{A}$.

2.1.2 *Potential dividers*

Potential dividers are used extensively as a method of obtaining reduced voltage, especially in low power circuits. The potential divider is a resistance which may be tapped at any intermediate point (Fig. 2.4). The input voltage is connected across the ends of the resistor and the output voltage is taken between one end of the resistor and the tapping point.

$$V_{\text{input}} = IR_A + IR_B$$

$$V_{\text{output}} = IR_B$$

$$\frac{V_{\text{output}}}{V_{\text{input}}} = \frac{R_B}{R_A + R_B} \tag{2.6}$$

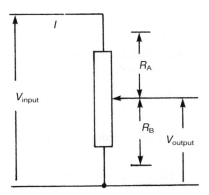

Fig. 2.4 Potential divider.

Example A student wishing to install a 9 V cassette player in a car decided to run the player from the 12 V car battery, via a potential divider. The potential divider that was constructed to give an output voltage of 9.2 V is shown in Fig. 2.5. However on connecting the cassette to the potential divider the measured output voltage was only 7.32 V. Why?

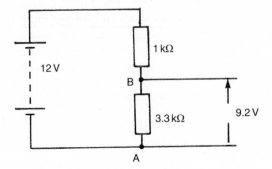

Fig. 2.5 Example.

The reason for the drop in voltage was that the cassette player, which had an effective resistance of about 3 kΩ, loaded the potential divider, as shown in Fig. 2.6. The resistance between A and B is thus

$$\frac{1}{R_{AB}} = \frac{1}{3} + \frac{1}{3.3}$$

Thus $R_{AB} = 1.572\,\Omega$. Thus

$$V_{AB} = \frac{1.572}{1 + 1.572} \times 12 = 7.32\,\text{V}$$

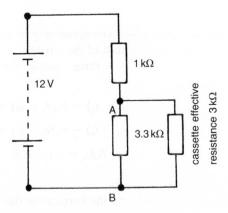

Fig. 2.6 Example.

2.1.3 Mesh analysis

The term *mesh* is used for a closed path in a circuit which does not contain any other closed paths within it. Meshes can be thought of as like the meshes of a fisherman's net. Figure 2.7 shows a circuit with two meshes. This form of analysis requires a mesh current to be defined, such a current being considered to circulate round the mesh. For Fig. 2.7 these currents are i_a and i_b. In considering the voltage sources and voltage drops in a mesh, the conventions adopted are that for voltage sources the voltage is positive in going from negative terminal to positive terminal and a voltage drop is positive in going in the opposite direction to the current (see Section 1.1.1). Using Kirchoff's voltage law for the first loop and proceeding round the loop in a clockwise direction gives

$$V_a - R_1 i_a - R_2(i_a - i_b) = 0$$

and for the second loop

$$-V_b - R_2 i_b - R_3(i_b - i_a) = 0$$

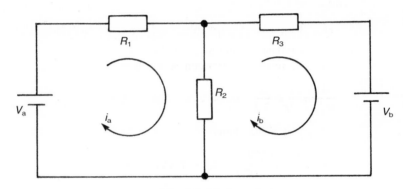

Fig. 2.7 Mesh analysis.

These equations can be solved simultaneously to give the mesh currents and hence the currents in each branch of the circuit.

Figure 2.8 shows a circuit with three meshes. For these, Kirchoff's voltage law gives

$$V_a - R_1 i_a - R_3(i_a - i_c) - R_2(i_a - i_b) = 0$$
$$V_b - R_4 i_b - R_2(i_b - i_a) - R_5(i_b - i_c) = 0$$
$$-V_c - R_7 i_c - R_5(i_c - i_b) - R_3(i_c - i_a) - R_6 i_c = 0$$

Example Use mesh analysis to determine the currents in the branches of the circuit in Fig. 2.2.

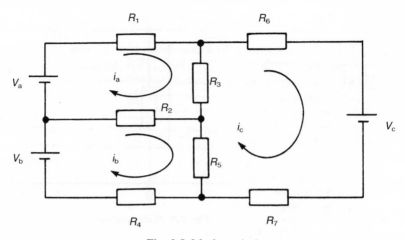

Fig. 2.8 Mesh analysis.

With a mesh current of i_a in the left-hand mesh and i_b in the right-hand mesh, then for the left-hand mesh

$$12 - 1i_a - 2(i_a - i_b) = 0$$

and for the right-hand mesh

$$-6 - 1i_b - 2(i_b - i_a) = 0$$

These two equations can be written as

$$12 = 3i_a - 2i_b$$

$$-6 = 3i_b - 2i_a g$$

Thus $i_b = 1.2\,\text{A}$ and $i_a = 4.8\,\text{A}$. Hence $I_1 = 4.8\,\text{A}$, $I_2 = -1.2\,\text{A}$ and $I_3 = i_a - i_b = 3.6\,\text{A}$.

2.1.4   Node analysis

Whereas mesh analysis uses Kirchoff's voltage law for obtaining mesh voltage equations, node analysis utilizes Kirchoff's current law to obtain node current equations. Consider the circuit shown in Fig. 2.9.

One of the nodes is taken as a reference node and the voltages at all other nodes considered relative to that node. Thus with node B as the reference node, node A is considered to be at a potential of V_A relative to it. Applying Kirchoff's current law to node A gives

$$\frac{V_a - V_A}{R_1} = \frac{V_A}{R_2} + \frac{V_A - V_b}{R_3}$$

Since there is only one node, other than the reference node, there is just one node equation.

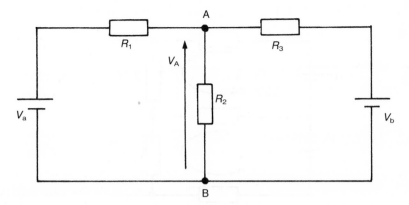

Fig. 2.9 Node analysis.

Figure 2.10 shows a circuit with more nodes. Nodes C and D are at the same potential and can be considered therefore to be a single node. Using Kirchoff's current law at node A gives

$$\frac{V_a - V_A}{R_1} = \frac{V_A}{R_2} + \frac{V_A - V_B}{R_3}$$

and for node B

$$\frac{V_b - V_B}{R_5} = \frac{V_B}{R_4} + \frac{V_B - V_A}{R_3}$$

These two simultaneous equations can be solved to obtain the currents in the branches of the circuit.

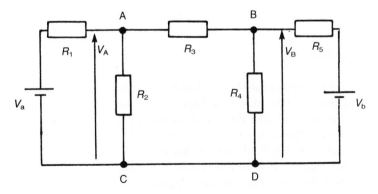

Fig. 2.10 Node analysis.

Example Use node analysis to determine the currents in the branches of the circuit shown in Fig. 2.2.

For the node at B, with E being the reference node,

$$\frac{12 - V_B}{1} = \frac{V_B}{2} + \frac{V_B - 6}{1}$$

Hence $V_B = 7.2\,\text{V}$. Thus $I_1 = (12 - 7.2)/1 = 4.8\,\text{A}$, $I_2 = -(7.2 - 6)/1 = -1.2\,\text{A}$, $I_3 = 7.2/2 = 3.6\,\text{A}$.

2.2 Thévenin's theorem

Thévenin's theorem states that any network consisting of resistors and voltage sources between a pair of terminals can be replaced by a voltage source in series with a resistance. The voltage of the replacement source is the open-circuit voltage of the network and the resistance is that measured at the terminals of the network when all independent sources are reduced to zero, i.e. voltage sources being replaced by a short circuit and current sources by an open circuit.

Thus, for example, the network between terminals A and B in Fig. 2.11(a) can be replaced by an equivalent network of the form shown in Fig. 2.11(b), where R_{sc} is the short circuit resistance obtained by considering all the voltage sources short circuited and calculating the resulting resistance of the circuit *looking in* at the terminals AB and V_{oc} is the open circuit voltage at AB with the $2\,\Omega$ resistor removed. R_{sc} is thus $0.5\,\Omega$ (Fig. 2.12(a)). Since with the open circuit (Fig. 2.12(b)) the circuit current is $(12 - 6)/2 = 3\,\text{A}$, then the voltage drop across a $1\,\Omega$ resistor is $3\,\text{V}$ and so the voltage between A and B, V_{oc}, is $9\,\text{V}$. Thus the Thévenin equivalent circuit is as shown in Fig. 2.12(c). The current through the $2\,\Omega$ resistor connected across AB can then be determined as $9/2.5 = 3.6\,\text{A}$, which checks with the current found in the same circuit in earlier examples in this chapter.

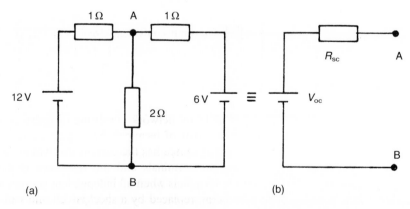

Fig. 2.11 (a) Original network, (b) Thévenin equivalent.

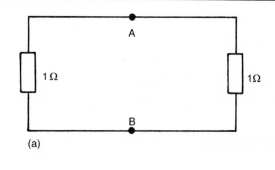

(a)

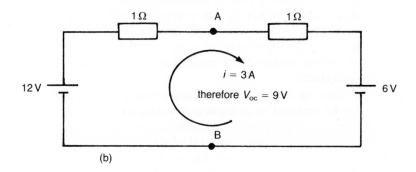

(b)

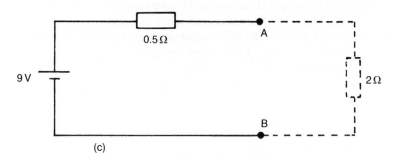

(c)

Fig. 2.12 Determining the Thévenin equivalent circuit, (a) R_{sc}, (b) V_{oc}, (c) the equivalent circuit.

2.3   Norton's theorem

Norton's theorem states that any network consisting of resistors and voltage sources can be replaced at a pair of terminals by a current source in parallel with a resistance. The current source has the current that would flow if a short circuit was placed across the terminals and the resistance is that which is measured looking into the terminals when all independent sources are reduced to zero, voltage sources being replaced by a short circuit and current sources by an open circuit. The Norton equivalent circuit for Fig. 2.11(a) is thus of the

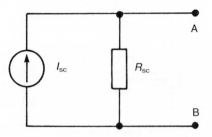

Fig. 2.13 Norton equivalent circuit.

form shown in Fig. 2.13, where R_{sc} is the short circuit resistance and I_{sc} the short circuit current. For Fig. 2.11(a), placing a short circuit between A and B means that A is at 0 V, the same voltage as at B, and thus the current between A and B is $12/1 + 6/1 = 18$ A. The short circuit resistance when the voltage sources are both short circuited is $1\,\Omega$ in parallel with $1\,\Omega$ and so is $0.5\,\Omega$. Thus the equivalent circuit is as shown in Fig. 2.14. The current through the $2\,\Omega$ resistor between terminals A and B is thus $I = (0.5/2 + 0.5)18 = 3.6$ A.

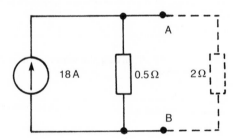

Fig. 2.14 Norton equivalent circuit.

2.4 Superposition theorem

The principle of superposition states that the response of a linear network containing several voltage or current sources is the sum of the responses of each source taken in turn, with the other voltage sources short circuited and current sources open circuited. Using the network of Fig. 2.11(a) as the example, the current through the $2\,\Omega$ resistor due to the 12 V source, with the 6 V source short circuited, is (Fig. 2.15(a))

$$I_a = \frac{12}{1 + \frac{2}{3}} \times \frac{1}{3} = 2.4\,\text{A}$$

The current through the $2\,\Omega$ resistor due to the 6 V source, with the 12 V source short circuited, is (Fig. 2.15(b))

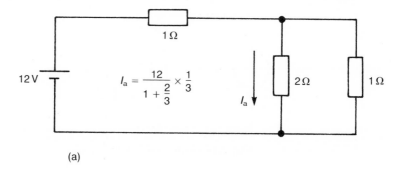

(a)

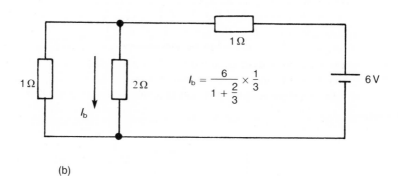

(b)

Fig. 2.15 Applying the superposition theorem.

$$I_b = \frac{6}{1 + \dfrac{2}{3}} \times \frac{1}{3} = 1.2\,\text{A}$$

Thus the total current through the $2\,\Omega$ resistor due to both sources is $2.4 + 1.2$ $= 3.6\,\text{A}$.

2.5 Ideal voltage and current sources

With an ideal voltage source, the voltage across its terminals will be independent of the current drawn from it. With real sources, this is not the case. A real source is thus considered as being equivalent to an ideal source in series with a resistance, this resistance being called the *internal resistance* (Fig. 2.16(a)). Thus for a source of e.m.f. *e* and internal resistance *r*, the potential difference v_o between its terminals is

$$v_o = e - ir \tag{2.7}$$

where *ir* is the potential drop across the internal resistance, *i* being the current taken by the external circuit.

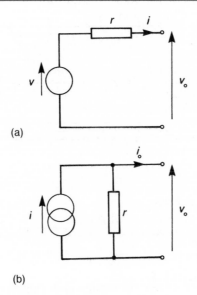

Fig. 2.16 (a) Real voltage source, (b) real current source.

With an ideal current source, the current taken from it is unaffected by the circuit into which the current is flowing. With real current sources, this is not the case. A real current source may be considered to be an ideal current source connected in parallel with a resistance, this resistance being called the internal resistance (Fig. 2.16(b)). Thus for a current source of i the output current i_o is i minus that part of the current that passes through the parallel resistance of v/r, where v is the potential difference between the output terminals. Thus

$$i_o = i - (v/r) \tag{2.8}$$

Example A voltage source has an e.m.f. of 12 V and an internal resistance of 0.5 Ω. What will be the potential difference between its terminals when the current drawn from it is 2 A?

Using Equation (2.7)

$$v_o = 12 - 0.5 \times 2 = 11 \text{ V}$$

2.6 Maximum power transfer

Consider the circuit shown in Fig. 2.17 in which a d.c. source of e.m.f. E and internal resistance r supplies a load of resistance R. The power supplied to the load is I^2R, where $I = E/(R + r)$. Therefore

$$P = \frac{E^2 R}{(R + r)^2} \tag{2.9}$$

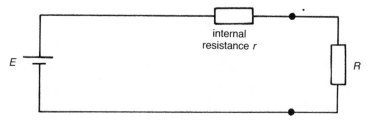

Fig. 2.17 Power transfer.

Differentiating with respect to R gives

$$\frac{dP}{dR} = -\frac{2E^2R}{(R + r)^3} + \frac{E^2}{(R + r)^2}$$

By equating dP/dR to zero the maximum or minimum value of R can be found.

$$\frac{2E^2R}{(R + r)^3} = \frac{E^2}{(R + r)^2}$$

Thus $R = r$.

$$\frac{d^2P}{dR^2} = \frac{6E^2R}{(R + r)^4} - \frac{4E^2}{(R + r)^3}$$

Substituting for $R = r$ in the above equation makes d^2P/dR^2 negative, hence showing that $R = r$ is the condition for maximum power transfer. The above discussion may be extended to any piece of electrical equipment, since by Thévenin's theorem any source may be transformed into an electrical circuit like Fig. 2.17. Maximum power transfer occurs when the loading resistance is equal to the internal resistance.

Example   A circuit has a Thévenin equivalent of a 10 V supply in series with an internal resistance of 5.0 Ω. What value of load resistance will lead to maximum power transfer and what will be the power transferred?

Maximum power will be transferred when the load resistance is equal to the internal resistance of the source. Thus the load resistance is 5.0 Ω. With this load the total circuit resistance is 10 Ω and so the circuit current is $V/R = 10/10 = 1.0$ A. The power transferred to the load is thus $I^2R_L = 1.0^2 \times 5 = 5.0$ W.

Problems

1.   For the circuits shown in Fig. 2.18 calculate the current I by (a) deriving simultaneous equations using Kirchoff's laws, (b) solving for series and parallel combinations of resistors.

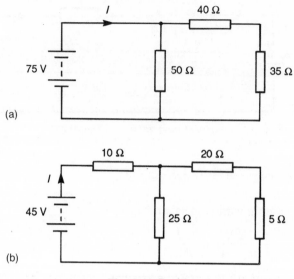

(a)

(b)

Fig. 2.18 Problem 1.

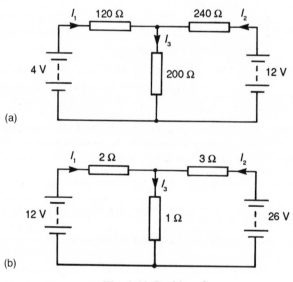

(a)

(b)

Fig. 2.19 Problem 2.

2. For the circuits shown in Fig. 2.19 calculate the currents I_1, I_2 and I_3 by (a) deriving simultaneous equations using Kirchoff's laws, (b) mesh analysis, (c) node analysis, (d) superposition.

3. Show for a linear potentiometer used as a voltage divider that the output voltage V_{out} is related to the linear displacement x of the sliding tapped point from one end when there is a load of R_L across the output by

$$V_L = \frac{V_{in}x}{(R_p/R_L) \times (1 - x) + 1}$$

where R_p is the total resistance of the potentiometer and V_{in} is the input voltage across the potentiometer.

4. For the circuit shown in Fig. 2.20, calculate the current in the $3\,\Omega$ resistor by (a) mesh analysis, (b) superposition theorem, (c) using the Thévenin equivalent circuit.

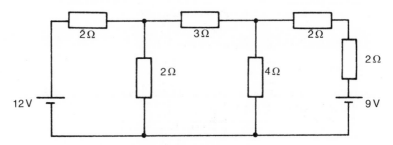

Fig. 2.20 Problem 4.

5. Using Norton's equivalent circuit, convert the circuit shown in Fig. 2.20 into one involving current sources and conductances, and hence find the current in the $3\,\Omega$ resistor using node analysis.

6. A voltage source has an e.m.f. of $20\,\text{V}$ and an internal resistance of $1.5\,\Omega$. What will be the potential difference between its terminals when the current drawn from it is $2\,\text{A}$?

7. A current source gives an output of $2.0\,\text{A}$ when the potential difference between its terminals is $8\,\text{V}$ and $1.7\,\text{A}$ when it is $10\,\text{V}$. What is the internal resistance of the source?

8. Six cells are connected (a) in series positive terminal to negative terminal, (b) in parallel positive terminal to positive terminal. If each cell has an e.m.f. of $2\,\text{V}$ and an internal resistance of $0.5\,\Omega$, what will be the e.m.f. and internal resistance of the combined cells?

9. Two batteries are connected in parallel, positive terminal to positive terminal, the e.m.f. of one being $12\,\text{V}$ and internal resistance $1.0\,\Omega$ and of the other being $15\,\text{V}$ and $2.0\,\Omega$. A resistor of $5\,\Omega$ is connected across the battery terminals. What is the current through the $5\,\Omega$ resistor?

10. A circuit has a Thévenin equivalent of a $100\,\text{V}$ supply in series with an internal resistance of $50\,\Omega$. What will be the load resistance for maximum power transfer and the amount of power thus transferred?

11. Four cells, each of e.m.f. $1.5\,\text{V}$ and internal resistance $0.5\,\Omega$, are connected in series with positive terminal to negative terminal. What will be the load resistance which can be connected across the arrangement in order to give maximum power transfer and what will be the power thus transferred?

Chapter 3
Transients

3.1 Circuits with resistors, capacitors and inductors

Consider the circuit shown in Fig. 3.1.

$$V_s = v_R + v_L + v_C \tag{3.1}$$

From Section 1.9 the voltage across the resistance, v_R, is iR, the voltage across the inductance v_L is given by

$$v_L = L\frac{di}{dt}$$

and across the capacitor

$$v_C = \frac{1}{C}\int i\,dt$$

Therefore Equation (3.1) can be written as

$$V_s = Ri + L\frac{di}{dt} + \frac{1}{C}\int i\,dt \tag{3.2}$$

The method used to determine the current from Equation (3.2) depends to a great extent on the form of the supply voltage. If the supply voltage varies with time, Equation (3.2) is solved as a differential equation. If the supply voltage varies sinusoidally with time and we are only concerned with the steady state values then certain specialized techniques can be used, these being discussed in Chapter 4. If we are only concerned with steady state values with constant supply voltages then because the current is not changing with time the inductance and capacitance terms have no significance and the concern is only with the circuit as a purely resistive circuit. The term steady state is used for the currents and voltages occurring in a circuit after all transient effects have subsided as a consequence of the supply being switched on, or off. This chapter is about these transient effects.

3.2 Series RL and RC circuits

For the series RL circuit in Fig. 3.2(a), consider the switch to be moved into position M. At the instant just preceding the switch closure the current is zero.

50

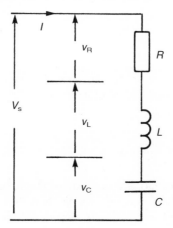

Fig. 3.1 RLC circuit.

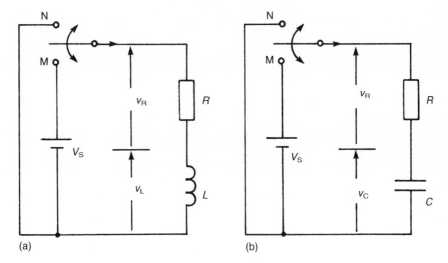

Fig. 3.2 (a) RL circuit, (b) RC circuit.

After the switch is closed, the current will rise in a transient manner to the steady state value eventually of V_s/R. Thus, at any instant

$$V_s = v_R + v_L$$

and so

$$V_s = Ri + L\frac{di}{dt} \tag{3.3}$$

For the series RC circuit in Fig. 3.2(b), consider the switch to be moved into position M. At the instant just preceding the switch closure the voltage across the capacitor is zero. After the switch is closed the voltage v_C will rise in a transient manner, as the capacitor charges up, until the capacitor is fully charged to the steady state value of V_s.

$$V_s = v_R + v_C$$

Since the charge on the capacitor at some instant q is Cv_C then as $i = dq/dt$ we have $i = C \, dv_C/dt$. Therefore

$$V_s = Ri + v_C$$

$$V_s = RC\frac{dv_C}{dt} + v_C \tag{3.4}$$

Equations (3.4) and (3.5) are of the same form, being first order differential equations. One way of solving them is by the technique of separating the variables and integrating. Consider this technique with Equation (3.3). The steady state current I is V_s/R and so the equation can be written as

$$I - i = \frac{L}{R}\frac{di}{dt}$$

$$\frac{R}{L}dt = \frac{1}{I - i}di$$

Integrating both sides with $i = 0$ at $t = 0$ and I at t

$$\int_0^t \frac{R}{L}dt = \int_0^I \frac{1}{I - i}di$$

$$\frac{Rt}{L} = \ln\frac{I}{I - i}$$

$$\frac{I - i}{I} = e^{-Rt/L}$$

$$i = I(1 - e^{-Rt/L}) \tag{3.5}$$

Since

$$v_L = L\frac{di}{dt}$$

then, using Equation (3.5),

$$v_L = RIe^{-Rt/L} = V_s e^{-Rt/L} \tag{3.6}$$

In a similar manner, Equation (3.4) can be written as

$$V_s - v_C = RC\frac{dv_C}{dt}$$

$$\frac{1}{RC}dt = \frac{1}{V_s - v_C}dv_C$$

At $t = 0$ then $v_C = 0$, with $v_C = V_s$ at time t. Thus integrating the above equation gives

$$\frac{t}{RC} = \ln \frac{V_s}{V_s - v_C}$$

$$\frac{V_s - v_C}{V_s} = e^{-t/RC}$$

$$v_C = V_s(1 - e^{-t/RC}) \tag{3.7}$$

Since $i = C\, dv_C/dt$ then, using Equation (3.7),

$$i = \frac{V_s}{R}e^{-t/RC} = Ie^{-t/RC} \tag{3.8}$$

Figure 3.3 shows graphs indicating how the voltage across the inductor or capacitor and the circuit current vary with time. If the initial rate of rise or decay of current or voltage were maintained then the steady state value would be reached in a time T. This interval of time is called the *time constant* of the circuit. In the case of the RL circuit, the initial current is zero and, since the current rises to I in a time T, the initial rate of growth of current is I/T. Therefore, initially,

$$v_L = L\frac{di}{dt} = \frac{LI}{T}$$

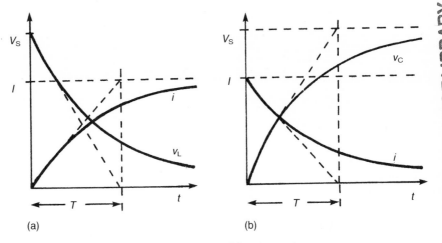

(a)   (b)

Fig. 3.3 (a) RL circuit, (b) RC circuit.

But initially $v_L = V_s$ and so, as $V_s = IR$,

$$T = \frac{L}{R} \tag{3.9}$$

In a time T the actual value of the current will have risen to

$$i = I(1 - e^{-Rt/L}) = I(1 - e^{-1}) = 0.632I \tag{3.10}$$

In the case of the RC circuit, the initial voltage across the capacitor is zero and the initial rate of growth of this voltage is V_s/T. But $i = C\, dv_C/dt$ and since initially the current is $I = V_s/R$, then

$$\frac{V_s}{R} = C\frac{V_s}{T}$$

$$T = RC \tag{3.11}$$

In a time T the actual value of the voltage across the capacitor will have risen to

$$v_C = V_s(1 - e^{-T/RC}) = V_s(1 - e^{-1}) = 0.632V_s \tag{3.12}$$

Now consider the circuits in Fig. 3.2 again. For the RL circuit, if the switch position after being at M for some time is changed to N, then

$$0 = i + L\frac{di}{dt}$$

$$-\frac{1}{i}di = \frac{R}{L}dt$$

At $t = 0$ then $i = I$ and at t we have i, thus integrating the above equation gives

$$-\ln\frac{i}{I} = \frac{Rt}{L}$$

$$i = Ie^{-Rt/L} \tag{3.13}$$

Since $v_L = L\, di/dt$ then, since $V_s = IR$,

$$v_L = -V_s e^{-Rt/L} \tag{3.14}$$

For the RC circuit in Fig. 3.2, then

$$0 = RC\frac{dv_C}{dt} + v_C$$

$$-\frac{1}{v_C}dv_C = \frac{1}{RC}dt$$

At $t = 0$ then $v_C = V_s$ and at t is v_C. Thus integrating the above equation gives

$$-\ln\frac{v_C}{V_s} = \frac{t}{RC}$$

$$v_C = V_s e^{-t/RC} \tag{3.15}$$

Since $i = C\, dv_C/dt$ with $I = V_s/R$,

$$i = -Ie^{-t/RC} \tag{3.16}$$

Figure 3.4 shows graphs illustrating how the current and voltages change with time for the RL and RC circuits. As with Fig. 3.3, if the initial rate of decay of current or voltage with time were maintained then the steady state

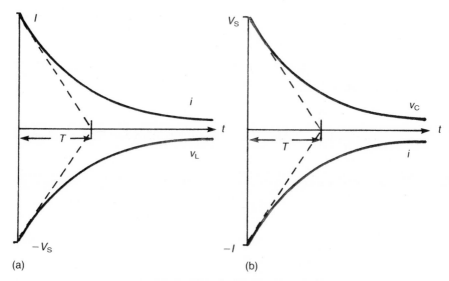

Fig. 3.4 (a) RL circuit, (b) RC circuit.

value of 0 would be reached in a time T. This time T is the time constant of the circuit. As before, it can be shown that $T = L/R$ or RC.

All real electronic instruments have some capacitance. For example, the input capacitance of an oscilloscope is about 10 to 25 pF and the coaxial cable used to connect an input to the oscilloscope is likely to have a capacitance of perhaps 150 pF per metre. A consequence of this is that when there is a rapidly changing input signal the instrument is not able to respond instantly but time is taken to charge the *capacitor* up to the input voltage. As small a capacitance as is feasible is desirable if the time constant is to be small and the instrument able to respond quickly.

Example   A circuit consists of an inductor of inductance 2 H in series with a resistor of resistance 1 kΩ. What is the current in the circuit (a) 2 ms, (b) 4 ms after a d.c. supply of 10 V is connected across the series arrangement?

(a) Using Equation (3.5), with the steady state current $I = V_s/R = 10/10^3 = 0.10$ A,

$$i = I(1 - e^{-Rt/L}) = 0.10(1 - e^{-10^3 \times 2 \times 10^{-3}/2}) = 0.063 \text{ A}$$

(b) With $t = 4$ ms

$$i = I(1 - e^{-Rt/L}) = 0.10(1 - e^{-10^3 \times 4 \times 10^{-3}/2}) = 0.086 \text{ A}$$

Example   A circuit consists of a 2 μF capacitor in series with a 1 MΩ resistor. What is (a) the time constant T of the circuit and (b) the voltage across the capacitor (i) $1T$, (ii) $2T$ after a 10 V supply is connected across the series arrangement?

(a) The time constant T is given by

$$T = CR = 2 \times 10^{-6} \times 1 \times 10^6 = 2\,\text{s}$$

(b) After $1T$ the potential difference across the capacitor will have risen to (Equation (3.12))

$$v_C = V_s(1 - e^{-t/RC}) = V_s(1 - e^{-1}) = 0.632\,V_s$$

and thus 6.32 V. After $2T$ the potential difference will have risen to

$$v_C = V_s(1 - e^{-t/RC}) = V_s(1 - e^{-2}) = 0.865\,V_s$$

and thus 8.65 V.

3.2.1 RC differentiator

For the circuit shown in Fig. 3.5(a), if the voltage drop across the capacitor v_C is very much greater than the voltage drop across the resistor v_R, then the source voltage v_s is approximately equal to v_C. Thus

$$i \approx C\frac{dv_s}{dt}$$

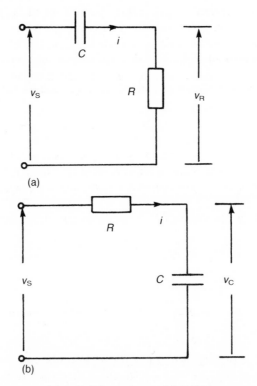

(a)

(b)

Fig. 3.5 (a) RC differentiator, (b) RC integrator.

If the output from the circuit is taken across the resistor, then

$$v_R = Ri \approx RC\frac{dv_s}{dt} \tag{3.17}$$

The output voltage is thus the differential of the input voltage multiplied by the time constant of the circuit.

The condition for the voltage drop across the capacitor to be greater than that across the resistor occurs when the time constant CR of the circuit is small so that the capaitor charges up or discharges very rapidly. Figure 3.6(a) shows the form of the output for such a circuit when the input is a rectangular pulse. The output approximates to a series of narrow spikes.

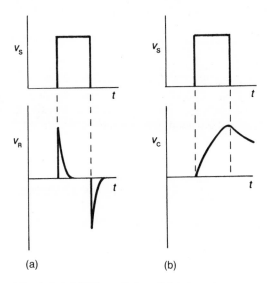

Fig. 3.6 (a) Differentiator, (b) integrator.

3.2.2 *RC integrator*

For the circuit shown in Fig. 3.5(b), if the voltage drop across the resistor v_R is very much greater than the voltage drop across the capacitor v_C, then the source voltage v_s is approximately equal to v_R. Thus, since $i = v_s/R$,

$$v_C = \frac{1}{C}\int i\ dt \approx \frac{1}{RC}\int v_s\ dt \tag{3.18}$$

The output voltage v_C is thus the integral of the input voltage v_s divided by the time constant of the circuit.

For the voltage drop across the resistor to be greater than that across the capacitor then the time constant of the circuit must be large. Only then will the capacitor take a long time to charge up or discharge. Figure 3.6(b) shows the form of the output from such a circuit when the input is a rectangular pulse. The output approximates to a signal which is proportional to the area under the input pulse.

3.3   RLC circuit

For a series connection of resistance, inductance and capacitance (as in Fig. 3.1), then as given by Equation (3.2)

$$V_s = Ri + L\frac{di}{dt} + \frac{1}{C}\int i\,dt$$

An alternative way of expressing this differential equation is in terms of the potential difference v_C across the capacitor. Since $i = dq/dt = C\,dv_C/dt$, then

$$V_s = RC\frac{dv_C}{dt} + LC\frac{d^2v_C}{dt^2} + v_C \tag{3.19}$$

This is a second order differential equation.

There are a number of ways this differential equation can be solved. One way is to consider v_C to be made up of two elements, a transient or natural part v_n and a forced part v_f which is responsible for the situation when all the transients have died away. Thus

$$v_C = v_n + v_f$$

Thus, substituting for v_C in Equation (3.19) gives, with some rearrangement of the terms,

$$V = \left(LC\frac{d^2v_f}{dt^2} + RC\frac{dv_f}{dt} + v_f\right) + \left(LC\frac{d^2v_n}{dt^2} + RC\frac{dv_n}{dt} + v_n\right)$$

We can thus have for the forcing differential equation

$$V = \left(LC\frac{d^2v_f}{dt^2} + RC\frac{dv_f}{dt} + v_f\right) \tag{3.20}$$

and for the transient differential equation

$$0 = \left(LC\frac{d^2v_n}{dt^2} + RC\frac{dv_n}{dt} + v_n\right) \tag{3.21}$$

To solve the transient differential equation we can try a solution of the form

$$v_n = Ae^{st}$$

With such a solution then Equation (3.21) becomes

$$0 = LCAs^2e^{st} + RCAse^{st} + Ae^{st}$$

$$0 = LCs^2 + RCs + 1$$

The roots of this quadratic equation are thus

$$s = \frac{-RC \pm \surd(R^2C^2 - 4LC)}{2LC} \tag{3.22}$$

There are thus two roots s_1 and s_2 and hence

$$v_n = Ae^{s_1 t} + Be^{s_2 t}$$

To solve the forcing differential Equation (3.20), since we expect there to be a steady state constant value for v_C then we can try the solution $v_f = k$, where k is a constant. Thus Equation (3.20) becomes

$$V = 0 + 0 + k$$

Thus the solution is $v_f = V$. Hence the solution for Equation (3.19) is

$$v_C = v_n + v_f = Ae^{s_1 t} + Be^{s_2 t} + V \qquad (3.23)$$

The values of A and B depend on the initial conditions in the circuit. Thus if the capacitor is initially uncharged then $v_C = 0$ when $t = 0$ and so $0 = A + B + V$. There is also zero current at $t = 0$ and so since $i = C\,dv_C/dt$ we must have $dv_C/dt = 0$ at $t = 0$. Thus differentiating Equation (3.22) gives $0 = As_1 + Bs_2$.

The way in which v_C varies with time depends on the values of s_1 and s_2 given by Equation (3.22). When we have $(R/L)^2 > (1/LC)$ then the square root is that of a positive number and so there are two real roots. This then results in the transient part of the voltage just slowly decaying with time. The circuit is then said to be *over damped*. With $(R/L)^2 = (1/LC)$ then the square root term is zero and so there are two equal real roots. The circuit is said to be *critically damped*. With $(R/L)^2 < (1/LC)$ then the square root is of a negative number and yields two complex roots. The circuit is then said to be *under damped* because the transient part of the voltage oscillates about the final steady value with an amplitude which decays with time. Figure 3.7 illustrates the effect of such damping on the way v_C varies with time.

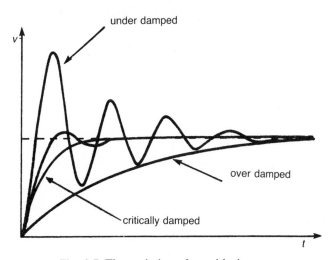

Fig. 3.7 The variation of v_C with time.

Example   A circuit consists of a $1\,\mu F$ capacitor, a $0.1\,H$ inductor and a $1\,k\Omega$ resistor in series. When a d.c. supply is connected to the circuit, what will be the form of the transient voltage across the capacitor?

The factor determining the form of the transient response is whether $(R/L)^2$ is less than, equal to, or greater than $(1/LC)$. Since $(R/L)^2 = 10^8$ and $(1/LC)$ is 10^7 then the circuit is over damped. The potential difference across the capacitor will thus slowly rise until it becomes equal to the steady state value.

3.3.1 Damping factor

An oscillation which has a constant amplitude A and does not decay with time can be described by the equation

$$v_n = A \sin \omega_n t$$

where ω_n is the angular frequency of the natural oscillations, i.e. $\omega_n = 2\pi f_n$ with f_n being the natural frequency. Differentiating this equation gives

$$\frac{d^2 v_n}{dt^2} = -\omega_n^2 A \sin \omega_n t = -\omega_n^2 v_n$$

$$\frac{d^2 v_n}{dt^2} + \omega_n^2 v_n = 0 \tag{3.24}$$

But this is just the transient differential equation (3.21) for the RLC circuit with $R = 0$. The resistance in the circuit is responsible for energy dissipation and so the damping. Rearrangement of Equation (3.21) with $R = 0$ gives

$$\frac{d^2 v_n}{dt^2} + \frac{1}{LC} v_n = 0$$

Comparison between this equation and (3.24) indicates that the natural angular frequency of the oscillation is

$$\omega_n = \frac{1}{\sqrt{(LC)}} \tag{3.25}$$

If we define a term ζ as being given by

$$\zeta \omega_n = (R/2L) \tag{3.26}$$

then the roots, Equation (3.22), are

$$s = -\zeta \omega_n \pm \sqrt{(\zeta^2 \omega_n^2 - \omega_n^2)} = -\zeta \omega_n \pm \omega_n \sqrt{(\zeta^2 - 1)} \tag{3.27}$$

Thus the value of ζ determines whether the square root term yields real or complex roots and so the form of the damping. For this reason it is called the *damping factor*. When $\zeta > 1$ there is over damping, when $\zeta = 1$ critical damping and when $\zeta < 1$ under damping. When $R = 0$ then $\zeta = 0$ and there is no damping.

Example   What is the natural angular frequency of oscillation of a circuit consisting of a $2\,\mu F$ capacitor in series with a $10\,mH$ inductor?

Using Equation (3.25)

$$\omega_n = \frac{1}{\sqrt{(LC)}} = \frac{1}{\sqrt{(0.010 \times 2 \times 10^{-6})}} = 7.07 \text{ krad/s}$$

Problems

1. A circuit consists of a 4 H inductor in series with a 1 kΩ resistor. What will be (a) the current, (b) the potential difference across the inductor, after 2 ms when a 10 V supply is connected across the series arrangement?

2. A circuit consists of a 2 H inductor in series with a 1 kΩ resistor. What is (a) the time constant T of the circuit, (b) the current in the circuit $1T$ after a 10 V d.c. supply was connected across the circuit?

3. An inductor is to be included in series with a 500 Ω resistor to limit the rate of growth of current in the circuit when the supply is switched on. If the current is to reach 50% of its value in 2 ms, what will be the required inductance?

4. A 0.5 H inductor is in series with a 2 kΩ resistor and a 10 V d.c. supply. What will be the current through the resistor and inductor 1 ms after the d.c. supply is short circuited?

5. A 2 μF capacitor is connected in series with a 1 kΩ resistor. What will be (a) the current in the circuit, (b) the potential difference across the capacitor, (c) the potential difference across the resistor 2 ms after a 10 V d.c. supply is connected across the series arrangement?

6. A 1 μF capacitor is in series with a 5 kΩ resistor and a d.c. supply of 10 V is connected across the series arrangement. What is (a) the time constant T of the circuit, (b) the current in the circuit after (i) $1T$, (ii) $2T$?

7. A capacitor may be represented by a capacitance C in parallel with a resistance R. The capacitor is charged to a voltage of 1000 V. When the charging source is removed the voltage drops to 500 V in 20 s. A pure capacitance of 60 μF is then connected in parallel with the first, and the combination again charged to 1000 V. When the source is removed, the voltage drops to 500 V in 60 s. What are the values of C and R?

8. What is the value of the resistance that should be included in series with a 1 μF capacitor if the series arrangement is to be connected across a 10 V supply and the potential difference across the capacitor is to become 5 V in 20 ms?

9. What is the time required for a 4 μF capacitor to discharge from 10 V to 2 V if the discharge is through a 1 MΩ resistance?

10. What resistance should be connected in series with a 1 μF capacitor if the initial discharge current when the capacitor has been charged to a potential difference of 10 V is to be limited to 1 mA?

11. Predict the form of the output that will occur when a sawtooth pulse of duration t is the input to a series capacitor–resistor circuit and the output is

taken from across the capacitor. The time constant of the circuit is much greater than t.

12. What is the form of the transient response of the following series circuits when a d.c. supply is connected to them?
    (a) $R = 1\,\mathrm{k}\Omega$, $L = 2\,\mathrm{H}$, $C = 2\,\mu\mathrm{F}$
    (b) $R = 10\,\Omega$, $L = 20\,\mathrm{mH}$, $C = 4\,\mu\mathrm{F}$
    (c) $R = 100\,\Omega$, $L = 100\,\mathrm{mH}$, $C = 0.1\,\mu\mathrm{F}$

13. What is the maximum resistance that should be in series with a $1\,\mu\mathrm{F}$ capacitor and a $100\,\mathrm{mH}$ inductor if the transient response when a d.c. supply is connected to them is to be critically damped?

14. What is the natural frequency of the transient oscillation in a circuit containing a $2\,\mu\mathrm{F}$ capacitor in series with a $10\,\mathrm{mH}$ inductor?

Chapter 4
Steady State Alternating Currents

4.1 Alternating currents

A large proportion of the applications in electrical and electronic engineering involve voltages and currents which are alternating and which have a sinusoidal form. This is not surprising since the electricity authorities generate and transmit sinusoidal voltages. It is important, therefore, to consider as a special case the response of circuits to sinusoidal inputs. In this chapter we are only concerned with the steady state response and not the transient responses that occur when a sinusoidal voltage is applied to a circuit.

4.1.1 Terminology

Figure 4.1 shows the waveform of a sinusoidal voltage. We can ideally consider such a waveform as being generated by the rotation of a coil in a magnetic field so that the magnetic flux linked by the coil is varied and the resulting induced e.m.f. varies sinusoidally with time (see Chapter 16). One rotation of the coil then results in one cycle of the voltage in Fig. 4.1, one cycle being the time taken for the waveform to repeat itself and hence every 360° (2π radians). The *amplitude*, *peak* or *maximum* value is the maximum value of the waveform and occurs when the coil is at 90° (π/2 radians), 270° (3π/2 radians), 450° (5π/2 radians), etc., and is designated as plus or minus V_m. The *instantaneous value* of the waveform is the voltage v at some instant of time and can be written as

$$v = V_m \sin \theta \qquad (4.1)$$

The *frequency f* is the number of cycles occurring in one second. The unit of frequency is the hertz (Hz), 1 cycle per second being 1 Hz. The *periodic time T* is the time taken for one cycle and is thus $1/f$. The rate at which the coil is rotated in the magnetic field to generate the waveform, i.e. the angular velocity ω, is such that 2π radians (360°) are covered in the periodic time T. Thus

$$\omega = \frac{2\pi}{T} = 2\pi f \qquad (4.2)$$

It is often more convenient to describe alternating waveforms by ω and since ω is proportional to the frequency f it is referred to as the *angular frequency*. It has the unit of radians/second but since the radian unit is a ratio the unit is often just written as /s or Hz.

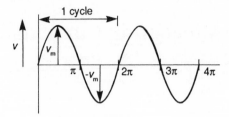

Fig. 4.1 Sinusoidal waveform.

Since the angular velocity is the angle θ covered per unit time, then $\theta = \omega t$ and so Equation (4.1) for the instantaneous voltage can be written as

$$v = V_m \sin \omega t \tag{4.3}$$

Consider now two sinusoidally varying signals, as shown in Fig. 4.2. Sinusoid b is displaced from sinusoid a by angle ϕ. This displacement is termed the *phase difference*, and the angle ϕ is termed the *phase angle*. If the instantaneous value of sinusoid a is given by

$$a = A \sin \omega t$$

then that of b is given by

$$b = B \sin(\omega t - \phi)$$

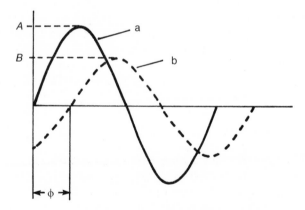

Fig. 4.2 Phase difference.

Because the peak of b occurs later (to the right on the figure) than that of a it is said to *lag* a and the phase angle is given a negative sign. In terms of the generation of the waveforms by rotating coils, then the coil for b is always rotating an angle ϕ behind that for a. In this discussion sinusoid a was taken as the reference, i.e. sinusoid b was defined with respect to a. We could however have taken sinusoid b as the reference and thus obtained

$$b = B \sin \omega t$$

$$a = A \sin(\omega t + \phi)$$

We can then say that a leads b by angle ϕ.

4.1.2 Root mean square values

The *root mean square (rms) value* of an alternating current is that current which when passed through a resistance will produce the same heating effect as a direct current passed through the same resistance. The heating effect of a current passed through a resistor is $i^2 R$. The average heating effect of a sinusoidal current, $i = I_m \sin \theta$, passed through a resistor R, is the average value of $(I_m \sin \theta)^2 R$. Over 1 cycle, when θ changes from 0 to 2π, then the average value is

$$\frac{1}{2\pi} \int_0^{2\pi} RI_m^2 \sin^2\theta \; d\theta = \frac{1}{2\pi} \int_0^{2\pi} \frac{RI_m^2}{2} (1 - \cos 2\theta) \; d\theta$$

$$= \frac{RI_m^2}{4\pi} \left[\theta - \frac{\sin 2\theta}{2} \right]_0^{2\pi} = \frac{I_m^2 R}{2}$$

Let I be the direct current which, when passed through the same resistor, will produce the same heating effect. Thus

$$I^2 R = \frac{I_m^2 R}{2}$$

$$I = \frac{I_m}{\sqrt{2}}$$

Thus the rms value of a sinusoidal current, or voltage, is 0.707 times the maximum value.

Example The mains alternating voltage is specified as being 240 V rms. What is the maximum value of this voltage if it is sinusoidal?

The maximum value is $\sqrt{2} \times V = \sqrt{2} \times 240 = 339\,V$

4.2 Sinusoidal response of RLC circuits

For the series RLC circuit in Fig. 4.3, let the instantaneous current be

$$i = I_m \sin \omega t \tag{4.4}$$

For the resistor, since $v = Ri$, then the voltage drop across it is

$$v_R = RI_m \sin \omega t \tag{4.5}$$

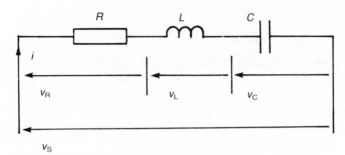

Fig. 4.3 Series RLC circuit.

For the inductor, since $v_L = L\, di/dt$, then the voltage drop across it is

$$v_L = L\frac{d(I_m \sin \omega t)}{dt}$$

$$= \omega L I_m \cos \omega t = \omega L I_m \sin(\omega t + 90°) \tag{4.6}$$

For the capacitor

$$v_C = \frac{1}{C}\int i\, dt = \frac{1}{C}\int I_m \sin \omega t\, dt$$

$$= -\frac{1}{\omega C} I_m \cos \omega t = \frac{1}{\omega C} I_m \sin(\omega t - 90°) \tag{4.7}$$

For the resistor, it can be seen from Equations (4.4) and (4.5) that the voltage drop across it is in phase with the current flowing through it. For the inductor, it can be seen from Equations (4.4) and (4.6) that the voltage drop across the inductor leads the current flowing through it by 90°. For the capacitor, it can be seen from Equations (4.4) and (4.7) that the voltage drop across it lags the current flowing through it by 90°.

For the resistor, the maximum value of the voltage drop will be when $\sin \omega t = 1$. Thus

$$V_{Rm} = RI_m$$

The ratio of the maximum voltage to the maximum current is the resistance, both the maximum voltage and current occurring at the same time because they are in phase. The resistance is independent of the frequency.

For the inductor, the maximum value of the voltage drop will be when $\cos \omega t = 1$. Thus

$$V_{Lm} = \omega L I_m$$

The ratio of the maximum voltage to the maximum current for the inductor is called the *inductive reactance* X_L, with the unit of ohms. Thus

$$X_L = \frac{V_{Lm}}{I_m} = \omega L \tag{4.8}$$

The maximum voltage across the inductor does not occur at the same time as the maximum current. The inductive reactance is thus proportional to the frequency, the higher the frequency the greater the reactance.

For the capacitor, the maximum value of the voltage drop will be when $\cos \omega t = 1$. Thus

$$V_{Cm} = \frac{1}{\omega C} I_m$$

The ratio of the maximum voltage to the maximum current for the capacitor is called the *capacitive reactance* X_C, with the unit of ohms. Thus

$$X_C = \frac{V_{Cm}}{I_m} = \frac{1}{\omega C} \tag{4.9}$$

The maximum voltage across the capacitor does not occur at the same time as the maximum current. The capacitive reactance is thus inversely proportional to the frequency, the higher the frequency the lower the reactance.

Example What is the reactance of a $1\,\mu F$ capacitor at a frequency of (a) 50 Hz, (b) 50 kHz?

(a) Using Equation (4.9),

$$X_C = \frac{1}{\omega C} = \frac{1}{2\pi \times 50 \times 1 \times 10^{-6}} = 3.18\,k\Omega$$

(b) Using Equation (4.9),

$$X_C = \frac{1}{\omega C} = \frac{1}{2\pi \times 50 \times 10^3 \times 1 \times 10^{-6}} = 3.18\,\Omega$$

Example A 500 mH inductor has an alternating voltage of maximum voltage 100 V and frequency 50 Hz connected across it. What will be the maximum value of the current in the circuit?

The inductive reactance is given by Equation (4.8) as

$$X_L = \omega L = 2\pi f L = 2\pi \times 50 \times 0.500 = 157\,\Omega$$

Thus the maximum current is

$$I_m = \frac{V_m}{X_L} = \frac{100}{157} = 0.64\,A$$

4.2.1   Phasors

It is often more convenient than dealing with instantaneous values, as above, to represent alternating voltages or currents graphically by a line which rotates

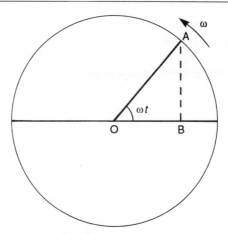

Fig. 4.4 Phasor representation.

in a circular path with an angular velocity ω (Fig. 4.4). The line OA makes an angle ωt with OX, the reference datum, and the length OA is V_m or I_m. The vertical projection AB of the rotating line is thus $V_m \sin \omega t$. The rotating line is called a *phasor* and the diagram on which it is drawn is called a *phasor diagram*. If there are two phasors with a phase difference between them then their lines on the phasor diagram will be separated by the phase angle. A phasor has both magnitude and a phase angle with respect to some reference datum.

One way of representing a phasor is by what is termed *polar notation*. This involves specifying the magnitude of the phasor and its angle in the form

$$\mathbf{V} = V_m \underline{/\phi} \tag{4.10}$$

Phasors are often written in bold type, as $\mathbf{V}$ is above.

Consider the addition of two voltages

$$v_A = V_{Am} \sin(\omega t + \phi_A)$$

$$v_B = V_{Bm} \sin(\omega t + \phi_B)$$

Their sum is

$$v_A + v_B = V_{Am} \sin(\omega t + \phi_A) + V_{Bm} \sin(\omega t + \phi_B)$$

Figure 4.5 shows the phasors representing these two voltages and how the sum is represented by the line OP which has the vertical projection of $v_A + v_B$. This line is the phasor obtained by the addition of the two phasors using vectorial means (the parallelogram of vectors).

Consider the RLC circuit in Fig. 4.3. The current is common to all the circuit elements and thus its phasor is used as the reference datum against which the phases of the voltages are represented. With circuits connected in parallel the source voltage is common to all the circuit elements and is generally used as the reference datum. The voltage drops across the three elements, i.e. v_R, v_L

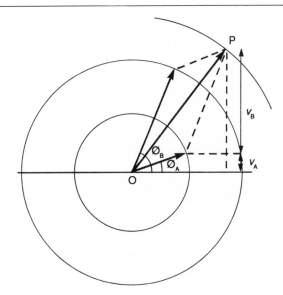

Fig. 4.5 Addition of two phasors.

and v_C, are thus represented by phasors specified in magnitude and phase, the phases being with respect to the current. Figure 4.6 shows the phasors. The source voltage is the vectorial addition of the phasors $\mathbf{V_R}$, $\mathbf{V_L}$ and $\mathbf{V_C}$. The source voltage phasor $\mathbf{V_s}$ has thus the magnitude, obtained using the Pythagoras theorem, of

$$V_s = \sqrt{[V_R^2 + (V_L - V_C)^2]} \tag{4.11}$$

and a phase angle ϕ of

$$\phi = \tan^{-1}\frac{V_L - V_C}{V_R} \tag{4.12}$$

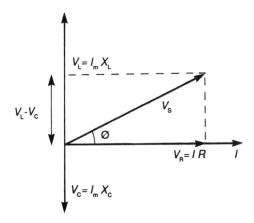

Fig. 4.6 Addition of phasors.

Since $V_R = IR$, $V_L = LX_L$ and $V_C = IX_C$ then the magnitude can be written as

$$V_s = I \, \sqrt{[R^2 + (X_L - X_C)^2]} \tag{4.13}$$

and the phase angle as

$$\phi = \tan^{-1}\frac{X_L - X_C}{R} \tag{4.14}$$

In the phasor diagram in Fig. 4.6, the magnitude of the sinusoidal current and voltages was represented by their maximum or peak values. It is more usual, however, to use the root mean square values of a sinusoid, these being just the maximum values divided by $\sqrt{2}$ (see Section 4.1.2).

Example   A circuit consists of an inductance of 10 mH in series with a 5 Ω resistance. An alternating current of frequency 100 Hz and maximum value 200 mA flows through the circuit. What is the voltage across (a) the inductance, (b) the resistance, (c) the series combination?

(a) The inductive reactance is

$$X_L = \omega L = 2\pi \times 100 \times 10 \times 10^{-3} = 6.28 \, \Omega$$

The voltage across the inductance will thus have a maximum value of $0.200 \times 6.28 = 1.26$ V. Its phasor will lead the current by 90°. Thus in polar notation the voltage phasor is $1.26/\underline{90°}$ V.

(b) The voltage across the resistance will have a maximum value of $0.200 \times 5 = 1.00$ V. Its phasor will be in phase with the current. Thus in polar notation the voltage phasor is $1.00/\underline{0°}$ V.

(c) The voltage across the combination will be represented by a phasor which is the sum of the voltage phasors for the inductor and the resistor. Thus, since the two phasors are at 90°, the Pythagoras theorem gives for the magnitude

$$\sqrt{(1.26^2 + 1.00^2)} = 1.61 \text{ V}$$

and for its phase

$$\tan^{-1}\frac{1.26}{1.00} = 51.6°$$

In polar notation the voltage is $1.61/\underline{51.6°}$ V.

4.2.2   *Impedance*

The term *impedance Z* is used to describe the relationship between the voltage phasor **V** and the current phasor **I** for a circuit or component.

$$Z = \frac{\mathbf{V}}{\mathbf{I}} \tag{4.15}$$

The units of impedance are ohms. Thus for just resistance then, in polar notation, $Z = R\underline{/0°}$. For just inductance $Z = X_L\underline{/90°}$ and for just capacitance $Z = X_C\underline{/90°}$.

For the series circuit in Fig. 4.3, the current phasor is $I\underline{/0°}$ and the voltage phasor $\mathbf{V_s}$ has a magnitude given by Equation (4.13) and a phase angle by Equation (4.14). Thus

$$Z = \sqrt{[R^2 + (X_L - X_C)^2]}\underline{/\tan^{-1}(X_L - X_C)/R} \tag{4.16}$$

The term *admittance* Y is used for the reciprocal of the impedance and has the unit of siemen (S).

Example   A circuit consists of a $20\,\Omega$ resistance in series with an inductance of $100\,\text{mH}$. An alternating supply of frequency $50\,\text{Hz}$ and maximum voltage $20\,\text{V}$ is connected across it. What is (a) the circuit impedance and (b) the current?

(a) The circuit impedance has a magnitude of

$$\sqrt{[R^2 + X_L^2]} = \sqrt{[20^2 + (2\pi \times 50 \times 0.100)^2]} = 37.2\,\Omega$$

It has a phase angle of

$$\phi = \tan^{-1}\frac{X_L}{R} = \tan^{-1}\frac{2\pi \times 50 \times 0.100}{20} = 57.5°$$

Thus in polar notation $Z = 37.2\underline{/57.5°}\,\Omega$. The phase angle of the impedance is the angle by which the voltage leads the current.

(b) The current in the circuit can be determined by the use of Equation (4.15).

$$\mathbf{I} = \frac{\mathbf{V}}{Z} = \frac{20\underline{/0°}}{37.2\underline{/57.5°}} = 0.538\underline{/-57.5°}\,\text{A}$$

The current has thus a magnitude of $0.538\,\text{A}$ and lags the voltage by $57.5°$.

4.3   j notation

Consider two phasors which have the same magnitude but just differ in phase by $90°$, i.e. $\mathbf{V_1} = V\underline{/\phi}$ and $\mathbf{V_2} = V\underline{/\phi + 90°}$. $\mathbf{V_2}$ is just $\mathbf{V_1}$ rotated through $90°$. We can represent this relationship be writing

$$\mathbf{V_2} = j\mathbf{V_1}$$

where the j indicates the rotation through $90°$. The operator j, when multiplying a phasor, shifts it by $90°$ in an anticlockwise direction.

Consider Fig. 4.7. Let $\mathbf{A}$ be a phasor along the horizontal axis. Multiplying $\mathbf{A}$ by j gives the phasor $j\mathbf{A}$ along the vertical axis. Multiplying $j\mathbf{A}$ by j gives a phasor $j^2\mathbf{A}$ along the horizontal axis in the opposite direction to $\mathbf{A}$, i.e. a rotation through $180°$. Thus

$$j^2 A = -A$$

$$j^2 = -1$$

Multiplying j^2A by j gives a phasor along the vertical axis in the opposite direction to jA. Therefore

$$j^3 A = -jA$$

$$j^3 = -j$$

j is the square root of -1 and is thus termed an imaginary or complex number. Thus with this notation the vertical axis in Fig. 4.7 is usually termed the imaginary axis and the horizontal axis the real axis.

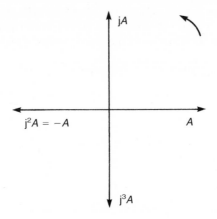

Fig. 4.7 j notation.

4.3.1   Manipulation of complex numbers

The addition of two complex numbers $(a + jb)$ and $(c + jd)$ is performed by adding the real and imaginary parts separately.

$$(a + jb) + (c + jd) = (a + c) + j(b + d) \tag{4.17}$$

Subtraction is by subtracting the real and imaginary parts separately.

$$(a + jb) - (c + jd) = (a - c) + j(b - d) \tag{4.18}$$

To multiply two complex numbers $(a + jb)$ and $(c + jd)$, each term of the first is multiplied by each term of the second.

$$(a + jb)(c + jd) = ac + jad + jbc + j^2bd$$

$$= (ac - bd) + j(ad + bc) \tag{4.19}$$

Division of two complex numbers is performed by multiplying both the numerator and denominator by the conjugate of the denominator. The conjugate of a complex number is obtained by changing the sign of the imaginary term. Thus the conjugate of $(c + jd)$ is $(c - jd)$.

$$\frac{(a + jb)}{(c + jd)} = \frac{(a + jb)(c - jd)}{(c + jd)(c - jd)} = \frac{(ac + bd) + j(bc - ad)}{c^2 + d^2}$$

$$= \frac{(ac + bc)}{(c^2 + d^2)} + j\frac{(bc - ad)}{(c^2 + d^2)} \tag{4.20}$$

A complex number may be converted into polar form. Thus the complex number $(a + jb)$, see Fig. 4.8, represents a phasor of magnitude $\sqrt{(a^2 + b^2)}$ at a phase angle of $\tan^{-1} b/a$.

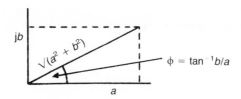

Fig. 4.8 Complex number $(a + jb)$.

$$\mathbf{V} = V\underline{/\phi} = \sqrt{(a^2 + b^2)}\underline{/\tan^{-1}(b/a)} \tag{4.21}$$

A phasor polar form can be converted into one in complex form, since for Fig. 4.8 we have $a/V = \cos \phi$ and $b/V = \sin \phi$ if V is the magnitude, i.e. length, of the phasor. Thus

$$\mathbf{V} = a + jb = V(\cos \phi + j \sin \phi) \tag{4.22}$$

The addition and subtraction of phasors written in polar form has to be by vectorial means and thus it is generally easier to convert them into complex form for addition and subtraction. Phasors in polar form can however be easily multiplied or divided.

$$A\underline{/\theta} \times B\underline{/\phi} = AB\underline{/\theta + \phi} \tag{4.23}$$

This can be shown by putting the phasors in the complex form given in Equation (4.22) and then multiplying them, i.e.

$$A(\cos \theta + j \sin \phi)B(\cos \phi + j \sin \phi)$$

$$= AB[(\cos \theta \cos \phi - \sin \theta \sin \phi) + j(\sin \theta \cos \phi + \cos \theta \sin \phi)]$$

$$= AB[\cos(\theta + \phi) + j \sin(\theta + \phi)]$$

Division of phasors in polar form, can similarly be shown to be

$$\frac{A\underline{/\theta}}{B\underline{/\phi}} = \frac{A}{B}\underline{/(\theta - \phi)} \tag{4.24}$$

4.3.2 Application to circuits

Consider now the application of this notation to the RLC circuit of Fig. 4.3 and the phasor diagram of Fig. 4.6. Taking the current as the reference datum, then since V_R is in phase with I we have $V = RI$, since V_L leads I by 90° we have $V_L = jX_LI$, and since V_C lags I by 90° we have $V_C = -jX_CI$. Thus

$$V_s = RI + jX_LI - jX_CI$$

$$V = I[R + j(X_L - X_C)] \tag{4.25}$$

The term in the square brackets is the impedance Z, i.e. $V = IZ$.

$$Z = R + j(X_L - X_C) \tag{4.26}$$

If we take the impedances of the resistor to be R, of the inductor to be jX_L and of the capacitor to be $-jX_C$, then Equation (4.26) is merely stating that the total impedance of the components in series is the sum of the impedances of each component.

$$Z = Z_1 + Z_2 + Z_3 \tag{4.27}$$

Consider the parallel RLC circuit shown in Fig. 4.9. Using the supply voltage as the reference and remembering that the current through a resistor is in phase with the voltage, the current through an inductor lags the voltage by 90°, and the current through a capacitor leads the voltage by 90°, then since the total current is given by the sum of the phasors of the branch currents, i.e.

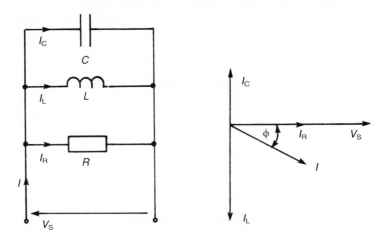

Fig. 4.9 Parallel RLC circuit.

$$I = I_R + I_L + I_C$$

$$= \frac{V_s}{R} - j\frac{V_s}{X_L} + j\frac{V_s}{X_C}$$

$$= V_s\left[\frac{1}{R} + j\left(\frac{1}{X_C} - \frac{1}{X_L}\right)\right] \tag{4.28}$$

the total impedance Z is given by

$$\frac{1}{Z} = \frac{1}{R} + j\left(\frac{1}{X_C} - \frac{1}{X_L}\right)$$

(4.29)

or alternatively

$$\frac{1}{Z} = \frac{1}{R} + \frac{1}{-jX_C} + \frac{1}{jX_L}$$

The reciprocal of the total impedance is equal to the sum of the reciprocals of the individual parallel impedances.

$$\frac{1}{Z} = \frac{1}{Z_1} + \frac{1}{Z_2} + \frac{1}{Z_3}$$

(4.30)

The reciprocal of impedance is admittance Y, unit siemen (S). Thus for impedances in parallel we can write

$$Y = Y_1 + Y_2 + Y_3$$

(4.31)

Example   For the circuit shown in Fig. 4.10, determine the magnitude and phase of the current drawn from the supply. The supply voltage is 20 V at 1000 Hz.

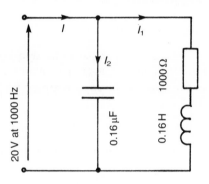

Fig. 4.10 Example.

$$X_C = \frac{1}{2\pi f C} = \frac{1}{2\pi \times 1000 \times 0.16 \times 10^{-6}} = 995\,\Omega$$

Hence the impedance for the capacitor arm is $-j995\,\Omega$. Thus, since $\mathbf{I} = \mathbf{V}/Z$,

$$\mathbf{I}_2 = \frac{20}{-j995} = j0.020\,\text{A}$$

The inductor arm has a resistance in series with an inductance. For the inductance

$$X_L = 2\pi f L = 2\pi \times 1000 \times 0.16 = 1005\,\Omega$$

The impedance of that arm is thus

$$Z = 1000 + j1005\,\Omega$$

Thus, since $\mathbf{I} = \mathbf{V}/Z$,

$$\mathbf{I_1} = \frac{20}{1000 + j1005} = \frac{20(1000 - j1005)}{(1000 + j1005)(1000 - j1005)}$$

$$= \frac{20(1000 - j1005)}{1000^2 + 1005^2} = 0.010 - j0.010 \text{ A}$$

Thus, by Kirchoff's current law,

$$\mathbf{I} = \mathbf{I_1} + \mathbf{I_2}$$

$$= 0.010 - j0.010 + j0.020 = 0.010 + j0.010 \text{ A}$$

Thus, in polar notation,

$$\mathbf{I} = \sqrt{(0.010^2 + 0.010^2)}/\tan^{-1} 0.010/0.010 \text{ A}$$

$$= 0.014\underline{/45°} \text{ A}$$

An alternative way of obtaining this answer would have been to determine the overall impedance Z of the circuit by using Equation (4.30) for impedances in parallel and hence obtain the current by $\mathbf{I} = \mathbf{V_s}/Z$.

4.3.3 Circuit theorems

Kirchoff's current law has been stated in Section 2.1 for d.c. circuits as: the algebraic sum of the currents at a node in a circuit is zero. This must also apply in a.c. circuits to the currents at any instant of time. But the instantaneous currents are of the form

$$i = I_m \sin(\omega t + \phi)$$

But we can represent such currents by a phasor $\mathbf{I}$. Thus

$$\Sigma i = \Sigma[I_m \sin(\omega t + \phi)] = \Sigma\mathbf{I} = 0$$

Thus the law can be stated as: the sum of the phasor currents at a node in a circuit is zero. Likewise, Kirchoff's voltage law can be stated as: the sum of the voltage phasor drops and voltage phasor sources around any closed circuit is zero. The techniques of node analysis and mesh analysis thus also apply with a.c. circuits.

Thévenin's theorem, Norton's theorem and the superposition theorem all apply to a.c. networks (see Chapter 2 for their use with d.c. networks). Thévenin's theorem becomes: a network consisting of voltage sources and impedances between a pair of terminals may be replaced by an equivalent network consisting of an ideal independent voltage source in series with an internal impedance, the voltage phasor of the source being the open circuit voltage between the terminals and the internal impedance being that measured looking into the terminals when all voltage sources are replaced by a short circuit and all current sources by an open circuit. Norton's theorem becomes: a network consisting of voltage sources and impedances between a pair of terminals may be replaced by an equivalent network consisting of an ideal independent current source in parallel with an internal impedance, the current phasor of the source being the current that would flow if the terminals were short circuited and the internal impedance that which is measured looking into the terminals when all voltage sources are replaced by a short circuit and all current sources by an open circuit.

Example   Use (a) node analysis, (b) mesh analysis, to determine the current through the capacitor in Fig. 4.11.

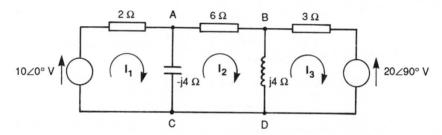

Fig. 4.11 Example.

(a) Nodes C and D are at the same potential and thus may be considered as a single node, in this case the reference node. Applying Kirchoff's current law to node A gives

$$\frac{(10 - \mathbf{V_A})}{2} - \frac{(\mathbf{V_A} - \mathbf{V_B})}{6} - \frac{\mathbf{V_A}}{(-j4)} = 0$$

On rearrangement this becomes

$$60 - \mathbf{V_A}(8 + j3) + 2\mathbf{V_B} = 0 \tag{4.32}$$

Applying Kirchoff's current law to node B gives

$$\frac{(\mathbf{V_A} - \mathbf{V_B})}{6} + \frac{(j20 - \mathbf{V_B})}{3} - \frac{\mathbf{V_B}}{j4} = 0$$

On rearrangement this becomes

$$2\mathbf{V_A} - \mathbf{V_B}(6 - j3) + j80 = 0 \tag{4.33}$$

Thus Equation (4.33) gives for $\mathbf{V_A}$

$$\mathbf{V_A} = 0.5\mathbf{V_B}(6 - j3) + j40$$

Hence, substituting this in Equation (4.32) gives

$$60 - [0.5\mathbf{V_B}(6 - j3) + j40](8 + j3) + 2\mathbf{V_B} = 0$$

$$60 - j40(8 + j3) - \mathbf{V_B}[0.5(6 - j3)(8 + j3) - 2] = 0$$

$$V_B = \frac{180 - j320}{26.5 - j3} = \frac{(180 - j320)(26.5 + j3)}{(26.5 - j3)(26.5 + j3)} = \frac{5730 - j7940}{711.25}$$

$$= 8.06 - j11.16 \, V$$

Substitution of this value in Equation (4.33) gives

$$V_A = 0.5(8.06 - j11.16)(6 - j3) + j40 = 7.44 - j5.57 \, V$$

The current through the capacitor is

$$I_C = \frac{V_A}{-j4} = \frac{7.44 - j5.57}{-j4} = j1.86 + 1.39 \, A$$

In polar form this is $2.32\underline{/53.2°}\, A$.

(b) Applying mesh analysis to Fig. 4.11, Kirchoff's voltage law gives for the mesh with the current I_1

$$10 - 2I_1 - (-j4)(I_1 - I_2) = 0$$

$$10 - I_1(2 - j4) - j4I_2 = 0 \tag{4.34}$$

For the mesh with current I_2

$$-6I_2 - j4(I_2 - I_3) - (-j4)(I_2 - I_1) = 0$$

$$-j4I_1 - 6I_2 + j4I_3 = 0 \tag{4.35}$$

For the mesh with current I_3

$$-3I_3 - j20 - j4(I_3 - I_2) = 0$$

$$-j20 + j4I_2 - I_3(3 + j4) = 0 \tag{4.36}$$

Solving the three simultaneous Equations (4.34), (4.35) and (4.36) leads to the currents being obtained and hence the current through the capacitor $(I_1 - I_2) = 1.39 + j1.86 \, A$, or in polar form $2.32\underline{/53.2°}\, A$.

Example   Determine the current through the $50\,\Omega$ resistor in Fig. 4.12 by the use of Thévenin's theorem.

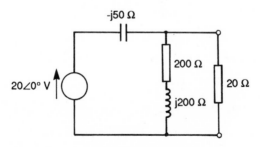

Fig. 4.12 Example.

Consider the terminals of the Thévenin circuit to be as shown in the figure. The open circuit voltage is the voltage drop across $(200 + j200)\,\Omega$ in a circuit of total impedance $(200 + j150)\,\Omega$. Thus

$$V_{Th} = \frac{20(200 + j200)}{200 + j150} = \frac{20(200 + j200)(200 - j150)}{(200 + j150)(200 - j150)}$$

$$= 22.4 + j3.2\,V$$

The Thévenin impedance is obtained by replacing the voltage source by a short circuit. Thus this is $(200 + j200)\,\Omega$ in parallel with $-j50\,\Omega$.

$$\frac{1}{Z_{Th}} = \frac{1}{-j50} + \frac{1}{200 + j200} = \frac{200 + j150}{-j50(200 + j200)}$$

$$Z_{Th} = \frac{200 - j200}{4 + j3} = \frac{(200 - j200)(4 - j3)}{(4 + j3)(4 - j3)}$$

$$= 8 - j56\,\Omega$$

Thus the Thévenin equivalent circuit with the load of $20\,\Omega$ is a source of $(22.4 + j3.2)\,V$ passing a current through a total circuit impedance of $(38 - j56)\,\Omega$. Hence the current is

$$I = \frac{22.4 + j3.2}{28 - j56} = \frac{(22.4 + j3.2)(28 + j56)}{(28 - j56)(28 + j56)}$$

$$= 0.11 + j0.34\,A$$

In polar form this is $0.36\underline{/72°}\,A$.

4.4 Power in a.c. circuits

Let the voltage across and the current flowing through a particular a.c. circuit be

$$v = V_m \sin(\omega t + \alpha) \tag{4.37}$$

$$i = I_m \sin(\omega t + \beta) \tag{4.38}$$

The instantaneous power p is

$$p = vi = V_m I_m[\sin(\omega t + \alpha)\sin(\omega t + \beta)]$$

$$= \tfrac{1}{2}V_m I_m[\cos(\alpha - \beta) - \cos(2\omega t + \alpha + \beta)]$$

The average power in the circuit, since the average over one cycle of the cosine term involving ωt is 0, is given by

$$P_{av} = \tfrac{1}{2}V_m I_m \cos(\alpha - \beta)$$

Let $\phi = (\alpha - \beta) = $ the phase angle between the current and the ·· Then (4.39)

$$P_{av} = \tfrac{1}{2}V_m I_m \cos\phi$$

since the rms value for the voltage is $V_m/\sqrt{2}$ and that for the current $I_m/\sqrt{2}$, then

$$P_{av} = V_{rms}I_{rms} \cos \phi \qquad (4.40)$$

The term $\cos \phi$ is known as the *power factor*. Thus the average power in an a.c. circuit is equal to the rms voltage multiplied by the rms current multiplied by the power factor. For a capacitor with just capacitance or an inductor with just inductance since the phase difference is 90° then $\cos \phi = 0$ and so the average power is 0. The power factor is said to be lagging when the current lags the supply voltage and leading when the current leads the supply voltage.

The higher the power factor of a load the greater the power delivered to it for a given voltage and current. A low power factor means that if a particular power is to be delivered, a higher value of IV is required than if the power factor were higher. There are thus obvious advantages in having a high power factor. Most domestic and industrial loads tend to be inductive and so have a lagging power factor. Hence one method of improving the power factor is to connect a capacitor in parallel with the inductive load.

Example   What is the average power consumed by a circuit if there is a voltage input of 100 V rms and a current of 4 A rms is produced which lags the voltage by 30°?

Using Equation (4.40),

$$P_{av} = V_{rms}I_{rms} \cos \phi = 100 \times 4 \times \cos 30° = 346 \text{ W}$$

4.4.1   Calculation of power using j notation

The phasors of the voltage and current described by Equations (4.39) and (4.40) may be drawn as shown in Fig. 4.13. In complex form the voltage and current are

$$\mathbf{V} = V \cos \alpha + jV \sin \alpha = a + jb$$

$$\mathbf{I} = I \cos \beta + jI \sin \beta = c + jd$$

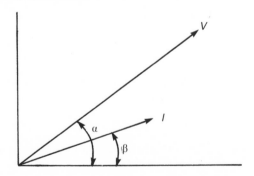

Fig. 4.13 Voltage and current phasors.

with $V \cos \alpha = a$, $V \sin \alpha = b$, $I \cos \beta = c$ and $I \sin \beta = d$. Since the average power (see Section 4.4 and Equation (4.40)), when V and I are rms values, is given by

$$P_{av} = VI \cos (\alpha - \beta) = VI[\cos \alpha \cos \beta + \sin \alpha \sin \beta]$$
$$= ac + bd$$

However, if we multiply the voltage $(a + jb)$ by the conjugate of the current $(c - jd)$

$$(a + jb)(c - jd) = (ac + bd) + j(bc - ad)$$

Thus the average power is the real part of the product of the voltage and the conjugate of the current, denoted by I^*.

$$P_{av} = \text{Re}(VI^*) \qquad (4.41)$$

The same result could have been obtained by taking the real part of the conjugate of voltage multiplied by the current, i.e.

$$P_{av} = \text{Re}(V^*I) \qquad (4.42)$$

4.4.2 *Power triangle*

When dealing with electrical transformers and generators it is found that their rating is given in kVA (kilovolt amperes), which is a measure of their current carrying capacity at a given voltage. The power that can be drawn from these devices depends on their kVA rating and the power factor of the load connected to them.

Equation (4.40) gives the average power as $IV \cos \phi$. Thus if we draw a right angled triangle with IV as the hypotenuse and ϕ as the angle between it and the horizontal, as in Fig. 4.14, then the power is the horizontal line. The hypotenuse IV is said to be the *apparent power* in units of VA. The horizontal is the *real power* in W. The vertical of the triangle is $IV \sin \phi$ and is called the *reactive power*. It has the unit VAr.

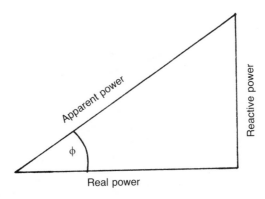

Fig. 4.14 Power triangle.

$$\text{Apparent power} = IV \tag{4.43}$$

$$\text{Real power} = IV \cos \phi \tag{4.44}$$

$$\text{Reactive power} = IV \sin \phi \tag{4.45}$$

Example A coil with a resistance of $5\,\Omega$ and inductance $20\,\text{mH}$ is connected across a $50\,\text{V}$, $50\,\text{Hz}$ supply. What is (a) the current and its phase angle with the applied voltage, (b) the power factor, (c) the apparent power, (d) the real power, (e) the reactive power?

(a) The impedance of the coil is

$$Z = R + j\omega L = 5 + j2\pi \times 50 \times 0.020 = 5 + j6.28\,\Omega$$

In polar notation this is $8.03\underline{/51.5°}\,\Omega$. The current is thus

$$I = \frac{50}{8.03\underline{/51.5°}} = 6.23\underline{/-51.5°}\,\text{A}$$

(b) The power factor $= \cos \phi = \cos(-51.5°) = -0.623$.

(c) The apparent power $= IV = 6.23 \times 50 = 311.5\,\text{VA}$

(d) The real power $=$ apparent power $\times \cos \phi = 311.5 \times 0.623 = 194.1\,\text{W}$.

(e) The reactive power $=$ apparent power $\times \sin \phi = 311.5 \times \sin 51.5° = 243.8\,\text{VAr}$

Example For the previous example, calculate the capacitance needed in parallel with the coil to raise the power factor to 0.9.

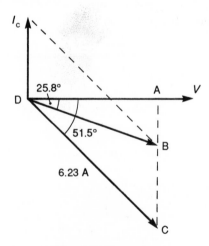

Fig. 4.15 Example.

The current taken by the capacitor must be such that when combined with the current taken by the coil, the resultant current lags the voltage by ϕ, where $\cos \phi = 0.9$ and so $\phi = 25.8°$. Figure 4.15 shows the phasor diagram for the currents.

Initially ϕ is 51.5°. Thus $AC/6.23 = \sin 51.5°$ and so $AC = 4.88$ A. This is sometimes referred to as the reactive component of the current. When ϕ is 25.8° then $AB/AD = \tan 25.8°$. But $AD = 6.23 \cos 51.5°$ and so $AB = 1.87$ A. Thus I_C which is equal to BC is $AC - AB = 4.88 - 1.87 = 3.01$ A. For the capacitor we have $V = I_C X_C$ and so $X_C = 50/3.01 = 16.6\,\Omega$. Hence, since $X_C = 1/2\pi fC$, then $C = 192\,\mu$F.

4.5 Frequency response of circuits

When analysing the operation of equipment which is designed for use over a range of frequencies, it is necessary to determine the performance at frequencies within that range. The *frequency response* of a circuit is often specified by plotting (a) the ratio of the magnitude of the output voltage, current or power to the magnitude of the input voltage, current or power against frequency, and (b) the phase of the ratio against frequency. Such a pair of graphs is known as a *Bode plot*.

The frequency range for many circuits is very large. For instance the response of an audio amplifier ranges from 30 Hz to 15 000 Hz. On a linear frequency scale the bass range of 30 to 200 Hz will be confined to about 1 per cent of the frequency spectrum. To overcome this cramping of information, frequency response curves are normally plotted on logarithmic graph paper so that the space allocated to each decade is the same, a decade being a factor of 10 change in frequency.

Normally the ratio of the magnitudes of the output voltage, current or power to the input voltage, current or power is also plotted in logarithmic form. The unit used for such a ratio is the bel.

$$\text{power ratio in bels} = \lg \frac{|P_{out}|}{|P_{in}|}$$

The vertical lines either side of the quantities are to indicate that we are only concerned with the magnitudes of the voltages. Generally the unit used is the decibel (dB), with 10 dB = 1 bel. Thus

$$\text{power ratio in dB} = 10 \lg \frac{|P_{out}|}{|P_{in}|} \tag{4.46}$$

Since $P = V^2/R$, then

$$\text{voltage ratio in dB} = 20 \lg \frac{|V_{out}|}{|V_{in}|} \tag{4.47}$$

Likewise, since $P = I^2 R$, then

$$\text{voltage ratio in dB} = 20 \lg \frac{|I_{out}|}{|I_{in}|} \tag{4.48}$$

The term *tuned circuit* is used for a circuit which is designed to transmit signals at some frequencies better than at others, i.e. is frequency selective, and so can be used to filter out signals in unwanted frequency ranges.

4.5.1 Series RLC circuit

Consider the frequency response of the series RLC network shown in Fig. 4.16.

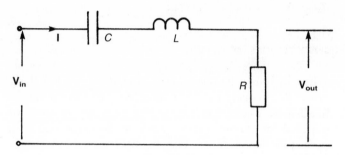

Fig. 4.16 RLC circuit.

$$\mathbf{V_{out}} = \mathbf{I}R$$

$$\mathbf{V_{in}} = \mathbf{I}\left[R + j\left(\omega L - \frac{1}{\omega C}\right)\right]$$

$$\frac{\mathbf{V_{out}}}{\mathbf{V_{in}}} = \frac{R}{R + j(\omega L - 1/\omega C)} = \frac{R[R - j(\omega L - 1/\omega C)]}{R^2 + (\omega - 1/\omega C)^2} \tag{4.49}$$

Thus considering the ratio of the magnitudes

$$\frac{V_{out}}{V_{in}} = \frac{R}{\sqrt{[R^2 + (\omega L - 1/\omega C)^2]}} \tag{4.50}$$

and the phase

$$\phi = \tan^{-1}\frac{\omega L - 1/\omega C}{R} \tag{4.51}$$

The maximum value of V_{out}/V_{in} is 1 and occurs when $\omega L = 1/\omega C$. At this condition $\phi = 0$. This condition is termed the *resonant* condition and the resonant angular frequency ω_0 is thus

$$\omega_0 = \frac{1}{\sqrt{(LC)}} \tag{4.52}$$

The ratio V_{out}/V_{in} is only 1 if the inductor is pure, i.e. does not possess any resistance. This is never the case in practice and thus the ratio at resonance is slightly less than 1. Figure 4.17 shows typical frequency plots for the RLC circuit.

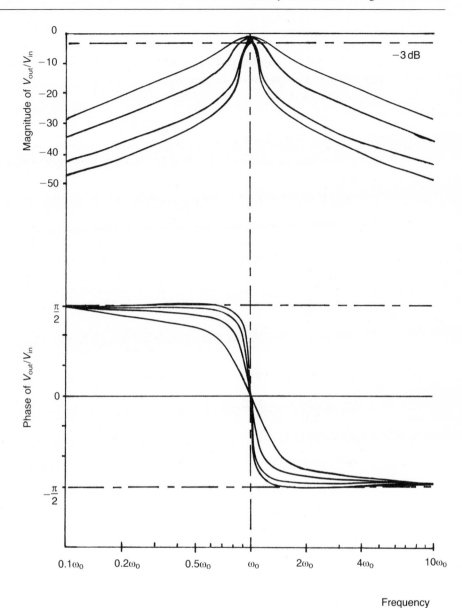

Fig. 4.17 Frequency plots for RLC circuit.

The circuit impedance is

$$Z = \sqrt{[R^2 + (\omega L - 1/\omega C)^2]}$$

Thus at resonance, when $\omega L = 1/\omega C$, the circuit impedance is just R.

4.5.2   Parallel RCL circuit

Consider a circuit consisting of a pure resistance, a pure inductance and a pure capacitance all in parallel with each other (as in Fig. 4.9). For such a circuit, the circuit impedance is given by (see Equation (4.29))

$$\frac{1}{Z} = \frac{1}{R} + j\left(\frac{1}{X_C} - \frac{1}{X_L}\right) = \frac{1}{R} + j\left(\omega C - \frac{1}{\omega L}\right)$$

For a current input to such a circuit of $\mathbf{I_{in}}$ then if $\mathbf{V}$ is the potential difference across the circuit

$$\mathbf{I_{in}} = \frac{\mathbf{V}}{Z}$$

Consider the current through the resistor as the output current $\mathbf{I_{out}}$. Then

$$\mathbf{I_{out}} = \frac{\mathbf{V}}{R}$$

Thus

$$\frac{\mathbf{I_{out}}}{\mathbf{I_{in}}} = \frac{1/R}{(1/R) + j(\omega C - 1/\omega L)} = \frac{1 - jR(\omega C - 1/\omega L)}{1 + R^2(\omega C - 1/\omega L)^2} \tag{4.53}$$

Thus, considering the ratio of the magnitudes,

$$\frac{I_{out}}{I_{in}} = \frac{1}{\sqrt{[1 + R^2(\omega C - 1/\omega L)^2]}} \tag{4.54}$$

and the phase

$$\phi = \tan^{-1} R(\omega C - 1/\omega L) \tag{4.55}$$

When $\omega C = 1/\omega L$ then $\phi = 0$ and $I_{out}/I_{in} = 1$. This is the resonance condition for the parallel circuit. The resonant angular frequency ω_0 is thus

$$\omega_0 = \frac{1}{\sqrt{(LC)}} \tag{4.56}$$

The circuit impedance at resonance is just R.

4.5.3   Bandwidth

In considering the frequency response of, say, an amplifier it is useful to know the range of frequencies over which the response can be considered flat, i.e. an essentially constant ratio V_{out}/V_{in}. A measure of this is given by the *bandwidth*. The bandwidth is defined as the range of frequencies between those at which the voltage ratio drops by 3 dB. The 3 dB point is chosen arbitrarily, because it represents a halving in the power. Thus if V_2 gives half the power of V_1 then

$$\frac{V_2^2/R}{V_1^2/R} = \frac{1}{2}$$

$$V_2 = (1/\sqrt{2})V_1$$

Thus the voltage ratio V_2/V_1 in dB is

$$20 \lg(1/\sqrt{2}) = -3\,\text{dB}$$

4.5.4 Q factor

The Q factor or quality factor is defined in general as being, for the resonance condition,

$$Q = 2\pi \times \frac{\text{maximum energy stored in a component}}{\text{energy dissipated per cycle}} \tag{4.57}$$

For a series RLC circuit, the average power dissipated at resonance is I^2R, since the circuit impedance is just the resistance. Thus the power dissipated during one cycle is I^2RT, where T is the periodic time. But $T = 1/f_0$, where f_0 is the resonant frequency. Thus the power dissipated during one cycle $I^2R/f_0 = (I_m/\sqrt{2})^2R/f_0$. For the capacitor in the circuit the maximum energy stored is $\frac{1}{2}CV_m^2$, where V_m is the maximum voltage across it. But $V_m = I_mX_C$ and so

$$Q = \frac{2\pi\frac{1}{2}C(I_mX_C)^2}{I_m^2R/2f_0} = \frac{X_C}{R} \tag{4.58}$$

The current through the capacitor is the same as the current through the resistance. But $V_C = IX_C$ and $V_R = IR$, and since at resonance the total circuit impedance is just R then the total voltage drop V across the circuit is equal to V_R then

$$Q = \frac{V_C}{V} \tag{4.59}$$

Hence Q is the voltage magnification produced by the component.

A similar equation can be developed for the inductor,

$$Q = \frac{X_L}{R} = \frac{V_L}{V} \tag{4.60}$$

The Q factor is thus the voltage magnification produced by a component. At resonance, since $\omega_0 L = 1/\omega_0 C$, then the Q factor for the capacitor is the same as that for the inductor.

At resonance the power dissipated in the series RLC circuit is I_0^2R, where I_0 is the current at resonance and equal to V/R. At the cut-off frequencies of the bandwidth the power has dropped to a half. Thus the power at these frequencies is $I_0^2R/2$. The current I at these frequencies is thus $I_0/\sqrt{2} = V/R\sqrt{2}$. But for the series RLC circuit, the current I at a frequency ω is given by

$$I = \frac{V}{Z} = \frac{V}{\sqrt{[R^2 + (\omega L - 1/\omega C)^2]}}$$

Thus

$$R\sqrt{2} = \sqrt{[R^2 + (\omega L - 1/\omega C)]}$$

$$R = \pm(\omega L - 1/\omega C)$$

Since $Q = 1/\omega_0 CR$ (Equation (4.58)) and $Q = \omega_0 L/R$ (Equation (4.60)), then

$$1 = \pm\left[\frac{\omega Q}{\omega_0} - \frac{\omega_0 Q}{\omega}\right]$$

$$\omega^2 \pm \frac{\omega_0 \omega}{Q} - \omega_0^2 = 0$$

This quadratic equation in ω gives the roots

$$\omega = \frac{\pm(\omega_0/Q) \pm \sqrt{[(\omega_0/Q)^2 + 4\omega_0^2]}}{2} = \pm\frac{\omega_0}{2Q} \pm \frac{\omega_0}{2Q}\sqrt{(1 + 4Q^2)}$$

Since generally $4Q^2$ is much greater than 1 and only positive values for ω have significance, this gives for the cut-off frequencies to the bandwidth the frequencies $\omega_0 \pm \omega_0/2Q$. Thus the bandwidth, which is the difference between these two frequencies, is given by

$$\text{bandwidth} = \frac{\omega_0}{Q} \qquad\qquad (4.61)$$

Thus the Q factor is the bandwidth divided by the resonant angular frequency and is thus a measure of the sharpness of the frequency response.

A similar analysis can be carried out for a parallel RLC circuit (Fig. 4.9). The definition of the Q factor in Equation (4.57) leads to

$$Q = \omega_0 CR = \frac{R}{\omega_0 L} \qquad\qquad (4.62)$$

and

$$Q = \frac{I_C}{I} = \frac{I_L}{I} \qquad\qquad (4.63)$$

and hence is the current magnification produced by a component. The bandwidth is the same as with Equation (4.61) for the series RLC circuit.

Example   A series RLC circuit has $R = 12\,\Omega$, $L = 100\,\text{mH}$ and $C = 0.2\,\mu\text{F}$. What is (a) the resonant frequency, (b) the Q factor, (c) the bandwidth, (d) the impedance at the resonant frequency?

(a) The resonant frequency is given by

$$f_0 = \frac{1}{2\pi\sqrt{(LC)}} = \frac{1}{2\pi\sqrt{(0.100 \times 0.2 \times 10^{-6})}} = 1125\,\text{Hz}$$

(b) The Q factor is

$$Q = \frac{\omega_0 L}{R} = \frac{2\pi \times 1125 \times 0.100}{12} = 58.9$$

(c) The bandwidth is given by Equation (4.61) as

$$\text{bandwidth} = \frac{f_0}{Q} = \frac{1125}{58.9} = 19.1\,\text{Hz}$$

(d) The impedance at resonance is just the resistance of $12\,\Omega$.

Problems

1. What is the reactance of a $2\,\mu\text{F}$ capacitor at (a) 100 Hz, (b) 1 MHz?

2. What is the reactance of a 100 mH inductor at (a) 100 Hz, (b) 1 MHz?

3. A 0.2 H inductor is connected across an a.c. supply having a maximum voltage of 200 V at a frequency of 50 Hz. What is (a) the inductive reactance and (b) the maximum current in the circuit?

4. A $10\,\mu\text{F}$ capacitor is connected across an alternating voltage of frequency 50 Hz and maximum value 250 V. What is (a) the capacitive reactance and (b) the maximum current in the circuit?

5. Express in complex form and polar form the impedances of each of the following circuits at a frequency of 50 Hz, (a) a resistance of $50\,\Omega$ in series with an inductance of 100 mH, (b) a resistance of $50\,\Omega$ in series with a capacitance of $10\,\mu\text{F}$.

6. A circuit consists of a $20\,\Omega$ resistance in series with an inductance of 20 mH. An alternating voltage of 50 V at a frequency of 100 Hz is applied across the series arrangement. What is (a) the total circuit impedance, (b) the magnitude of the current and its phase with respect to the voltage?

7. A circuit consists of a $50\,\Omega$ resistance in series with a capacitance of $10\,\mu\text{F}$. An alternating voltage of 50 V at a frequency of 100 Hz is applied across the series arrangement. What is (a) the total circuit impedance, (b) the magnitude of the current and its phase with respect to the voltage?

8. A series RLC circuit has a resistance of $20\,\Omega$, an inductance of 50 mH and a capacitance of $10\,\mu\text{F}$. An alternating voltage of 50 V at a frequency of 300 Hz is applied across the series arrangement. What is (a) the total circuit impedance, (b) the magnitude of the current and its phase with respect to the voltage?

9. A circuit consists of a $20\,\Omega$ resistance in parallel with an inductance having an inductive reactance of $10\,\Omega$. What is the impedance of the arrangement in (a) complex notation, (b) polar notation?

10. What is the voltage across a coil of inductance 50 mH and resistance $10\,\Omega$ if a current of 0.50 A and angular frequency 100 rad/s flows through it?

11. A circuit consists of a 100 Ω resistance in parallel with a 5 μF capacitance. What is the current supplied when a voltage of 25 V and angular frequency 1000 rad/s is connected across the arrangement?

12. A parallel RLC circuit has a resistance of 100 Ω, an inductance of 20 mH and a capacitance of 10 μF in parallel with an alternating voltage of 20 V at a frequency of 500 Hz. What will be the current taken by the circuit?

13. A network consists of a 20 Ω resistance in parallel with a capacitor having a reactance of 10 Ω. What is the current taken by the network when a 10 V alternating supply is connected across it?

14. A network consists of a 8 Ω resistor in parallel with a capacitor of reactance 6 Ω, the parallel arrangement then being connected in series with a 5 Ω resistor to an alternating supply of 10 V. What is the current taken from the supply?

15. A circuit consists of a capacitor C in series with a resistance R_1, both then being in parallel with a resistance R_2. What is the impedance of the circuit?

16. By the use of (i) node analysis, (ii) mesh analysis and (iii) by considering impedances in series and parallel, determine the current through the capacitors in Fig. 4.18(a) and (b).

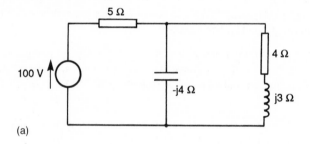

(a)

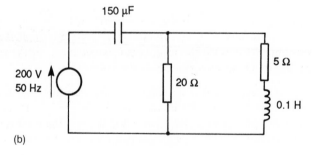

(b)

Fig. 4.18 Problem 16.

17. By the use of (i) node analysis, (ii) mesh analysis, (iii) superposition, (iv) Thévenin's theorem, determine the current through the capacitor in Fig. 4.19.

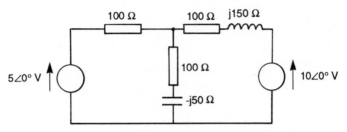

Fig. 4.19 Problem 17.

18. A circuit consists of a source of alternating voltage V with an internal impedance Z_1 connected in parallel with an impedance Z_2 and also a load of impedance Z_L. Use (i) Thévenin's theorem and (ii) Norton's theorem to determine the current through the load.

19. A coil of resistance $12\,\Omega$ and inductance $0.12\,H$ is connected in parallel with a capacitor of $60\,\mu F$ and a variable frequency voltage supply connected across the parallel arrangement. Calculate the frequency at which the circuit becomes purely resistive, and the value of the effective resistance.

20. What is the average power dissipated in a circuit if for a voltage input of $20\,V$ rms there is a current of $1.0\,A$ which leads the voltage by (a) $0°$, (b) $60°$, (c) $90°$?

21. What are the power factors for circuits consisting of (a) just a pure capacitor, (b) just a pure inductance, (c) an inductance in series with a resistance?

22. In order to use three $110\,V$, $60\,W$ lamps on a $230\,V$, $50\,Hz$ supply, they are connected in parallel and a capacitor put in series with the group. What is (a) the capacitance required to give the correct voltage across the lamps, (b) the power factor of the circuit?

23. A factory has a load of $150\,kVA$ at a power factor of 0.8 lagging. Calculate the value of the capacitance which, when connected in parallel with the load, will improve the power factor to 0.9 lagging. The supply voltage is $500\,V$ at $50\,Hz$.

24. An inductive load when connected across a $240\,V$, $50\,Hz$ supply takes a current of $5\,A$ at a lagging power factor of 0.7. What capacitance should be connected in parallel with the load to increase the power factor to 0.9 lagging?

25. What is (a) the resonant frequency, (b) the impedance at resonance for a series RLC circuit consisting of an inductor with inductance $50\,mH$ and resistance $10\,\Omega$ and a capacitor of $10\,\mu F$ capacitance?

26. Derive an equation for the resonant frequency and the impedance at that frequency of a circuit consisting of a coil having inductance and resistance in parallel with a capacitor.

27. What is (a) the resonant frequency, (b) the Q factor and (c) the bandwidth for a series RLC circuit with $R = 20\,\Omega$, $L = 0.10\,H$, and $C = 0.5\,\mu F$?

28.   A series RLC circuit has $R = 5\,\Omega$, $L = 0.5\,\text{H}$, and C a variable capacitor. A 10 V, 50 Hz supply is connected across the arrangement. What is (a) the capacitance needed to give resonance, (b) the current at resonance, and (c) the Q factor?

Chapter 5
Three-phase Circuits

5.1 Three-phase systems

Historically, the first electricity was supplied by d.c. generators and used mainly for lighting in the immediate locality. As the demand increased more powerful generators were required, operating at higher voltages so that power could be distributed over a wider locality. The power loss in passing a current I through a cable is proportional to the square of the current. Since for a given power input IV is a constant, then the power loss is inversely proportional to V^2. Thus the higher the transmission voltage the small the losses. It was found that the commutator, an integral part of any d.c. machine, severely limited the maximum voltage that could be generated, and so attention was turned to a.c. generators or alternators where transformers could be readily used to step up voltages. These alternators initially were single-phase, again mainly for lighting loads. Over the years, however, the development of alternators tended towards three-phase machines because these were more efficient and because three-phase a.c. motors supplied by these alternators performed much better than their single-phase counterparts. All a.c. motors greater than about 4 kW are of the three-phase type. Gradually the three-phase distribution system was built up, and is now accepted world-wide as the optimum a.c. distribution system.

5.1.1 Generation of three-phase voltages

Consider three coils, labelled red, yellow and blue and designated by R_1R_0, Y_1Y_0 and B_1B_0 in Fig. 5.1. The three coils are physically displaced by 120° from each other and rotate anticlockwise in the magnetic field produced between the N and S poles. By Faraday's law, Equation (1.27), the voltage induced in a coil is

$$e = N\frac{d\phi}{dt}$$

Assuming that the flux ϕ produced by the magnet system and linked by a coil is proportional to the sine of the angular displacement (practical alternators approach this condition very closely), then we have

$$\phi = \phi_m \sin \omega t$$

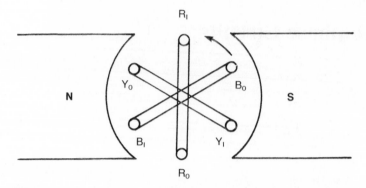

Fig. 5.1 Three-phase generator.

where ϕ_m is the maximum flux linked by a coil and ω is its angular velocity. The voltages induced in the three coils are thus

$$e_R = \omega N\ \phi_m\ \cos \omega t \tag{5.1}$$

$$e_Y = \omega N\ \phi_m\ \cos(\omega t - 120°) \tag{5.2}$$

$$e_B = \omega N\ \phi_m\ \cos(\omega t - 240°) \tag{5.3}$$

and are shown in Fig. 5.2.

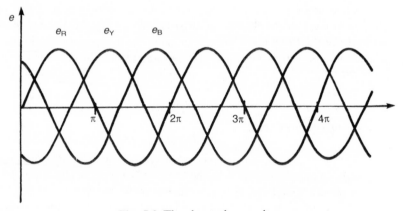

Fig. 5.2 The three-phase voltages.

5.1.2 Star and delta connections

It might appear that to transmit the voltages from the three coils would require six conductors, a costly and physically undesirable condition. It is possible, however, to reduce the number of conductors required to four, or in some cases three, by connecting the coils in a symmetric fashion as shown below.

Basically the three coils shown in Fig. 5.1 can be connected symmetrically in two ways, *star* (or *Y*) connected as shown in Fig. 5.3(a), or *delta* (or *mesh*) connected as shown in Fig. 5.3(b). The star connection may be made by connecting R_0, Y_0 and B_0 together and the delta connection by connecting R_0 to Y_1, Y_0 to B_1, and B_0 to R_1. With the star connection, the point at which R_0, Y_0 and B_0 is connected is called the *neutral point*.

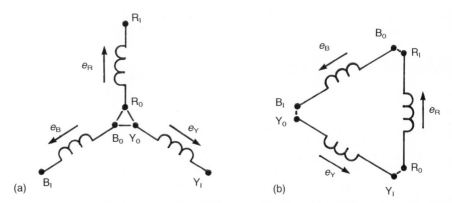

Fig. 5.3 (a) Star connection, (b) delta connection.

5.2   Voltage and current relationships

Figure 5.4 shows a three-phase star connected alternator supplying currents I_R, I_Y and I_B to a balanced resistive–inductive load. The term balanced means that the three elements of the load are identical. Figure 5.5 shows the phasor diagram for the circuit. The phase voltages, i.e. the voltages with respect to the

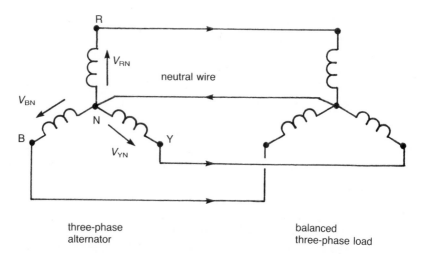

Fig. 5.4 Star-connected alternator and balanced load.

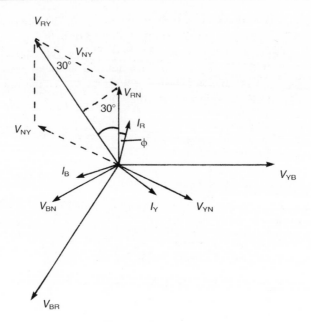

Fig. 5.5 Phasors.

neutral point, are $\mathbf{V_{RN}}$, $\mathbf{V_{YN}}$ and $\mathbf{V_{BN}}$. With a balanced alternator the three-phase voltages are equal in magnitude and displaced by 120°. The voltages between the lines are called the *line voltages* and are the voltages between terminals R and Y, R and B, and B and Y. Denoting the voltage of the red line with respect to the yellow line by $\mathbf{V_{RY}}$, then

$$\mathbf{V_{RY}} = \mathbf{V_{RN}} - \mathbf{V_{YN}}$$

$$\mathbf{V_{RY}} = \mathbf{V_{RN}} + \mathbf{V_{NY}}$$

This phasor addition is shown on the phasor diagram of Fig. 5.5. The line voltage has a magnitude of

$$V_{RY} = V_{RN} \cos 30° + V_{NY} \cos 30°$$

Hence if the phase voltages are equal and since $\cos 30° = \sqrt{3}/2$, the line voltage V_L is related to the phase voltage V_P by

$$V_L = \sqrt{3}V_P \tag{5.4}$$

It can be seen from Fig. 5.4 that, for the star connection, the magnitudes of the currents in the lines I_L are equal to those of the currents in the phases I_P, i.e.

$$I_L = I_P \tag{5.5}$$

The currents supplied to the load along the lines lag the respective phase voltages by some phase angle ϕ. The neutral line current $\mathbf{I_N}$ is, by Kirchoff's current law, equal to the phasor addition of the phase currents, i.e.

$$\mathbf{I_N} = \mathbf{I_R} + \mathbf{I_Y} + \mathbf{I_B} \tag{5.6}$$

With the star form of connection we can have either a three-wire or a four-wire system. The four-wire system includes a wire for the neutral line current. However, when there is a balanced load the neutral wire current is zero and so is often dispensed with.

Now consider the alternator connected in delta mode, as in Fig. 5.6. The voltage of line 1 with respect to line 2 is equal to the red phase voltage, the voltage of line 2 with respect to line 3 is equal to the yellow phase voltage, etc. Thus for a delta connection

$$V_L = V_P \tag{5.7}$$

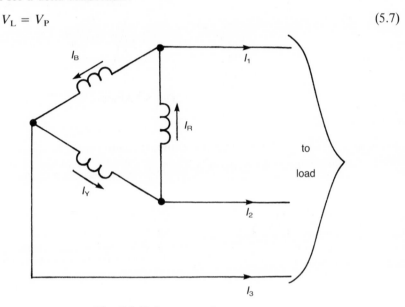

Fig. 5.6 Delta connection.

The line current $\mathbf{I_1}$ is obtained by Kirchoff's current law as

$$\mathbf{I_1} = \mathbf{I_R} - \mathbf{I_B}$$

$$\mathbf{I_1} = \mathbf{I_R} + (-\mathbf{I_B})$$

This phasor addition is shown in the phasor diagram of Fig. 5.7. The magnitude of the line current is thus

$$I_1 = I_R \cos 30° + I_B \cos 30°$$

With a balanced load the phase currents are equal and thus, since $\cos 30° = \sqrt{3}/2$, the line current magnitude I_L is related to the phase current I_p by

$$I_L = \sqrt{3}I_P \tag{5.8}$$

To summarize:

Star connection: $V_L = \sqrt{3}V_P$ $I_L = I_P$
Delta connection: $V_L = V_P$ $I_L = \sqrt{3}I_P$

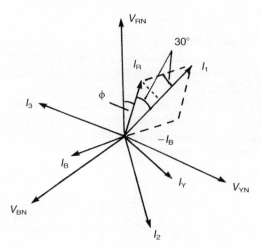

Fig. 5.7 Phasors.

Example What are the line voltages and currents for a three-phase star connected alternator with a phase voltage of 240 V and connected to a balanced star connected load of three 100 Ω resistances?

Figure 5.4 shows the circuit. Using Equation (5.4),

$$V_L = \sqrt{3} \times 240 = 416\,\text{V}$$

Each phase voltage is connected across a 100 Ω resistance. Thus the current is 240/100 = 2.40 A.

5.3   Power in three-phase circuits

If V_P is the rms phase voltage magnitude, I_P the rms phase current and ϕ the phase angle between V_P and I_P, then the power per phase is given by Equation (4.40) as

$$P/\text{phase} = V_P I_P \cos\phi$$

Thus the total power for a three-phase circuit is

$$P = 3V_P I_P \cos\phi \tag{5.9}$$

Thus for a star connection

$$P = 3(V_L/\sqrt{3})I_L \cos\phi = \sqrt{3}V_L I_L \cos\phi$$

and for a delta connection

$$P = 3V_L(I_L/\sqrt{3}) \cos\phi = \sqrt{3}V_L I_L \cos\phi$$

Therefore, in terms of the line values, the power in either a star or delta connected system is

$$P = \sqrt{3}V_\mathrm{L}I_\mathrm{L} \cos \phi \qquad\qquad (5.10)$$

Example A 500 kVA, 6600 V alternator is delta connected. Calculate the magnitude of the current in each phase of the alternator when it delivers full output to a balanced three-phase load at a power factor of 0.8 lagging.

Note: The rating of almost all three-phase electrical plant is given in terms of the line values.

The total power supplied by the alternator is $500 \times 0.8 = 400$ kW. But

$$P = \sqrt{3}V_\mathrm{L}I_\mathrm{L} \cos \phi$$

$$400\,000 = \sqrt{3} \times 6600 \times I_\mathrm{L} \cos \phi$$

Hence $I_\mathrm{L} = 43.7$ A.

For a delta connection

$$I_\mathrm{p} = \frac{I_\mathrm{L}}{\sqrt{3}} = \frac{43.7}{\sqrt{3}} = 25.3 \text{ A}$$

Example A four-wire cable supplies a star connected load consisting of red phase $(5 - \mathrm{j}10)\,\Omega$, yellow phase $(3 - \mathrm{j}6)\,\Omega$, and blue phase $(2 + \mathrm{j}4)\,\Omega$. If the cable voltage is balanced and equal to 415 V, calculate the magnitude and phase of the line and neutral currents, drawing the phasor diagrams, and the total power supplied to the load.

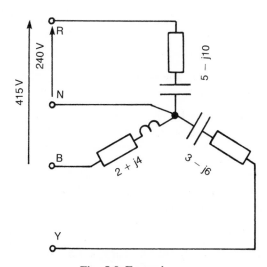

Fig. 5.8 Example.

Figure 5.8 shows the load system. The line-neutral voltage, i.e. the phase voltage V_p, has the magnitude

$$V_p = \frac{415}{\sqrt{3}} = 240\,\text{V}$$

Taking the phases in order and calculating the phase currents (and hence, since it is star connected, line currents) by $\mathbf{I} = \mathbf{V}/\mathbf{Z}$, using the red phase voltage as the reference phasor, then for the red phase

$$\mathbf{I_R} = \frac{240}{5 - \text{j}10} = \frac{240(5 + \text{j}10)}{25 + 100} = 9.6 + \text{j}19.2\,\text{A}$$

In polar notation, this is $21.5\underline{/63.4°}\,\text{A}$. For the yellow phase, the yellow phase voltage with respect to the red phase voltage is

$$\mathbf{V_{YN}} = 240\underline{/-120°} = 240[\cos(-120°) + \text{j}\sin(-120°)]$$

$$= -120 - \text{j}207.8\,\text{V}$$

Thus

$$\mathbf{I_Y} = \frac{-120 - \text{j}207.8}{3 - \text{j}6} = \frac{-(120 + \text{j}207.8)(3 + \text{j}6)}{9 + 36}$$

$$= 19.71 - \text{j}29.85\,\text{A} = 35.8\underline{/-56.6°}\,\text{A}.$$

For the blue phase, the blue phase voltage with respect to the red phase voltage is

$$\mathbf{V_{BN}} = 240\underline{/-240°} = 240[\cos(-240°) + \text{j}\sin(-240°)]$$

$$= -120 + \text{j}207.8\,\text{V}$$

Thus

$$\mathbf{I_B} = \frac{-120 + \text{j}207.8}{2 + \text{j}4} = \frac{(-120 + \text{j}207.8)(2 - \text{j}4)}{4 + 16}$$

$$= 29.56 + \text{j}44.78\,\text{A} = 53.7\underline{/56.6°}\,\text{A}$$

Figure 5.9 shows the phasors for the voltages and currents.

By Kirchoff's current law, the neutral line current is given by

$$\mathbf{I_N} = \mathbf{I_R} + \mathbf{I_Y} + \mathbf{I_B}$$
$$= 9.6 + \text{j}19.2 + 19.71 - \text{j}29.85 + 29.56 + \text{j}44.78$$
$$= 58.87 + \text{j}34.13\,\text{A} = 68.0\underline{/30.1°}\,\text{A}$$

Because the currents are not balanced, the power cannot be calculated from Equation (5.10). The total power for an unbalanced three-phase system must be obtained by summing the powers in the individual phases.

$$\text{Power in red phase} = V_p I_R \cos \phi_R$$

The phase angle between the current and voltage in the red phase is $63.4°$. Therefore

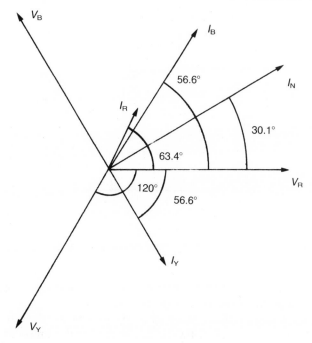

Fig. 5.9 Example.

$$P_R = 240 \times 21.5 \times \cos 63.4° = 2310 \, W$$

For the yellow phase

$$P_Y = V_P I_Y \cos \phi_Y$$

The phase angle between the current and voltage in the yellow phase is $120° - 56.6° = 63.4°$. Thus

$$P_Y = 240 \times 35.8 \times \cos 63.4° = 3847 \, W$$

For the blue phase

$$P_B = V_P I_B \cos \phi_B$$

The phase angle between the current and voltage in the blue phase is $120° - 56.6° = 63.4°$. Thus

$$P_B = 240 \times 53.7 \times \cos 63.4° = 5771 \, W$$

Thus the total power is $P_R + P_Y + P_B = 2310 + 3847 + 5771 = 11\,928 \, W$.

5.3.1 Measurement of three-phase power

Provided there is a neutral point available, the simplest way to measure the power supplied to a three-phase balanced load is to connect a watt meter (see

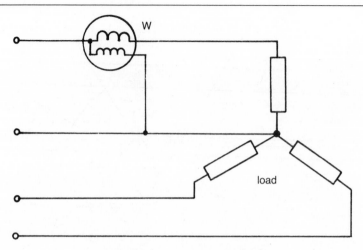

Fig. 5.10 Single watt meter method.

Chapter 19) as shown in Fig. 5.10. The total power supplied to the load is then three times the watt meter reading.

However, for a three-wire system with a balanced or unbalanced load the usual technique is by the use of two watt meters, as shown in Fig. 5.11. In instantaneous form, the total power supplied to a three-phase load, balanced or unbalanced, is

$$p = v_R i_R + v_Y i_Y + v_B i_B$$

For the star connected load shown in Fig. 5.10

$$i_R + i_Y + i_B = 0$$

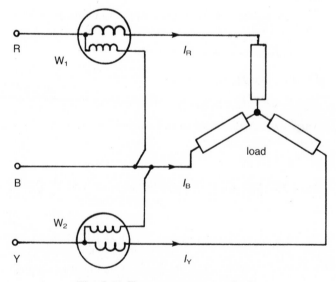

Fig. 5.11 Two watt meter method.

Thus $i_B = -i_R - i_Y$. Hence the total instantaneous power supplied is

$$p = v_R i_R + v_Y i_Y + v_B(- i_R - i_Y)$$
$$= (v_R - v_B)i_R + (v_Y - v_B)i_Y$$

The average value of $(v_R - v_B)i_R$ is the reading on W_1 and the average value of $(v_Y - v_B)i_Y$ is the reading on W_2. Therefore the sum of the two watt meter readings is the total average power supplied to the load.

Problems

1. A delta connected balanced load consists of three identical coils as the load, each coil having a resistance of $15\,\Omega$ and a reactance of $20\,\Omega$. The load is connected across a three-phase supply with a 415 V rms line voltage. What is the line current magnitude and its phase relationship to the line voltage?

2. A balanced star connected load of three identical resistors has a line voltage of 200 V and a line current of 3.5 A. What are the values of the load resistances?

3. A balanced delta connected load of three identical resistors has a line voltage of 200 V and a line current of 3.5 A. What are the values of the load resistances?

4. A three-wire, three-phase supply of 415 V, 50 Hz, is star connected to three identical $20\,\mu F$ capacitors. What is the line current?

5. Three coils, each of 48 mH inductance and $20\,\Omega$ resistance, are connected in star to a 400 V, three-phase, 50 Hz supply. Calculate (a) the line current, (b) the power factor, (c) the total power supplied.

6. Three coils, each of inductance 20 mH and resistance $4\,\Omega$, are connected (a) in star form, (b) in delta form, as the balanced load for a three-phase 415 V, 50 Hz system. What is the total power in each case?

7. A three-phase 240 V supply is connected with ABC phase rotation to a delta connected load consisting of $Z_{AB} = 10\,\Omega$, $Z_{BC} = 8.66 + j5\,\Omega$, and $Z_{CA} = 15\underline{/-30°}\,\Omega$. Calculate the line currents and draw a phasor diagram showing voltages and currents. Take V to be $240\underline{/0°}$ V.

8. Two watt meters are used to measure the power in a three-phase, three-wire system supplying a balanced load and give readings of 12.2 kW and -3.0 kW. What is the total power?

9. A balanced three-phase load consists of a three-phase induction motor operating at a power factor of 0.866 lagging and delivering 59.68 kW at an efficiency of 0.89, together with a three-phase synchronous motor taking an input of 20 kVA at a power factor of 0.3 leading. Find the reading on each of two watt meters connected to measure the total power.

Chapter 6
Semiconductor Devices

6.1 Semiconductors

Materials may be roughly classified by their electrical resistivities. The metals are, in general, good conductors with resistivities typically being of the order of $10^{-8}\,\Omega\,\text{m}$. At the other extreme are insulators, a good insulator having a resistivity of the order of 10^{10} to $10^{12}\,\Omega\,\text{m}$. *Semiconductors* are materials with resistivities which fall between these two extremes. There are a number of materials which may be classified as semiconductors, these including elements and intermetallic compounds. The elements germanium and silicon have however found the widest use.

Each individual atom of germanium, or silicon, comprises a positively charged nucleus and a number of negatively charged orbiting electrons. In the normal state, the atom has no net charge and so the total negative charge of the electrons is equal in magnitude to the positive charge on the nucleus. Both germanium and silicon are in Group IV of the Periodic Table, having four electrons in their outer orbit. These are termed the valence electrons. The atomic structure of a pure germanium or silicon crystal can be represented schematically as shown in Fig. 6.1. The four valence electrons of each atom form covalent bonds with one electron of each of four neighbouring atoms, so forming a crystal structure. The term covalent bond is used where the bonding is as a result of neighbouring atoms sharing electrons. At absolute zero, i.e. 0 K, all the atomic electrons would be bound in this manner into the crystal lattice and there would be no free electrons. At room temperature, however, some electrons have acquired sufficient energy to be able to break free from their bound positions and are then free to move throughout the crystal lattice. A material is only able to conduct electricity if there are some free charge carriers which can be moved by the application of an electric field. Thus, at room temperature, because there are some free electrons the material has some conducting properties. The resistivity of the material is infinite at 0 K and its value decreases with an increase in temperature as more electrons break free from their bound positions in the crystal lattice.

This contrasts with the effect of temperature on the resistivity of metals, the resistivity normally increasing with increasing temperature. Metals, as good conductors, have large numbers of free electrons. When the temperature increases we can consider the situation to be rather like a crowd of people with all of them swaying back and forth with an increasing amplitude as the temperature increases, so rendering it more difficult for movement through the crowd.

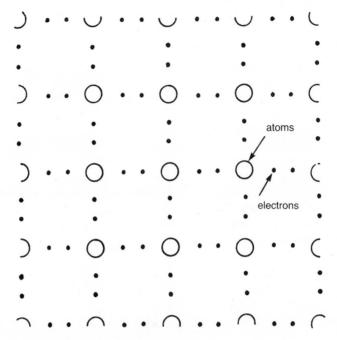

Fig. 6.1 Two dimensional schematic representation of the crystal lattice.

When an electron, charge $-e$, in a semiconductor is freed from its attachment to the nucleus, the nucleus is left with a net positive charge of $+e$. A vacancy now exists in the lattice which may be filled by any other available electron. It is useful to consider the vacancy in the lattice as a *hole* having a positive charge of $+e$, since it is able to *attract* electrons. The charge associated with the hole is the net positive charge of the atom after its loss of an electron. Thus a quantity of energy is able to produce an *electron–hole pair*, i.e. a free negatively charged electron and a hole with effectively a positive charge. The amount of energy required to generate an electron–hole pair is a function of the type of material.

For a given material, the number n of electrons per unit volume that have broken free from being bound to atoms, and so are available for conduction, at a temperature of T K is given by

$$n \approx \text{constant} \times e^{-E_g/kT} \tag{6.1}$$

where k = Boltzmann's constant (1.38×10^{-23} J/K) and E_g is the amount of energy required for an electron to break free from its bound condition, i.e. to go from being a valence electron to a conduction electron, and is about 0.7 eV for germanium and 1.1 eV for silicon. The electron volt (eV) is the energy gained by an electron in moving through a potential difference of 1 V and so, since the charge on an electron is 1.6×10^{-19} C, is 1.6×10^{-19} J. At any temperature, therefore, there are far more free electron–hole pairs in a germanium crystal than in a silicon crystal.

A useful way of representing the above situation is in terms of *energy bands*. In simplistic terms we can consider the valence electrons to have energies which are not high enough for them to escape and become conduction electrons. The valence electrons can thus be considered to lie within a valence band of energies which is separated from a conduction band of energies by E_g. For a valence electron to become a conduction electron it has to be moved into the conduction band, across the energy gap E between the bands of E_g (Fig. 6.2). It is rather like the evaporation of water. For a molecule in the liquid to escape and become a free molecule in the vapour then the latent heat has to be supplied. The latent heat thus represents the energy gap between the bound molecular state and the free state. With a semiconductor there is a small gap between the valence and conduction bands and so at room temperature some electrons have been able to cross the gap. With a metal there is no gap between the valence and conduction bands. With an insulator the gap is too large for electrons to move across the gap at room temperature.

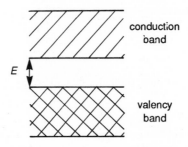

Fig. 6.2 Energy band picture of a semiconductor.

6.1.1 Intrinsic semiconductors

A pure semiconductor has equal numbers of free electrons and holes and is said be in the *intrinsic state*. Conduction in the semiconductor is due to both the movement of the free electrons and the holes. The concept of a movable hole may be better understood by considering a simple analogy. If a patron at a cinema wishes to sit in an empty seat which is in a row of occupied seats, then he or she can either walk along the row past all the full seats (analogous to a moving electron) or ask all the occupants of the seats to move along one seat, so effectively moving the empty seat along the row to him or her (analogous to the moving hole). The hole in the crystal lattice thus can be considered to move when in actual fact the movement is due to the original hole being re-occupied by another electron, so producing a further hole in a different place. In fact, free electrons and holes are continually being produced, and are continually recombining, the number of electrons and holes in the crystal at any instant being a balance being the generation and recombination effects. The actual life time of a free electron or hole before recombination takes place is dependent upon the concentration of the electrons and holes, and would normally be a very small fraction of a second. During this lifetime the electron or hole would be able to travel some distance through the lattice, the mean distance travelled being termed the *mean free path*.

If a potential difference is applied across such a semiconductor crystal, there will be a drift of the free electrons towards the positive terminal and of the holes towards the negative terminal. Under the action of the electric field E, established by the applied potential difference (E = potential gradient), the electrons acquire an average drift velocity v_e which is related to the applied electric field by

$$v_e = -\mu_e E \tag{6.2}$$

The quantity μ_e is known as the *electron mobility*. The minus sign indicates that the direction of the drift velocity is in the opposite direction to that of the electric field. Similarly, the average drift velocity v_h for the holes is given by

$$v_h = \mu_h E \tag{6.3}$$

where μ_h is the *hole mobility*, no minus sign being required because the hole velocity is in the opposite direction to that of the electrons and so in the direction of the electric field. With the drift velocity in m/s, the field strength in V/m, then mobility is in units of $m^2 V^{-1} s^{-1}$.

A velocity of v_e for the electrons means that in a time t all the electrons in a volume $v_e t A$ will have moved through the cross-sectional area A of the material. With n electrons per unit volume each with a charge e, then in a time t the charge movement through the area A is $nev_e t A$. The current is the rate of movement of charge and so is $nev_e A$. Hence the current density J_e produced by the electron drift is

$$J_e = nev_e \tag{6.4}$$

Similarly for the holes, the current density is

$$J_h = pev_h \tag{6.5}$$

where p is the number of free holes per unit volume. As explained earlier, the holes and electrons drift in opposite directions under the action of the electric field. Their contributions to the total current density, however, add directly. The total current density is, therefore,

$$J = J_e + J_h = e(nv_e + pv_h) = eE(n\mu_e + p\mu_h) \tag{6.6}$$

The resistivity ρ of the crystal is given by

$$\rho = \frac{RA}{l} = \frac{VA}{Il} = \frac{E}{J}$$

Hence, using Equation (6.6),

$$\rho = \frac{E}{J} = \frac{1}{e(n\mu_e + p\mu_h)} \tag{6.7}$$

In an intrinsic material, the numbers of electrons and holes are equal, i.e. $n = p = n_i$, where n_i is called the *intrinsic carrier concentration*. Thus we may write

$$\rho = \frac{1}{en_i(\mu_e + \mu_h)} \tag{6.8}$$

Example   What is the resistivity of intrinsic silicon at $300\,\text{K}$ if the intrinsic carrier concentration is $1.4 \times 10^{16}/\text{m}^3$, the electron mobility is $0.15\,\text{m}^2\,\text{V}^{-1}\text{s}^{-1}$ and the hole mobility $0.048\,\text{m}^2\,\text{V}^{-1}\text{s}^{-1}$?

Using Equation (6.8), with $e = 1.6 \times 10^{-19}\,\text{C}$,

$$\rho = \frac{1}{en_i(\mu_e + \mu_h)} = \frac{1}{1.6 \times 10^{-19} \times 1.4 \times 10^{16}(0.15 + 0.048)}$$

$$= 2255\,\Omega\,\text{m}$$

6.2   Doped semiconductors

The production of very pure silicon forms the first part of the process used in the preparation of silicon transistors and other devices. The conduction property of the silicon is then markedly altered by the controlled addition of minute quantities of impurity. The impurity added is either from the group V elements of the periodic Table, such as antimony, phosphorus or arsenic, if a so-called *n type* semiconductor is required, or from the group III elements, perhaps boron or aluminium, if a *p type* semiconductor is required. The group V elements have five orbital electrons in their outer shell. When a trace of such an element is added to intrinsic silicon, and a crystal forms, the impurity atoms take the place of silicon atoms in the lattice, with four of their five outer electrons forming the required covalent bonds with the neighbouring silicon atoms. At $0\,\text{K}$ the extra electron is loosely held to its parent atom, but requires only a small addition of energy to release it into the conduction band. Compared with the energy gap of silicon of about $1.1\,\text{eV}$, the energy required for the extra impurity electron is of the order of $0.01\,\text{eV}$. At room temperature, therefore, the impurity atoms have provided extra conduction electrons and hence the resistivity of the doped silicon is reduced substantially from that of the intrinsic silicon. Doped semiconductors are called *extrinsic semiconductors*. An impurity that provides extra free conduction electrons is called a *donor*, the resulting doped semiconductor being called *n type*.

It is also possible to produce doped or extrinsic semiconductors with extra holes. Such a *p type* semiconductor requires the addition of a group III impurity element. Such an element has only three electrons in its outer shell. When the crystal lattice is formed, a hole is produced by the shortage of valence electrons at the impurity atom centres. Only a little additional energy, of the order of $0.005\,\text{eV}$, is required to move electrons from silicon atoms into these holes and so produce holes in the valence band of the silicon. An impurity that provides extra free holes is called an *acceptor*.

Whereas the intrinsic semiconductor contains equal numbers of electrons and holes, the addition of impurities upsets this balance. In the n type semiconductor the presence of the additional free electrons increases the probability that a hole will recombine with a free electron. The number of holes is thus much smaller in the doped n type semiconductor than in the intrinsic semiconductor. With the doped p type semiconductor the presence of the additional free holes increases the probability that an electron will recombine with a free hole. The number of electrons is thus much smaller in the doped p type semiconductor than in the intrinsic semiconductor. In doped semiconductors, therefore, conduction is due mainly to one type of charge carrier, electrons in n type material and holes in p type material. This carrier is referred to, in each case, as the *majority carrier*, while the other carrier is the *minority carrier*. Thus, in p type material, holes form the majority carriers while electrons form the minority carriers.

6.2.1   *Carrier diffusion*

Movement of charge carriers can occur as the result of an applied electric field, as discussed earlier in Section 6.1.1. This is referred to as current due to carrier drift. Movement of charge carriers can also occur due to carrier diffusion. If somebody releases a strong smelling substance in one corner of a room the smell will gradually spread throughout the room. This is because the smell molecules have kinetic energy, because they are at room temperature, and so bounce around between and are knocked around by the air molecules until they end up spread out over the room. There has thus been a movement of smell molecules from the initial region of high concentration to the other parts of the room which had a low concentration. This process is called *diffusion*. Movement of charge carriers can also occur due to carrier diffusion. This effect occurs where the concentration of a charge carrier in not the same throughout the material. A drift of carriers occurs from areas of high concentration towards areas of low concentration.

The rate at which diffusion occurs, whether for charge carriers or smell, is proportional to the concentration gradient. Thus, a general diffusion law can be expressed as

$$\text{rate of diffusion} = -D\frac{\mathrm{d}c}{\mathrm{d}x} \tag{6.9}$$

where the rate of diffusion is the number of particles passing through unit area per second, $\mathrm{d}c/\mathrm{d}x$ is the concentration gradient and D a constant called the *diffusion coefficient*. The minus sign indicates that the positive flow of particles goes in the direction of falling concentration. In the case of charge carriers, their motion constitutes a current, a so-called *diffusion current*. Thus where there is a concentration gradient of electrons, since each electron carries a charge $-e$, the diffusion current density J_e is

$$J_e = eD_e\frac{\mathrm{d}n}{\mathrm{d}x} \tag{6.10}$$

where dn/dx is the electron concentration gradient and D_e the *electron diffusion coefficient*. Where there is a concentration gradient of holes, the diffusion current density J_h is

$$J_h = -eD_h\frac{dp}{dx}$$

(6.11)

6.3   The pn junction diode

A *pn junction diode* is formed when the doping in a crystal is such that the semiconductor changes from p type to n type in a very small distance, typically of the order of 10^{-6} or 10^{-7} m. Figure 6.3(a) shows two isolated sections of semiconductor, one p type and one n type. Omitting, for the sake of clarity, the basic semiconductor lattice, the p type material is represented as fixed negatively charged impurity atom nuclei, together with free positively charged holes, these being the majority carriers in the p type material. There are also, of course, a relatively small number of free electrons, the minority carriers. Remember, however, that the piece of semiconductor is electrically neutral, that is it contains equal amounts of positive and negative charge. Similarly, the n type material is represented in the diagram by fixed positively charged impurity atom centres, together with free negatively charged electrons, the majority carrier. Again, a small number of free holes, the minority carriers, are also present.

A pn junction is not formed by merely bringing together the two sections of semiconductor. In order to realize the rectifying properties of a junction diode, it is necessary that the crystal structure is continuous across the transition from p type material to n type material. The junction is, therefore, produced by the addition of the necessary impurities to a single crystal of the basic semiconductor.

Consider the conditions at the transition between the n and p type materials in a correctly formed junction. In the p material, holes form the majority carriers, while in the n material, holes form the minority carriers. The concentration of holes is, therefore, much greater on the p side of the junction than on the n side. Because of the difference in hole concentration, a diffusion of holes occurs from the p material to the n material. This results in a transferral of positive charge from the p material to the n material. Consider now also the free electrons which form the majority carriers in the n type material, and the minority carriers in the p material. The concentration of electrons is much greater in the n material than in the p material; a diffusion of electrons occurs from the n to the p material, transferring negative charge to the p material. The result of the diffusion of both holes and electrons is, therefore, to produce a potential difference across the junction with the n side of the junction positive with respect to the p side. This potential difference is in such a sense as to oppose the diffusion of carriers which originally caused it; it is effectively a *potential barrier* across the junction.

The majority carriers which cross the junction become, of course, minority

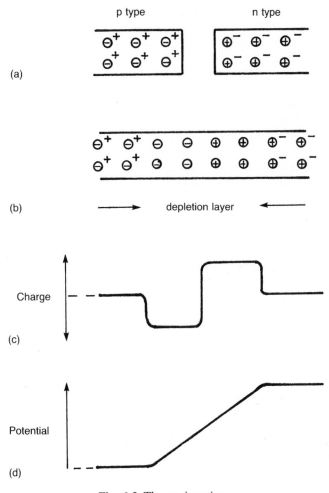

Fig. 6.3 The pn junction.

carriers in the semiconductor material of the opposite type. Once across the junction, the carriers take part in the continuous recombination and regeneration of holes and electrons which is normal in a semiconductor. The existence of the barrier potential and the resulting electric field across the junction produces a narrow region about the junction which is *depleted* of charge carriers. The charge distribution and the potential distribution are illustrated in Fig. 6.3(c) and 6.3(d).

Although the electric field in the depletion layer is in the sense to oppose the continued crossing of the junction by majority carriers, it is, of course, in the correct sense to accelerate minority carriers across the junction. Any minority carriers entering the depletion layer and coming under the influence of the electric field are, therefore, rapidly transferred across the junction. In an isolated pn junction, without external circuit connections, the net current, i.e. the mean rate of movement of electrical charge, must of course, be zero. A state of balance is, therefore, produced in the junction whereby the potential barrier is of the exact magnitude to allow sufficient majority carriers to cross the junction to equal and cancel the charge carried across the junction by the minority carriers.

The electric field in the depletion layer is produced as we have seen by the removal of free charge carriers across the junction, leaving the charged donor or acceptor impurity atom centres fixed in the crystal lattice. The width of the depletion layer is, therefore, dependent upon the actual concentration of the impurity atoms on each side of the junction. Thus, a relatively high concentration of impurity atoms will result in a narrow depletion layer. Again, if the impurity concentrations on each side of the junction are equal, the depletion layer will be equally spaced in both the p and the n types of material. However, if the impurity concentration is lower in, for example, the p material, than in the n material, the depletion layer will extend further into the p material than into the n material.

Outside the depletion region, however, the net charge is zero, i.e. the material is electrically neutral. The electric field in the material away from the depletion layer is, therefore, also zero.

6.3.1 The biased pn junction diode

If a voltage is applied across the junction via connections made to the p and n sides, the conditions in the junction are altered. If the negative connection is made to the p type material and the positive connection is made to the n type material, the applied voltage is in such a direction as to increase the potential barrier across the depletion layer. A small external applied voltage is found to be sufficient to increase the barrier potential to a level which effectively prevents majority carriers from crossing the junction. Charge flow across the junction is then due solely to minority carriers which continue to be accelerated across the junction. Under these conditions the junction is said to be *reverse biased*. All minority carriers which enter the depletion layer region are accelerated across the junction and the current in the reverse biassed junction is due only to the minority carriers. The current in the junction, and the external circuit, is, therefore, very small, and over a large range is independent of the magnitude of the applied reverse voltage and, because it is fixed by the number of thermally generated minority charge carriers, is temperature dependent. *Forward bias* is applied to a pn junction by connecting the external source in such a direction to make the p material positive with respect to the n material. The external voltage now has the effect of reducing the magnitude of the barrier potential, so allowing a great increase in the number of majority carriers crossing the junction. The forward current thus increases very rapidly for small applied forward bias voltages. The junction current–voltage relationship is found to be of the form

$$I = I_s(e^{eV/kT} - 1) \tag{6.12}$$

where I is the current across the junction, V the applied voltage, k Boltzmann's constant, T the absolute temperature, and I_s the reverse saturation current due to the thermally generated minority carriers.

Figure 6.4 shows a current/voltage characteristic typical of a silicon diode. The figure shows clearly the basic rectifying characteristic of the diode, i.e. the very high impedance presented by the diode to an applied voltage of the reverse polarity and the very low impedance presented by the diode to an applied voltage of the forward polarity. The diode is, in effect, a switch which is controlled by the polarity of the applied voltage, the switch being closed, or on, for forward voltages, and open, or off, for reverse voltages. Note that, in the figure, the voltage and current scales are different for the forward and reverse bias conditions. The reason for this is that the reverse current I_s, which may typically be of the order of a few nA, would be indistinguishable from zero if the reverse current scale was not magnified in relation to the forward current scale.

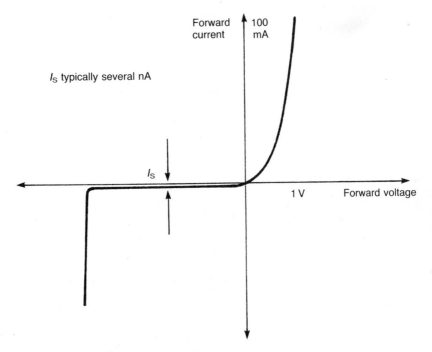

Fig. 6.4 Typical silicon diode characteristic.

At room temperature (approximately 300 K) the value of the term e/kT is approximately 40. With an applied reverse voltage of a little over 0.1 V, therefore, the reverse current has risen to within 1% of the saturation value I_s. The reverse current is then virtually independent of the applied reverse voltage until breakdown of the junction commences at a voltage V_{BO} which, depending upon the construction of the diode, may be as low as 1 V, or as high as several thousand volts. In the forward direction, a bias of a little over 0.1 V will result in a forward current of about $100I_s$. However, because I_s is so very small, a forward voltage of the order of 0.7–1.0 V is required to achieve the normal forward current range of the diode.

For a germanium diode the reverse saturation current is higher than that of a silicon diode, and the forward bias voltage is slightly lower (0.3–0.6 V). For both diodes the reverse current will double for a temperature rise of the order of 7 to 8°C.

6.4   The diode in a circuit

The term *linear* is used for a circuit element such as a resistor if the current through it is proportional to the voltage across it. This means that if the voltage is doubled then the current is doubled. There are, however, many circuit elements which are not linear. The junction diode is such an example.

Consider a typical diode characteristic, e.g. that in Fig. 6.4. If we define the resistance of the diode in the same way that was used for the linear resistor, i.e.

$$\text{resistance} = \frac{\text{voltage across device}}{\text{current flowing through the device}}$$

we immediately meet the difficulty that the value of resistance obtained varies with the value of the current flowing. As an example, consider the diode in its reverse biassed condition, where the application of a relatively large voltage results in a very small reverse current. The effective diode resistance is obviously very large, for silicon diodes it may be of the order of 100 MΩ. Conversely, in the forward direction the application of about 0.4 V will result in hardly any forward current at all, whereas about 0.8 V will result in a current of the order of perhaps 30 mA. The resistance in the forward direction thus varies. Ohm's law is not followed, the current is not proportional to the voltage. Thus to determine the current in a circuit it is necessary to know the full current–voltage characteristic.

Kirchoff's laws still apply to circuits containing non-linear elements. This is because they depend only on the fundamental laws of charge conservation and energy conservation. Problems may thus still be solved using node analysis or mesh analysis.

Figure 6.5(a) shows the diode symbol used in circuit diagrams. The arrow head indicates the direction of forward current flow. A very simple model of a diode, and one which is used frequently in practice, is as a device which is of zero resistance when forward biased, and infinite resistance when reverse biased. Figure 6.5(b) shows the characteristic for such a representation and the equivalent circuit. This is just a resistor.

By the use of more complex equivalent circuits, greater accuracy can be achieved. The non-linearity at low voltages can be accounted for by the equivalent circuit shown in Fig. 6.5(c). This considers that the diode does not begin to conduct until a particular value of forward voltage has been reached, about 0.6 V for a silicon diode. This model considers the diode when reverse biased or forward biased with a voltage across it of less than 0.6 V to have

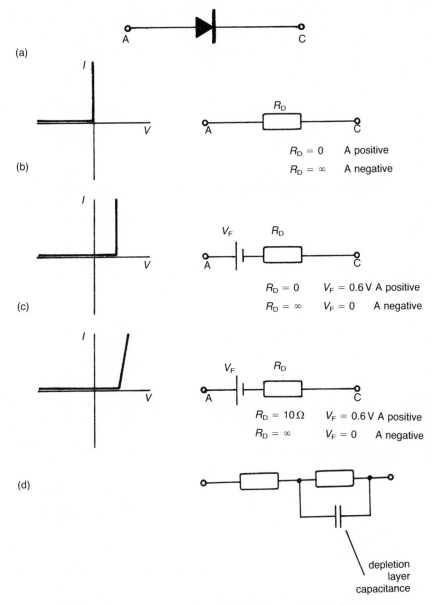

Fig. 6.5 (a) Diode symbol, (b), (c), (d) equivalent circuits.

infinite resistance and so there is no current. When forward biassed with more than 0.6 V then the diode can be considered to have a constant voltage drop across it of 0.6 V with the series resistance being zero. This equivalent circuit can be made more accurate by considering the series resistance, when the diode is forward biased at more than 0.6 V, to have a finite value. A typical value might be $10\,\Omega$ (Fig. 6.5(d)).

The junction depletion layer consists of a region of semiconductor without free charge carriers, sandwiched between two sections of conducting material, the p and n type semiconductors. A capacitor therefore exists across the junction. This can be represented in an equivalent circuit in the way shown in Fig. 6.5(d). A typical value for the capacitance is about 100 pF.

Example A sinusoidal voltage source of peak voltage 10 V and zero internal impedance is applied to a series circuit of diode and a 100 Ω load resistance. Using the simple equivalent circuit of Fig. 6.5(b), determine the waveform of the current through the load resistor.

Figure 6.6(a) shows the circuit. The current through the load resistor is given by

$$i = \frac{v}{100 + R_D}$$

During the forward biased half cycle, when $R_D = 0$, the current i is $v/100$ and so

$$i = (1/100)10 \sin \omega t = 0.1 \sin \omega t$$

During the reverse biased half cycle, when $R_D = \infty$, the current is zero. The current waveform is thus as shown in Fig. 6.6(b).

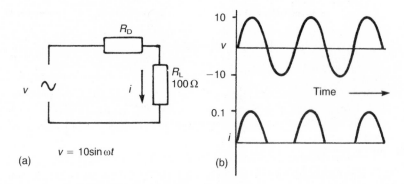

Fig. 6.6 Example.

6.4.1 *Small signal equivalent circuit*

Consider the case of an applied voltage which is varied by a small amount about its steady value. Figure 6.7(a) shows the characteristic for a diode with an applied steady forward bias voltage V_B which results in a diode current of I_B. Only the forward conduction quadrant of the characteristic has been shown.

The point A on the characteristic is the effective operating point for this

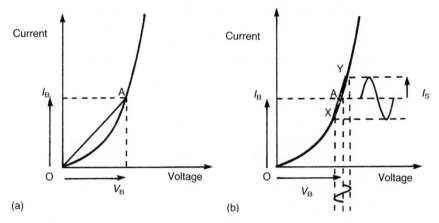

Fig. 6.7 (a) Direct voltage, (b) direct plus alternating voltage applied to a diode.

value of applied voltage. The effective resistance of the diode at this point, and remember this applies only at this point, is given by

$$R_{DC} = \frac{V_B}{I_B} \tag{6.13}$$

and is represented by the reciprocal of the slope of the chord OA of the characteristic.

Now consider the addition of a small alternating voltage to the steady diode voltage V_B such that the voltage applied to the diode can be expressed as

$$v = V_B + V_S \sin \omega t$$

The conditions with the diode are now represented by the characteristic shown in Fig. 6.7(b). The current produced in the diode is effectively the same current I_B as before, but with a varying current added to it. Although the varying voltage added to the steady voltage V_B was sinusoidal, the varying component of the diode current cannot be sinusoidal unless the diode characteristic between the points X and Y on the characteristic is exactly straight, i.e. unless a linear relationship exists between the alternating component of the applied voltage and the alternating component of the resulting current.

For all practical diodes, however, such a linear relationship does not exist and the curvature of the characteristic results in a distorted current waveform (such a waveform can be considered to be the result of the superposition of a number of harmonics of sinusoidal alternating voltages). The effect of such curvature is reduced if the magnitude of the alternating component of the voltage is reduced, and almost linear operation can be approached if the magnitude is reduced to a very low level. Such operation is called *small signal* operation.

Under small signal conditions, for an applied voltage of

$$v = V_B + V_S \sin \omega t \qquad (V_S \ll V_B)$$

the diode current will be very nearly

$$i = I_B + I_S \sin \omega t \qquad (I_S \ll I_B)$$

The direct components of current and voltage will be related to each other, as before, by

$$R_{DC} = \frac{V_B}{I_B}$$

where R_{DC} represents the effective *d.c. resistance* of the diode at the operating point A. The alternating components of the current and voltage will be related to each other by

$$R_{AC} = \frac{V_S}{I_S} \tag{6.14}$$

where R_{AC} is the effective *a.c. resistance* of the diode at the operating point A and is effectively the reciprocal of the gradient of the diode characteristic at the operating point. Alternative names for the a.c. resistance are the *incremental resistance* or *dynamic resistance*.

The use, in this manner, of a direct *bias* voltage with a small alternating voltage is very common. The function of the direct bias voltage is effectively to select the position of the operating point A on the diode characteristic, and hence effectively control the value of the a.c. resistance represented by the device to the alternating signal voltage.

A *small signal equivalent circuit* can be used to represent the diode operating in this way. Referring to Fig. 6.7(a), the a.c. resistance at the operating point A is equivalent to the reciprocal of the gradient of the characteristic at point A. But the current–voltage characteristic has the equation

$$I = I_S(e^{eV/kT} - 1)$$

Hence the gradient is

$$\frac{dI}{dV} = \frac{e}{kT} I_S \, e^{eV/kT} \approx \frac{e}{kT} I$$

The a.c. resistance is thus very nearly

$$\text{a.c. resistance} = \frac{dV}{dI} = \frac{e}{kT}\frac{1}{I}$$

At room temperature kT/e has a value of about 1/40. Hence

$$\text{a.c. resistance} \approx \frac{1}{40I} \tag{6.15}$$

The value of the a.c. resistance is thus fixed by the position of the operating point on the characteristic, and this is determined by the value of the biasing direct voltage.

A small signal equivalent circuit would, in its simplest form, merely consist of a resistor of value $1/40I$. At other than low frequencies, however, the small signal equivalent circuit should also include the effects of the depletion layer capacitance, this being considered to be in parallel with the $1/40I$ resistor.

6.4.2 *Load lines*

Determining the current in a series diode–resistor circuit (Fig. 6.8) is rather like the chicken and the egg situation, in that, in order to calculate the circuit current, the voltage across the diode must be known. But in order to obtain the diode voltage from the characteristic, the circuit current must be known. One way of graphically determining the d.c. current and voltage conditions in a circuit consisting of a diode in series with a resistor (Fig. 6.8) involves the use of *load lines*. Applying Kirchoff's voltage law to the circuit

$$V = V_D + V_R$$

and thus at some particular current

$$V = V_{DI} + RI$$

where V_{DI} is the value according to the diode characteristic of V_D at the current I. This equation can be written as

$$I = \frac{V}{R} - \frac{V_{DI}}{R}$$

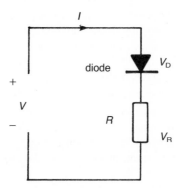

Fig. 6.8 Diode circuit.

A graph of I against V is a straight line called the *load line* with a gradient of $1/R$. Figure 6.9 shows such a line. It is most conveniently drawn by connecting two points with a straight line, namely the point on the voltage axis with $I = 0$ of V and the point on the current axis of V/R. Superimposed on the graph is the diode characteristic, i.e. I against V_D. The value of the current that satisfies both these graphs is where they intersect. This point is called the *operating point* and enables the values of V_{DI} and the current to be read off from the graph.

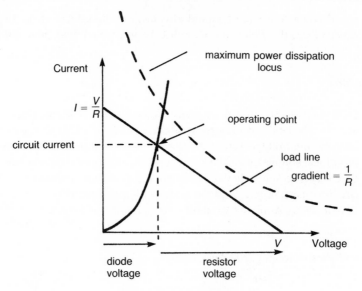

Fig. 6.9 Using load lines.

In a particular circuit, the choice of the operating point is often governed by the maximum allowed power dissipation in the device. The maximum power rating for a particular diode is specified by the manufacturer, as is the maximum allowable reverse voltage and maximum allowable forward current. At the operating point, the power dissipation in the diode is given by

$$\text{power} = V_{\text{DI}}I$$

Thus for any given diode voltage there is a maximum value of the allowable diode current. Using the manufacturer's specified maximum power then values of V_{DI} at different currents can be determined and thus the locus of the extreme limits of the operating point can be drawn on the characteristic (it is a rectangular hyperbola). This is shown in Fig. 6.9. The operating point must lie to the left of the maximum power dissipation curve.

6.5   The Zener diode

From the diode characteristic of Fig. 6.4 it can be seen that when the reverse voltage is increased beyond a certain value, V_{BO}, the reverse current increases rapidly. The diode is said to break down. The critical voltage V_{BO} is termed the *breakdown* or *Zener voltage*. The sudden and large increase in reverse current is caused by either or both of two different effects, depending upon the impurity doping levels in the semiconductor.

(1) The large value of reverse voltage applied to the diode results in an electric field in the depletion layer sufficiently high to break the covalent bonds, so

producing free electron-hole pairs. This effect is known as the *Zener effect*.

(2) Mobile charge carriers are accelerated by the electric field until they collide with atoms in the crystal lattice, moving, on average, a distance known as the *mean free path*. If, between collisions, they are accelerated so that their kinetic energy is sufficient to ionize atoms with which they collide, then free carriers will be produced, which in turn will be accelerated, and can cause ionizing collisions. This results in an *avalanche breakdown* effect.

The mechanism of breakdown in a given diode depends upon the doping levels in the semiconductor, and determines the actual value of the breakdown voltage V_{BO}. In fact, both mechanisms may occur at the same time.

The breakdown characteristic of correctly designed diodes is utilized in a range of diodes termed Zener diodes, which are used to provide stable reference voltages. They are manufactured to have breakdown voltages in the range from about 3.0 V to 20.0 V. The lower voltage reference diodes rely mainly upon the Zener effect, while the higher voltage diodes operate mainly due to the avalanche effect. The devices are used in practice to provide a voltage drop which is, to a large extent, independent of the current flowing through the diode. The symbol used to represent a Zener diode is shown in Fig. 6.10. Because the diode is used in a reverse biased condition, the voltage polarity is as indicated in the figure.

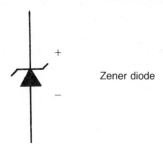

Fig. 6.10 Symbol for the Zener diode.

Example   A Zener diode has a breakdown voltage of 10 V and a maximum power dissipation of 400 mW. What is the maximum current the diode should pass?

The current $I = P/V = 0.400/10 = 40$ mA.

6.6   The varactor diode

As indicated earlier, a pn junction consists of the depletion layer, a region of semiconductor material without free charge carriers, sandwiched between two sections of conducting material, the p and n type semiconductor. The arrangement is thus a capacitor, the capacitance being $\varepsilon A/d$, where ε is the permittivity of the semiconductor material, A the area of the junction and d the width of the depletion layer. Semiconductor diodes are generally manufactured so that the capacitance is a minimum, however the varactor diode is designed to have specific capacitance values. The varactor diode is operated with a reverse bias voltage. For a junction between the p and n type materials which is abrupt the capacitance is inversely proportional to the square root of the reverse bias voltage, i.e.

$$C = \frac{k}{\sqrt{V}} \tag{6.16}$$

Where the junction is linearly graded then the capacitance is inversely proportional to the cube root of the reverse bias voltage. Such capacitors have a voltage-adjustable capacitance. Typically the capacitances vary between about 50 pF and 100 pF for an abrupt junction and between 5 pF and 50 pF for a graded junction.

6.7 The junction transistor

The development of the transistor, from its beginnings in 1948, to the cheap, mass-produced device of today, has revolutionized the field of electronics, and affected most aspects of life. The junction transistor is a bipolar device, in that current flow in the transistor is due to movement of both holes and electrons. The transistor consists, essentially, of two pn diodes formed with one common section, so giving a three element device, as illustrated in Fig. 6.11.

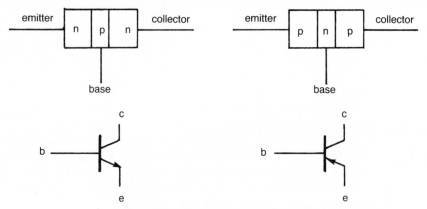

Fig. 6.11 The junction transistor.

The common section, called the *base*, may be of either p or n type semiconductor, resulting in two basic types of transistor known as the *npn* and the *pnp* types. The central base section, which is extremely thin, separates the *emitter* and *collector* sections of the transistor. Connecting leads are provided to all three sections. The figure also shows symbols which are commonly used to represent the two types of transistor in circuit diagrams. In describing the operation of the bipolar transistor, reference will be made to the npn transistor. The mode of operation of the pnp transistor is exactly the same as that of the npn transistor, with the exception that the polarities of all applied voltages and currents, and that of the charge carriers, are reversed.

In normal use, as a linear amplifying device, the transistor is used with the emitter to base junction forward biased, and with the collector to base junction reverse biased. In the case of the npn transistor, this means that the emitter is negative with respect to the base, while the collector is positive with respect to the base. Because the collector–base junction is reverse biased and assuming, for the moment, that the emitter connection is open circuit, then the only current flowing in the collector connection will be the small reverse current of the collector–base junction. This current is termed the collector–base leakage current, and is given the symbol L_{CBO}.

When the emitter–base diode is forward biased, the current flowing in the junction will be due to electrons crossing from the emitter to the base, and also to holes crossing from the base to the emitter. By deliberately using lightly doped p semiconductor for the base material, and relatively heavily doped n semiconductor for the emitter material, the junction current may be designed to be mainly due to electron flow from the emitter to the base region. (Similarly, in considering the pnp transistor, the emitter–base junction current is designed to be mainly due to holes crossing from the emitter to the base.) The majority carriers of the emitter region are said to be *injected* into the base region by the emitter–base bias.

The emitter current may thus be written

$$I_E = I_{electron} + I_{hole}$$

where $I_{electron}$ and I_{hole} are respectively the components due to electron flow and hole flow. Remember that the passage of negatively charged electrons from the emitter to the base is represented by a conventional current from base to emitter.

Once the electrons cross into the p type base region they become minority carriers and if they remained in the base for more than a very short time, they would disappear, due to recombination with the majority carrier holes. However, the increased concentration of electrons in the base region near the emitter–base junction causes a diffusion of the electrons across the base and, because the base region is very thin, most reach the collector side of the base region without recombination. The very small number of electrons which do recombine with holes in the base region gives rise to a small component of base current, because the positive charge removed in the recombination must be replaced in order to maintain steady conditions.

On reaching the collector–base junction, the electrons come under the influence of the electric field across the junction which, because the junction is reverse biased, is in such a direction to accelerate the carriers into the collector region. The carriers reaching the collector then form the collector current.

Figure 6.12 shows diagrammatically the transistor action in both npn and pnp transistors, omitting the effects of the collector base leakage current I_{CBO}.

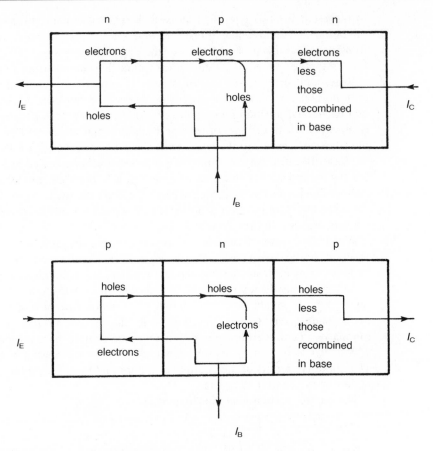

Fig. 6.12 Charge carrier movements.

Summarizing, the current due to the electrons injected from the emitter into the base of the npn transistor forms a fraction θ_1 of the total emitter current. A further fraction θ_2 of the electrons injected into the base actually reaches the collector base junction and crosses into the collector region. The steady collector current I_C is thus a fraction of the steady emitter current I_E. The fraction is known as the current gain between collector and emitter, and is given the symbol h_{FB} or, very commonly, α, where

$$h_{FB} = \alpha = \theta_1 \theta_2$$

In a well designed transistor, the value of h_{FB} will be very close to unity, typical values lying in the range 0.95 to 0.995. The transistor is, of course, a three terminal device, so that the difference between the emitter current and the collector current must form the base current.

The basic equations which characterize the transistor may thus be written

$$I_E = I_C + I_B \tag{6.17}$$

$$I_C = h_{FB} I_E \tag{6.18}$$

Equation (6.18) describes the basic transistor action, while Equation (6.17) relates the actual currents in the three transistor leads.

6.7.1 The common base characteristics

Figure 6.13 shows an npn transistor connected so that its common base characteristics may be determined. The emitter current is determined by the value of the resistor R_1, together with the value of the supply voltage. Notice that it would be impractical to vary the emitter current by means of a variable voltage supply alone. This is so because, as the base emitter junction is forward biased, an increase in emitter supply voltage of from about 0.6 to 0.7 V would increase the emitter current from a very low value almost through its full range. The use of a series resistor and a high supply voltage effectively makes the emitter current almost independent of the base emitter diode characteristic.

Thus

$$I_E = \frac{V_1 - V_{BE}}{R_1}$$

and if V_{BE} is small in comparison with V_1 then

$$I_E \simeq \frac{V_1}{R_1} \tag{6.19}$$

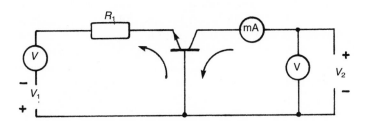

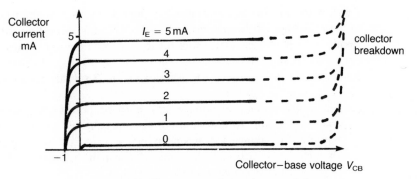

Fig. 6.13 The common base characteristics.

The common base collector current/voltage characteristics are also shown in Fig. 6.13.

The following points should be noted from the characteristics

(1) Over a wide voltage range, the collector current is very nearly independent of the collector voltage, i.e. the collector circuit may be said to be of very high impedance. The collector circuit is almost a constant current source.
(2) The value of the collector current is determined by the value of the emitter current.
(3) The collector base voltage must actually be reversed (collector negative with respect to the base) in order to reduce the collector current to zero.
(4) With zero emitter current the collector current is very small and is, in effect, the reverse current of the collector base junction I_{CBO}.
(5) If the collector base voltage is increased beyond a certain level collector junction breakdown begins, and the collector current increases rapidly.

The mechanism of collector breakdown is similar to that described in Section 6.5, and may be due to either or both avalanche and Zener effects. A further effect which occurs is that as the collector base voltage is increased, the width of the depletion layer also increases, so decreasing the width of the already very narrow base region. As the base width decreases, the time spent by the carriers in crossing the base also decreases, giving less chance of the minority carriers recombining. As a result, the current gain of the transistor increases. At relatively high collector–base voltages, the depletion layer width extends well into the base region, and may approach the emitter base junction. At excessive voltages, *punch through* occurs, whereby carriers injected from the emitter move directly across into the collector; the resulting current may cause permanent damage to the transistor.

The characteristics shown in Fig. 6.13 are termed the common base characteristics, because the base connection is common to both the emitter current circuit and to the collector current circuit.

6.7.2   *The common emitter characteristics*

Figure 6.14(a) shows an npn transistor connected with its emitter lead common to both the base current circuit and to the collector current circuit. Using this circuit the collector characteristics for the common emitter connection may be obtained. These characteristics are shown in Fig. 6.14(b). Comparison with the common base collector characteristics of Fig. 6.13 reveals several differences.

(1) The collector–emitter voltage V_{CE} must be positive to produce positive collector current. At low values of V_{CE} the collector current I_C is itself low, but after the voltage V_{CE} has been increased to exceed a *knee* voltage, the characteristics assume an almost constant current state, i.e. the characteristics become almost horizontal.
(2) The characteristics are not quite as near to horizontal as are those of the

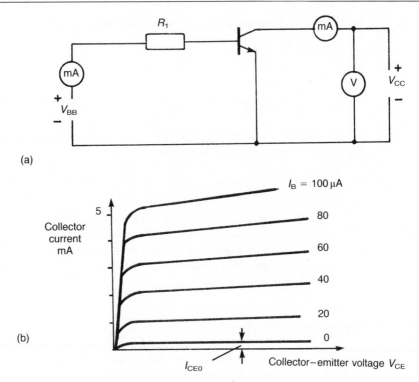

Fig. 6.14 The common emitter characteristics.

common base connection. The effective collector impedance, i.e. the reciprocal of the gradient of the characteristic, is therefore not as high in the common emitter connection as in the common base connection.

(3) With the base current I_B at zero, the collector current is not zero but has a value I_{CEO} which is larger than the equivalent leakage current in common base, I_{CBO}.

(4) The current gain in the common emitter connection, with the symbol h_{FE} or sometimes β is defined as

$$\beta = h_{FE} = \frac{I_C}{I_B} = \frac{\text{collector current}}{\text{base current}} \qquad (6.20)$$

From the figure it can be seen that the characteristics are not as evenly spaced nor as parallel as in the case of the common base connection. This shows that the current gain h_{FE} varies more with a variation of the collector voltage than was the case with the common base connection.

Let us now make an estimate of the value of the common emitter gain. The two basic equations (6.17) and (6.18) which characterize the transistor are

$$I_E = I_C + I_B$$

$$I_C = h_{FB} I_E$$

Eliminating I_E from these two equations, we have

$$I_C = h_{FB}(I_C + I_B)$$

whence

$$h_{FE} = \frac{I_C}{I_B} = \frac{h_{FB}}{1 - h_{FB}} \tag{6.21}$$

For a transistor whose steady state (d.c.) current gain in common base is 0.97, the current gain in common emitter is

$$h_{FE} = \frac{0.97}{1 - 0.97} = 32.3$$

Notice also that a small change in the value of h_{FB} from 0.97 to 0.98 gives a relatively large change in the value of h_{FE} from 32.3 to 49.0, i.e. the current gain h_{FE} is much more sensitive to second order effects, for example changes in collector voltage, collector current or perhaps changes in temperature, than is the common base current gain h_{FB}.

6.8 The transistor in a circuit

Like the diode, discussed earlier in this chapter, the transistor is a non-linear device. As with the diode, the equations describing the behaviour of the device can be used, a graphical technique involving a load line used or equivalent circuits used to solve circuit problems involving it.

6.8.1 Load lines

Figure 6.15(a) shows an npn transistor connected in the common emitter mode, and with a resistor R_C as its collector load. As with the diode resistor network of Section 6.4.2, various methods may be used to determine the d.c. conditions in the circuit. Figure 6.15(b) shows the common emitter collector characteristics with a load line for the resistor R_C superimposed. In the diode circuit the operating point is given by the intersection of the load line with the diode characteristic. In the transistor circuit we are presented with a choice of operating points, the point of intersection of the load line with the collector characteristic may be varied by varying the transistor base current. In the figure, the load line for the resistor is drawn using the method described in Section 6.4.2 by connecting the point V_{CC} on the voltage axis to the point $I_C = V_{CC}/R_C$ on the current axis. The operating point A is then selected by setting the base current I_B to be $40\,\mu A$. The maximum allowable power dissipation curve given by the locus of

$$V_{CE}I_C = W$$

is also drawn on the characteristics; the operating point must lie to the left of this curve.

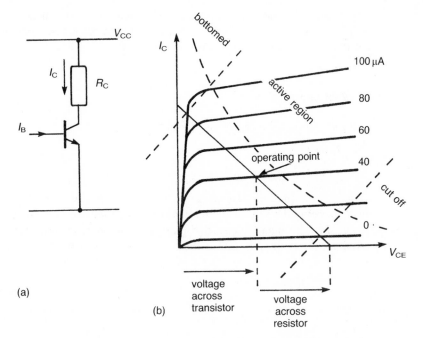

Fig. 6.15 An npn transistor in common emitter mode.

By varying the value of the base current I_B, the value of the collector current can be changed by a larger amount, the ratio of the two currents being, of course, the circuit current gain. Notice, however, that the extent of the change in the collector current is limited. Thus, as the base current I_B is reduced, eventually to zero, the operating point moves down the load line to reach the point B. At this point the transistor is said to be cut off; the only current flowing in the collector circuit is the collector emitter leakage current I_{CEO} which is a relatively small current of, at the most, several μA. There is, therefore, very little voltage drop across the load resistor R_C, and the collector emitter voltage V_{CE} is very nearly equal to the collector supply voltage V_{CC}.

Alternatively, if the base current I_B is increased, the operating point A moves upwards along the load line to reach the point C at which the collector current has increased to such an extent that the voltage drop in the load resistor R_C is almost equal to the supply voltage V_{CC}. The collector emitter voltage cannot be reduced any further, and the transistor is said to be *bottomed* or *saturated*. In a typical transistor, the bottoming voltage may be as low as 0.1 to 0.2 V. Notice that as the base emitter voltage in this heavily conducting condition will be of the order of 0.6 to 0.7 V, then the transistor collector has become more negative than the base; i.e. the collector base junction is now so saturated with carriers that it has become forward biased.

The *cut-off* and *bottomed* region of the characteristics are of great interest when the transistor is being used as an on–off switch, because the two conditions represent a very high impedance and a very low impedance state of the transistor, analogous with a switch open and a switch closed. However, when in use as a linear amplifier, the operating point is positioned in the centre of the active region of the characteristics.

6.8.2 Large signal equivalent circuits

Consider Equations (6.17) and (6.20).

$$I_E = I_C + I_B$$

$$I_C = h_{FE}I_B$$

These two equations can represent the circuit described by Fig. 6.16(a), i.e. with I_E being the sum of the base current and that from a current generator $I_C = h_{FE}I_B$. The figure shows two alternative ways of drawing the circuit. This equivalent circuit is useful for the common emitter form of connection.

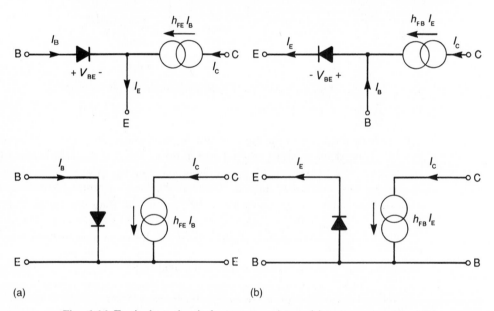

(a)   (b)

Fig. 6.16 Equivalent circuit for npn transistor, (a) common emitter, (b) common base.

Now consider Equations (6.17) and (6.18), i.e.

$$I_C = h_{FB}I_E$$

These two equations can represent the circuit described in Fig. 6.16(b). The figure shows two alternative ways of drawing the same circuit and is a useful form of circuit for the common base form of connection.

The equivalent circuits include a diode. This can be replaced by its equivalent circuit (see Fig. 6.5). Generally the equivalent circuit used, when the transistor is operating in active mode, involves just replacing the diode by a battery of voltage 0.7 V.

Example   In the circuit of Fig. 6.17(a), determine the transistor collector potential. Take the transistor current gain, h_{FB}, to be 0.97 and the emitter base voltage to be 0.7 V.

Figure 6.17(b) shows the same circuit with the transistor replaced by the equivalent circuit. The potential difference across the base resistor is $(10 - 0.7)$ V and thus the base current $I_B = 9.3/100 = 0.093$ mA. The current generator provides a current of $h_{FE}I_B$. But $h_{FE} = h_{FB}/(1 - h_{FB}) = 0.97/0.03 = 32.3$ (Equation (6.21)). Thus $I_C = 32.3 \times 0.093 = 3.0$ mA. The potential difference across the collector resistance is thus $3.0 \times 10^{-3} \times 2.2 \times 10^3 = 6.6$ V. Hence the collector voltage is $(10 - 6.6) = 3.4$ V.

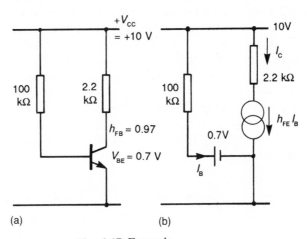

(a)   (b)

Fig. 6.17 Example.

6.8.3   *Small signal equivalent circuits*

Consider a transistor which has a d.c. bias component plus a small signals component, i.e.

$$i_B = I_B + i_b$$

$$i_C = I_C + i_c$$

$$v_{BE} = V_{BE} + v_{be}$$

$$v_{CE} = V_{CE} + v_{ce}$$

What is required is an equivalent circuit for the small signal components. For such small signals, since the biasing moves the operating point to what are essentially flat parts of the characteristics it is possible to regard the transistor as being a linear device.

Consequently, the transistor may be considered as a *two port system*, or *four terminal network* as it is sometimes called. Such a system has an input and an output and can be represented by a *black box* (Fig. 6.18). The performance of the box can be described in terms of the input voltage and current and the output voltage and current. Thus if v_1 is the input voltage, i_1 the input current, v_2 the output voltage and i_2 the output current, a general relationship can be specified as

$$v_1 = h_{11}i_1 + h_{12}v_2 \tag{6.22}$$

$$i_2 = h_{21}i_1 + h_{22}v_2 \tag{6.23}$$

where h_{11}, h_{12}, h_{21} and h_{22} are the *h parameters*. Each of these parameters is defined by some zero condition. Thus, when $v_2 = 0$, i.e. the output is short-circuited, then the two equations give $h_{11} = v_1/i_1$ and $h_{21} = i_2/i_1$. When $i_1 = 0$, i.e. the input is open-circuit, then $h_{12} = v_1/v_2$ and $h_{22} = i_2/v_2$. The subscripts used with the *h* parameters relate to the subscripts of the quantities in the short- or open-circuited conditions when the parameters are defined. These definitions mean that h_{11} is the short-circuit input impedance, h_{12} the reverse voltage gain with port 1 open-circuit, h_{21} the short-circuit current gain and h_{22} the admittance seen looking back into port 2 with port 1 open-circuit. When dealing with transistors the nomenclature used for the *h* parameters is modified so that the subscript refers to these terms. Thus $h_{11} = h_i$, the input impedance; $h_{12} = h_r$, the reverse gain; $h_{21} = h_f$, the forward gain; $h_{22} = h_o$, the output admittance.

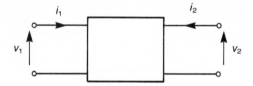

Fig. 6.18 Two port network.

If a common-emitter transistor is considered as a two port system, then the small signal input voltage is v_{be} and the input current i_b. The small signal output voltage is v_{ce} and the output current I_c. Lower case letters are used for the base, collector and emitter to indicate that we are dealing with small signal conditions superimposed on a d.c. bias condition. Thus Equations (6.22) and (6.23) can be written as

$$v_{be} = h_{ie}i_b + h_{re}v_{ce} \tag{6.24}$$

$$i_c = h_{fe}i_b + h_{oe}v_{ce} \tag{6.25}$$

The extra subscript e added to the *h* parameters is to indicate that they refer to the common-emitter condition. Similar sets of equations can be written for

Table 6.1 h parameters.

Common base	Common emitter	Common collector
$h_{ib} = \dfrac{v_{eb}}{i_e}$	$h_{ie} = \dfrac{v_{be}}{i_b}$	$h_{ic} = \dfrac{v_{bc}}{i_b}$
$h_{rb} = \dfrac{v_{eb}}{v_{cb}}$	$h_{re} = \dfrac{v_{be}}{v_{ce}}$	$h_{rc} = \dfrac{v_{bc}}{v_{ec}}$
$h_{fb} = \dfrac{i_c}{i_e}$	$h_{fe} = \dfrac{i_c}{i_b}$	$h_{fc} = \dfrac{i_e}{i_b}$
$h_{ob} = \dfrac{i_c}{v_{cb}}$	$h_{oe} = \dfrac{i_c}{v_{ce}}$	$h_{oc} = \dfrac{i_e}{v_{ec}}$

common-base and common-collector transistors. Table 6.1 shows the full range of h parameters.

The h parameter equations can be used to describe equivalent circuits for transistors. Thus, for example, Equation (6.23) can, in general, be considered to describe an input circuit which has an input voltage of v_{be} in series with an impedance of h_{ie} and a voltage source of $h_{re}v_{ce}$. Equation (6.24) can be rearranged as

$$v_{ce} = \frac{i_c - h_{fe}i_b}{h_{oe}}$$

and so describe an output circuit which has an input voltage source of v_{ce} across an impedance of h_{oe} through which flows current i_c and current from a current source of $h_{fe}i_b$. Figure 6.19 shows the equivalent circuit.

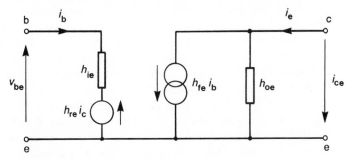

Fig. 6.19 Equivalent circuit for common emitter transistor.

Typical values of the h parameters are $h_{ie} = 1000\,\Omega$, $h_{re} = 3 \times 10^{-4}$, $h_{oe} = 300 \times 10^{-6}\,\text{S}$ and $h_{fe} = 250$. Because of the small values of h_{re} and h_{oe} a simplified version of the equivalent circuit is often used. Figure 6.20 shows the simplified circuit. This circuit is used further in Chapter 8.

Example Determine the values of the components in the simplified equivalent circuit for a common emitter transistor if $h_{fe} = 200$ and $h_{ie} = 750\,\Omega$.

The impedance in series with the input is $h_{ie} = 750\,\Omega$. The current generator in the output is $h_{fe}i_b = 200i_b$.

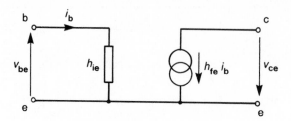

Fig. 6.20 Simplified equivalent circuit.

6.9   The field effect transistor (FET)

The operation of the *FET* is based upon completely different principles to those of the bipolar transistor. It is often described as a unipolar device in that conduction through it takes place predominantly by means of one type of charge carrier, i.e. either by electrons or by holes depending upon its type. This may be contrasted with the method of operation of the bipolar transistor in which conduction takes place due to the movement of both holes and electrons.

6.9.1   *The junction field effect transistor (JFET)*

The *JFET* consists, basically, of a short thin bar of semiconductor which forms a channel between its two end connections, termed the *source* and the *drain*. The device is known as an *n channel* device if n type semiconductor is used in its construction, or a *p channel* device if p type semiconductor is used.

Considering an n channel device, when a direct voltage source is connected between the end connections, conduction occurs by electron flow from the negative of the voltage supply along the channel from the source to the drain and back to the positive of the supply. Thinking in terms of conventional current flow, in the n channel JFET, current flow occurs from the drain to the source. The magnitude of the current flow is determined by the magnitude of the applied voltage, V_{DS}, and by the resistance of the channel. For any conductor, its resistance is given by

$$R = \frac{1}{\sigma} \times \frac{\text{length}}{\text{area}}$$

where σ is the conductivity of the material. If the channel is formed of uniformly doped semiconductor, then the conductivity σ will be constant throughout its length, and the channel will be, in effect, a linear resistor.

The operation of the device is based upon the control of the current flow through the device by effectively controlling the thickness of the channel. Control of the thickness is achieved by forming at both sides of the channel a p type region (n type region for a p channel FET). The basic construction is illustrated in Fig. 6.21. The two gate p regions, together with the n type channel, form pn junctions which, in normal use, are reverse biased by a d.c. potential V_{GS}. Because they are reverse biased, only a very small leakage current flows; the gate inputs are, therefore, extremely high impedance inputs. As in any other pn junction, the depletion layer formed at the junction is almost free of charge carriers and has, therefore, an extremely high resistivity. By using a lightly doped n-type semiconductor for the channel, the gate depletion layer may be made to extend well into the channel width. As the depletion layer width is also a function of the voltage applied across the junction, increasing with increasing voltage, the gate voltage has a controlling effect upon the extent of the depletion layer penetration into the channel, and hence upon the effective remaining channel thickness. The result is a device in which a control voltage, the gate–source voltage V_{GS}, effectively controls the resistance of the path between the source and the drain. Figure 6.21 shows the depletion layer at each of the gate–channel junctions. The voltage difference between the gate and the channel is greater at the drain end of the gate than at the source end of the gate, because of the voltage gradient along the channel, this resulting in a greater extension of the depletion layer into the channel at the drain end.

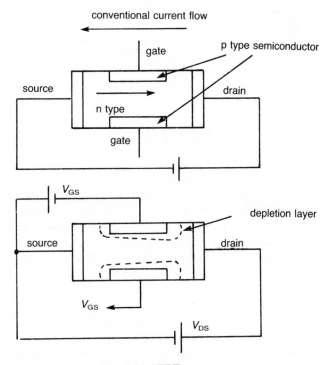

Fig. 6.21 JFET.

The characteristic curves for an n channel junction FET are shown in Fig. 6.22. For a given value of gate–source voltage an increase of drain-source voltage, from zero, initially produces a relatively linear rise in drain current. Because of the potential gradient along the channel, the channel width is a minimum near the drain end of the gate junction. As the drain voltage is increased, the depletion layer at the gate extends further into the channel and starts to *pinch off* the drain current. The drain current then turns the *knee* of the characteristic and becomes virtually independent of the drain voltage V_{DS}, until the voltage exceeds the value at which drain–gate junction breakdown commences.

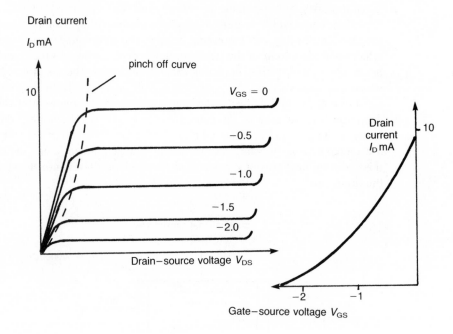

Fig. 6.22 Characteristic curves.

Figure 6.22 also shows the characteristic plotted in terms of the relationship between the drain current I_D and the controlling gate–source voltage V_{GS}. This curve applies only for the constant current region between the pinch off region and the breakdown region. Symbols used to represent junction field effect transistors are shown in Fig. 6.23. As in the case of npn and pnp transistors, p and n channel FETs are indicated by the different directions of the arrowhead used in the gate symbol.

6.9.2 The metal oxide semiconductor transistor (MOST)

The *metal oxide semiconductor transistor* (*MOST*), which is also known as the *insulated gate FET* (*IGFET*), or the *MOSFET*, utilizes a different mechanism

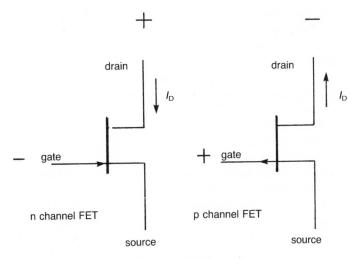

Fig. 6.23 JFET symbols.

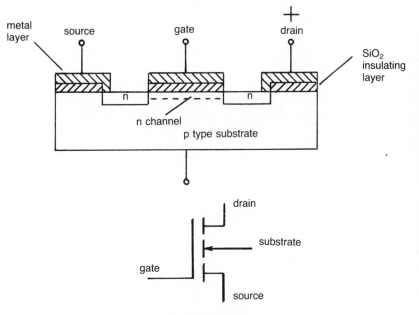

Fig. 6.24 MOST.

to control the resistance of the source to drain channel. Figure 6.24 shows a schematic diagram of the construction of an n channel MOST. A p type semiconductor is used for the basic substrate and has two strongly doped n type regions formed in it; these are the source and drain regions. The complete surface is then insulated with a diffused layer of SiO_2, and a metallized layer formed on top, this providing the gate connection, and also the source and drain connections via windows in the insulating layer. In normal use the drain is held positive with respect to the source, by the drain–source voltage, V_{DS}. Because the drain to substrate junction is reverse biased, no conducting path exists between the source and the drain.

When the gate electrode is made positive an electric field is produced in the insulating oxide layer, which induces negative charges in the substrate near the insulated surface. These charges consist of electrons drawn into the substrate. The negative charges form an induced conducting channel between the source and the drain; the number of charges and hence the conductivity of the channel is controlled by the voltage applied to the gate electrode. Because the gate electrode is insulated from the substrate, the impedance of the gate circuit is very high indeed, the gate current being typically of the order of 10^{-12} A. A MOST operating in this manner is said to be operating in the *enhancement mode*, because the action of the gate potential is to enhance the conducting properties of the channel. It is also possible to construct a MOST which, even with zero voltage at the gate, has conducting properties between the source and the drain. These conducting properties are due to a layer of positive surface charge at the interface between the substrate and the oxide layer, this resulting from a different manufacturing treatment of the substrate. The positive surface charge thus induces an n channel between the source and the drain. A positive gate potential will now increase the conductivity of the n channel by the inducement of more carriers, while a negative gate potential will decrease the conductivity of the channel. Such a device is said to work in either the enhancement mode, or the *depletion mode*. With either polarity of gate potential, however, the gate circuit impedance is very high; this contrasts with the operation of the gate circuit of the junction FET, which, being a pn junction, cannot be forward biased without becoming low impedance. The characteristics for a device of this type are shown in Fig. 6.25.

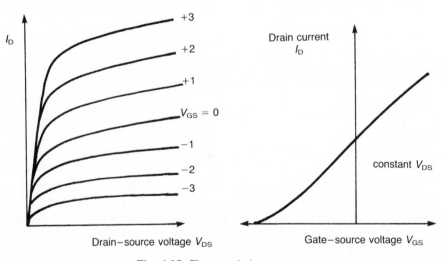

Fig. 6.25 Characteristic curves.

6.10 Switching devices

The desirable properties of an ideal switch may be summarized as follows:

(1) Infinite impedance when in the open circuit state.
(2) Zero impedance when in the closed state.
(3) Zero transition time between the two states.
(4) Zero power dissipation at the control input, i.e. at the input which initiates the change of state of the switch.

A transistor switch with a collector load resistor R_L is shown in Fig. 6.26(a), together with the load line for resistor R_L drawn on the collector characteristics. The diagram shows the effect of increasing the transistor base current I_B from zero to a value of 800 µA. With zero base current, the transistor is in its OFF state, the collector voltage V_{CE} being a little less than the supply voltage V_{CC}, because of the voltage drop across the load resistor R_L, due to the collector–emitter leakage current. The operating point in the off state is point A in the diagram; because this point is below the maximum power dissipation curve, the transistor may be operated continuously in this state.

A base current I_B of 800 µA will cause the operating point to move to point B, giving a very low value of collector voltage, represented in the figure by the ON voltage, $V_{CE(SAT)}$. The transistor is said to be *saturated*, or *bottomed*, and the collector voltage can be as low as 0.1 to 0.2 V. For a supply voltage of 20 V the collector voltage could be expected to change, therefore, from about 0.1 V to well over 19.9 V for all reasonable values of collector load resistor, and for a silicon transistor. Figure 6.26(c) shows the switching operation.

Switching operations of this sort may be performed using either bipolar transistors or field effect transistors. Specially designed bipolar switching transistors are available which will allow switching times as low as several nanoseconds.

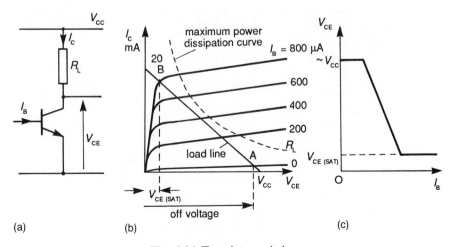

Fig. 6.26 Transistor switch.

Switching operations are of many sorts, and while some applications may require very low switching times, other applications may be more concerned with the switching of very large current and voltage levels rather than with the achievement of very high switching speeds. There are several devices available which are specially designed for switching applications.

6.10.1 The thyristor

This device, which is also known as the *silicon controlled rectifier*, is in very wide use as a power switch. Units are available with current ratings of several hundred amperes and voltage ratings of several thousand volts. The thyristor is a unidirectional device with three connections, known as the anode, cathode and the control gate. Current flow is from the anode to the cathode only. With the cathode positive with respect to the anode the device has a very high impedance. With the anode positive with respect to the cathode the device has two stable states:

(1) A high impedance state (order of megohms).
(2) A low impedance state (order of fraction of one ohm).

The operation of the thyristor can best be explained by considering it as consisting of two transistors, one an npn and one a pnp device, interconnected to form a four layer pnpn structure. This arrangement is illustrated in Fig. 6.27. In the figure, the three pn junctions of the four layer device are labelled A, B and C.

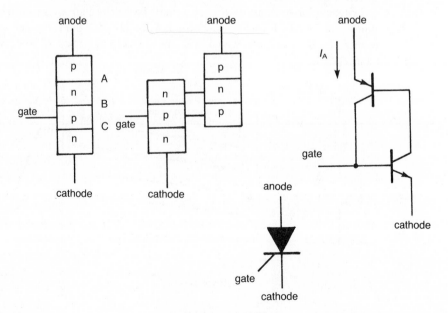

Fig. 6.27 The thyristor.

Reverse polarity: When the cathode is positive with respect to the anode, junction B is forward biased, but junctions A and C are reverse biased. The current flow is very small, being the leakage current of junction B.

Forward polarity: When the anode is positive with respect to the cathode, junctions A and C are forward biased, but junction B is reverse biased. Almost all the applied voltage is across junction B. The current flow is very small, being the leakage current of junction B. Let the leakage current across junction B be I_{SAT}. Consider, initially, only the hole component of the total current I. Let the hole component be $\gamma_1 I$. Thus, considering the pnp transistor section of the complete device, the hole current crossing junction A is $\gamma_1 I$. Of these holes, only a fraction β_1 reach junction B without recombining. The hole current reaching junction B is, therefore $\gamma_1 \beta_1 I$. At junction B, due to avalanche multiplication effect, the hole current crossing the junction is $M\gamma_1\beta_1 I$, where M is an avalanche multiplication factor whose value is 1 or greater. In addition, there will be a small leakage current across junction B due to holes, of value I_{SATp}. The total hole current across junction B is $M\beta_1\gamma_1 I + MI_{SATp}$. Similarly, the total electron current across junction B is $M\beta_2\gamma_2 I + MI_{SATn}$.

Putting $I_{SATp} + I_{SATn} = I_{SAT}$, the total current across junction B is

$$I = M\beta_1\gamma_1 I + M\beta_2\gamma_2 I + MI_{SAT}$$

or

$$I = \frac{MI_{SAT}}{1 - M(\beta_1\gamma_1 + \beta_2\gamma_2)}$$

$\gamma_1\beta_1$ and $\gamma_2\beta_2$ are, in effect, the current gains of the two transistors making up the thyristor.

Thus

$$I = \frac{MI_{SAT}}{1 - M(h_{FE1} + h_{FE2})} \tag{6.26}$$

If conditions in the thyristor are such that

$$M(h_{FE1} + h_{FE2}) = 1$$

then the denominator is zero, and the device current is limited only by the impedance of the external circuit. The thyristor can be switched from its forward blocking (high impedance) state into its forward conducting (low impedance) state by any effect which can increase the value of the factor $M(h_{FE1} + h_{FE2})$, called the *loop gain G*, to unity.

Once triggered into the conducting state, all the junctions in the device become forward biased, and the potential drop across the device is approximately equal to that of a single pn junction. As the thyristor is constructed, the current gains h_{FE1} and h_{FE2} are low at low emitter currents, but increase fairly rapidly as the current is increased. The thyristor may, therefore, be *turned on* by any mechanism which can cause a momentary increase in the effective emitter currents.

(1) Switch on by voltage breakdown. If the forward voltage is increased, so increasing the value of the avalanche multiplication factor M, the leakage current carriers are sufficient to cause an avalanche multiplication of the current. This effect occurs when the voltage is increased to the *breakover voltage*.

(2) Switch on by rate of change of voltage. In Section 6.4 the existence of a depletion layer capacitance across a pn junction was described. If a step voltage function is applied across the thyristor, when in its forward blocking condition, the charging current which flows in the capacitance of junction B effectively acts as a base current for the npn transistor. Due to the transistor action this allows the loop gain G to approach unity, switching on the device.

(3) Switch on by transistor action. The introduction of a short duration current into the gate connection (i.e. the base of the npn transistor) will switch the thyristor on.

(4) Switch on by radiation. Radiant energy within the correct spectral frequency can generate hole electron pairs. The consequent increase in the junction leakage current will cause switch on. This effect forms the basis of a range of optical pnpn switches, but is not applicable to normal thyristor devices.

Conduction in a thyristor will continue as long as the current in the device is maintained above the *holding current*. It is necessary, therefore, in order to turn off the device, to remove the forward voltage from the device. In alternating current circuits, therefore, conduction normally ceases at the end of the forward conduction half cycle of the supply voltage. For extremely rapid turn off of the device, it is advantageous to reverse the polarity of the voltage across the device. Figure 6.28 shows the current/voltage characteristic for a thyristor for different gate currents.

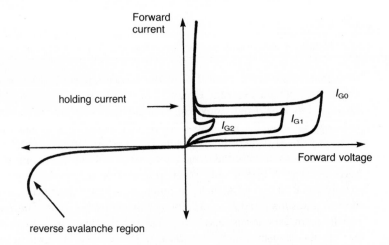

Fig. 6.28 Characteristics.

6.10.2 *The triac*

The *triac* (triode a.c. semiconductor switch) is similar in basic operation to a thyristor, but differs in that it can be switched into conduction in both directions. It is in essence equivalent to two thyristors mounted back to back, and finds application in systems requiring the control of conduction in both directions, for example, in full wave alternating current power systems. The commonly used circuit symbol for a triac is shown in Fig. 6.29.

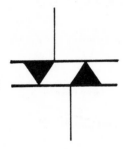

Fig. 6.29 The triac symbol.

6.10.3 *The unijunction transistor*

The *unijunction transistor* is a three terminal device which may be considered as a piece of n type silicon semiconductor, with connections at each end, and with a third connection made via a pn junction. The two end connections are known as the base connections, B_1 and B_2, while the third connection is the emitter E. Between B_1 and B_2, the unijunction has the characteristics of a resistance, called the interbase resistance R_{BB}, and whose value is dependent upon the doping levels used in the silicon; R_{BB} is normally in the range 5 to $9\,k\Omega$. Under normal conditions, base B_2 is biased positive with respect to B_1. The current flow from B_2 to B_1 is determined by the value of R_{BB}.

The potential at the pn junction is determined by the positioning of the junction on the body of the device, and is given by

$$V = \eta V_{BB} \tag{6.27}$$

where η is a fraction, normally of the order of 0.6 to 0.7. If the emitter potential is less positive than the potential V, the pn junction is reverse biased, and only a very small leakage current flows. If, however, the emitter potential is taken more positive than the voltage V, to a voltage V_p, where

$$V_p = \eta V_{BB} + V_D \tag{6.28}$$

then the emitter pn junction becomes forward biased, holes being injected into the body of the transistor. The voltage V_p is known as the *peak point voltage* and V_D is the emitter pn diode voltage and is of the order of 0.5 V. The injection of holes into the device causes the voltage between the emitter and base B_1 to fall rapidly to a low value, termed the valley point voltage V_V. The unijunction is thus a device with two stable states which finds many applications in switching circuits. The static emitter voltage/emitter current characteristic is shown in Fig. 6.30, together with the unijunction circuit symbol. The most important characteristic of the device is the peak point voltage V_p, this determining the critical switching voltage level in circuit applications. The fraction η is known as the *intrinsic standoff ratio*.

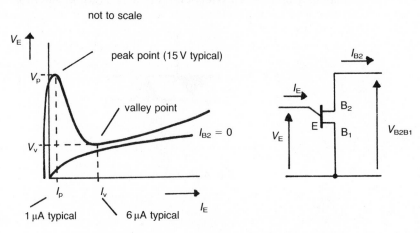

Fig. 6.30 Static characteristics of unijunction transistor.

6.11 Integrated circuits

In the same way that the invention of the transistor in 1948 revolutionized electronics, so the development of integrated circuits that started in the 1960s has been a similar revolution. Integration allows complex circuits, involving thousands of resistors, capacitors, diodes and transistors, to be included on a single chip of semiconductor. Compared with circuits involving discrete components and wired interconnections, integrated circuits offer the advantages of low cost (as a consequence of mass production), higher reliability, sophisticated circuitry more readily available, and all in a minute amount of space.

Most integrated circuits are formed on a substrate of p type silicon. The types of steps involved in producing, for example, a transistor are:

(1) Obtain the p type substrate.
(2) Grow an oxide layer on it.
(3) Form windows in the oxide layer for the base diffusion.
(4) Perform a boron diffusion through the windows to produce p type material.
(5) Grow a second oxide layer.
(6) Form windows for the emitter diffusion.
(7) Perform a phosphorus diffusion through the windows to produce n type material.

(8) Grow a third oxide layer.

(9) Form windows for the base and emitter contacts.

(10) Evaporate aluminium on to the surface.

(11) Remove the aluminium from all but the required conductor pattern.

The result of carrying out steps like those outlined above is that simultaneously large numbers of identical components and circuits are produced.

Integrated resistors can be obtained as a result of diffusion to produce a suitable doped channel. Capacitors may be formed by using the depletion layer capacitance between p and n type material or using the oxide layer as a dielectric between a metallized area and a suitably doped sub layer. Integrated diodes often are just a suitable connected transistor, often the emitter junction being the diode with the collector and base short-circuited.

Problems

1. In intrinsic germanium at room temperature the numbers of free electrons and holes are $n_i = 2.32 \times 10^{19}/m^3$. If the mobilities of electrons and holes are respectively $\mu_e = 0.36\,m^2\,V^{-1}\,s^{-1}$ and $\mu_h = 0.18\,m^2\,V^{-1}\,s^{-1}$, calculate the resistivity of the germanium.

2. What is the resistivity of silicon doped with boron at room temperature if the number of free holes is $10^{21}/m^3$ and the number of free electrons negligible in comparison. The holes have a mobility of $0.048\,m^2\,V^{-1}\,s^{-1}$.

3. What is the mobility of electrons in copper if it has a resistivity of $1.6 \times 10^{-8}\,\Omega m$ and 8.5×10^{28} electrons per cubic metre?

4. What is the diffusion current density in a p type semiconductor if the hole concentration changes from $2.0 \times 10^{10}/m^3$ to $1.4 \times 10^{10}/m^3$ in a distance of $10\,\mu m$? $D_h = 1.1 \times 10^{-3}\,m^2/s$.

5. With reverse bias a pn junction diode saturates at $2.5\,\mu A$ at a temperature of $27°C$. Calculate the current for a forward voltage of $0.22\,V$.

6. Using the diode equivalent circuit given in Fig. 6.5(c), determine the current through the $250\,\Omega$ resistance for the circuit in Fig. 6.31. What would be the value of the current if the battery connections were reversed?

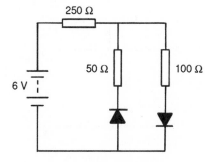

Fig. 6.31 Problem 6.

7. The following are the current–voltage data for the forward part of the characteristic of a diode. Plot the characteristic and hence determine the d.c. resistance and a.c. resistance when the applied voltage is (a) 0.25 V, (b) 0.275 V.

Voltage V	0	0.05	0.10	0.15	0.20	0.25	0.30
Current mA	0	0.2	0.4	0.6	4.0	30.0	200

8. The following are the current–voltage data for the forward part of the characteristic of a diode. What current will flow in a circuit with the diode in series with a 50 Ω resistor if the applied voltage is 4.0 V?

Voltage V	0	1.0	1.5	2.0	2.5	3.0
Current mA	0	4.0	9.7	18	30	50

9. The characteristic of a diode is described by the equation

    $$V = aI + bI^2$$

    where a and b are constants. What are the values of the d.c. resistance and the a.c. resistance at a current I?

10. A Zener diode has a breakdown voltage of 3.3 V with a maximum power dissipation of 400 mW. What is the maximum allowable current?

11. A varactor diode has a capacitance of 20 pF when the reverse bias voltage across it is 5 V. What will be the capacitance when the voltage is 6 V?

12. A transistor in the common emitter connection has $h_{fe} = 100$. Ignoring the small leakage current, what is the collector current when the base current is 0.1 mA?

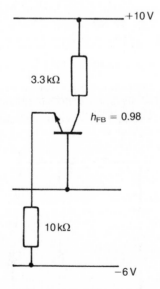

Fig. 6.32 Problem 15.

13. Determine the values of the components in the simplified equivalent circuit for a common emitter transistor if $h_{fe} = 400$ and $h_{ie} = 1200\,\Omega$.

14. A transistor in the common emitter configuration gave the following data for a base current of $20\,\mu A$. Plot the I_C–V_{CE} characteristic and hence determine the output impedance for a base current of $60\,\mu A$ at an operating point of $V_{CE} = 6\,V$.

V_{CE}	V	2	4	6	8	10
I_C	mA	3.5	3.7	3.9	4.1	4.3

15. For the circuit shown in Fig. 6.32, determine the transistor collector potential. Take the transistor current gain, h_{FB}, to be 0.98 and the emitter–base diode voltage to be 0.7 V.

16. For the circuit shown in Fig. 6.33, determine base current and the transistor collector potential. Take the value of h_{FE} to be 80 and the emitter–base diode voltage to be 0.7 V.

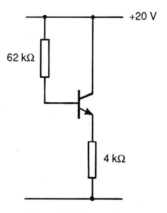

Fig. 6.33 Example 16.

17. The I_C/V_{CE} characteristics of a transistor are

V_{CE}	V	1	4	8	10	
I_C	mA	3.0	3.1	3.3	3.4	$I_B = 20\,\mu A$
		6.8	7.5	8.4	8.9	40
		10.8	11.9	13.1	13.9	60
		14.5	15.9	17.8	18.8	80
		18.6	20.8	23.3	24.8	100
		22.9	25.2	28.7	30.2	120
		26.0	29.0	33.0	35.0	140
		29.1	32.9	37.8	40.1	160
		32.5	36.7	42.0	44.7	180
		36.0	40.5	46.5	49.5	200

Plot the characteristics and draw the curve representing a dissipation in the transistor of 100 mW. What is the lowest value of collector load resistor which may be used with a collector supply voltage of 10 V if the dissipation in the transistor is not to exceed 100 mW for any value of base current? What base current is required with this value of collector load resistor to set the collector–emitter voltage to 5 V? Estimate the value of the d.c. current gain (h_{FE}) and the a.c. current gain (h_{fe}) at this operating point.

Chapter 7
Amplifiers

7.1 Amplifier gain

Throughout engineering practice, it is necessary to find the means whereby signals developed in a system can be increased in power to be eventually sufficiently powerful to perform their allotted tasks. Thus the mechanical force available from a man's arm can be used to raise from the ground a load weighing more than he does himself.

Consider a method by which a man may lift a very heavy weight. Figure 7.1 shows a four part pulley arrangement. Ignoring any frictional losses in the pulleys, the tension in the string in all parts is equal to the pull F. Since there are, in all, four strings supporting the lower block and load, the system will be in equilibrium if the downward force due to the weight of the load W and of the lower block is equal to $4F$. The pulley system is thus a *force amplifier*. Suppose now that, in order to raise the load, the pull F moves the string a distance of 1 m. Since there are four strings supporting the lower block, the load will be raised by a distance $\frac{1}{4}$ m. The work done by the pull is given by

$$\text{work done} = F \times 1$$

The work done on the load is given by

$$\text{work done} = 4F \times \tfrac{1}{4} = F$$

In practice, of course, no pulley arrangement, or indeed any other form of machine, can be constructed perfectly, and there would be some energy loss in overcoming frictional forces at the pulley bearings. Even if, however, these losses are negligible, we can see that in using such a force amplifier nothing is gained that is not put into the system. There is a *mechanical advantage* in amplifying the original available force of F to a magnitude of $4F$, but all energy used must be supplied by the applied pull.

Since the input power is equal to the rate of supplying energy, and considering a time interval during the movement of δt, then

$$\text{input power} = \frac{F \times 1}{\delta t}$$

and

$$\text{output power} = \frac{4F}{\delta t} \times \frac{1}{4} = \frac{F \times 1}{\delta t}$$

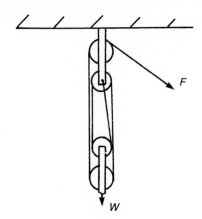

Fig. 7.1 Pulley system.

Thus even though we obtain a mechanical advantage by using the force amplifier, the output power can never exceed the input power and would, in practice, be less due to internal losses.

An analogy may be drawn between the pulley arrangement and the electrical transformer discussed in Chapter 1. Suppose that we have available an alternating voltage of rms value V, and that for a particular task a voltage of $4V$ is required. This could be arranged by using a transformer with a secondary winding consisting of four times the number of turns of the primary winding. Figure 7.2 shows the transformer connected to a load resistor of R, giving a current in the transformer secondary winding of

$$I_s = \frac{4V}{R}$$

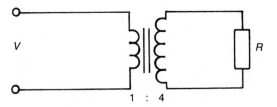

Fig. 7.2 Transformer system.

The current in the transformer primary, because of the 1:4 turns ratio, is equal to four times the secondary current and is, therefore,

$$I_p = \frac{16V}{R}$$

The power supplied to the load resistor is

$$P = V_s I_s = 4V \times \frac{4V}{R} = \frac{16V^2}{R}$$

The power input to the transformer is

$$P = V_p I_p = V \times \frac{16V}{R} = \frac{16V^2}{R}$$

The effect of the transformer is thus to amplify the voltage available, to the value required to supply the load. As in the pulley example, however, there is no gain of power, and in fact the input power to the transformer is exactly the same as that supplied to the load. Of course, we have again neglected the unavoidable small losses which would occur in the transformer due to winding resistance and losses in the iron core.

In the types of machine we have considered, the system is designed to give the operating conditions desired for a particular purpose, but no gain of power is possible. Consider, however, again, the problem of raising the heavy weight of W. If, instead of using the simple block and tackle, we obtain a motor driven crane, a different condition exists. All the energy used to raise the weight W will now be obtained from the electric mains supplying the crane motor. All that is required from the system *input signal*, i.e. from the crane operator, is a control signal to switch on the motor, perhaps to adjust the speed of winding, and to judge when to switch off. An energy balance between the system input, i.e. the crane operator, and the system output, i.e. the work done in raising the load, is now different in meaning from that applied in the previous examples. The function of the system input is only to control the transfer of energy from a power supply to the system output. The actual work done by the operator in pushing the control switches will be very small, and so the power gain of the system when defined as

$$\text{Power gain} = \frac{\text{Useful power output applied to the load}}{\text{Power input to the control switches}} \tag{7.1}$$

is very high indeed. The point to note is, of course, that as in the earlier examples, we have not succeeded in obtaining something for nothing. All power supplied to the load is taken from the motor main supply and must be paid for.

A similar condition exists in the electronic amplifiers which are the subject of this chapter. They consist in general of electronic circuits which are supplied with energy from a d.c. source, which may be a battery or a rectified alternating current source (see Chapter 11). A signal input to the circuit is used to control the transfer of energy from the power supply to the signal output. The normal requirement for an amplifier is that the output signal should be a higher power version of that applied to the input; thus the signal waveform at the output should be exactly the same shape as that at the input. In particular, if the input signal waveform is sinusoidal, the output waveform should also be sinusoidal. Any deviation from this linear relationship is caused by the presence in the output signal of frequencies not present in the input signal, and which must have been generated internally in the amplifier. Such internally generated output signals are termed *distortion* and the amplifier design should reduce such signals to a minimum.

The power gain of an amplifier is defined as

$$\text{power gain} = A_p = \frac{\text{output signal power}}{\text{input signal power}} \tag{7.2}$$

Consider, for example, an amplifier designed to drive a high fidelity loudspeaker system when driven by the output from a record player pick-up cartridge. The output from such a cartridge would be a signal containing frequencies in the range from about 20 Hz to perhaps 20 kHz. An input power level of about 1 microwatt (μW) would be available at a voltage of the order of 100 mV. A power output would be required of perhaps 30 W, which, with a loudspeaker load resistance nominally of 8 Ω, means an amplifier output voltage of the order of

$$V = \sqrt{(P_o R_L)} = \sqrt{(30 \times 8)} \simeq 15 \text{ V rms}$$

The amplifier would have an overall power gain of

$$A_p = \frac{30}{10^{-6}} = 30 \times 10^6$$

and a voltage gain of

$$A_v = \frac{15}{0.1} = 150$$

These values of power and voltage gain would be achieved, in practice, by a cascaded arrangement of amplifiers of lower gain.

7.1.1 Types of electronic amplifiers

Amplifiers may be broadly classified by considering the frequency range over which they are designed to operate.

Wideband amplifiers
These amplifiers are designed to provide power gain over a wide frequency range. In the example considered of an audio frequency amplifier, the requirement is for constant gain over the complete audio frequency spectrum, i.e. from approximately 20 Hz to 20 kHz. The operating bandwidth of an amplifier is normally specified as that frequency range over which the amplifier will give a power gain of not less than half of its maximum value. Figure 7.3(a) shows a power gain/frequency response curve typical for a wide band amplifier. Such a response is normally plotted on a logarithmic frequency scale because of the wide frequency range which must be considered. A further classification depends upon whether the amplifier bandwidth extends downward to include

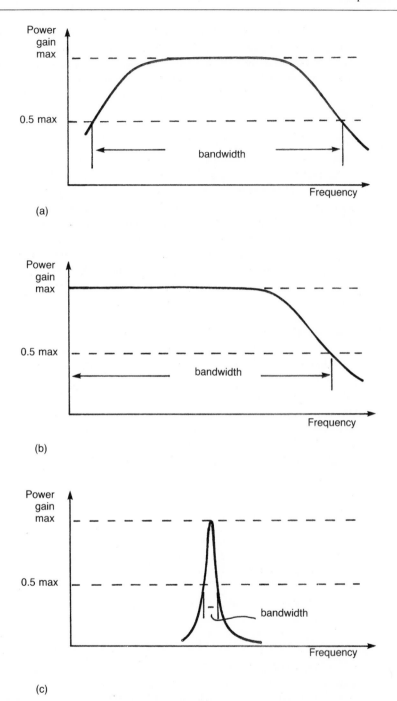

Fig. 7.3 Amplifiers, (a) wide band amplifier, (b) d.c. coupled wide band amplifier, (c) narrow band (tuned) amplifier.

0 Hz, i.e. whether the amplifier will respond to a d.c. input signal. In many applications a response to a d.c. signal is required; for example, a television receiver video amplifier requires a bandwidth of from 0 to about 5 MHz, while amplifiers designed for use with transducers in instrumentation systems often have a bandwidth of from 0 to 1.5 kHz. Figure 7.3(b) shows a response of this type.

Narrow band (tuned) amplifiers

These amplifiers are designed to provide power gain over a very narrow band of frequencies about a nominal centre frequency, and are used to select a desired signal frequency from a range of signals. A typical frequency response is shown in Fig. 7.3(c). Amplifiers of this type form the tuning arrangements in radio and television receivers. Another application of narrow band amplifiers is as null detector amplifiers to increase the sensitivity of alternating current bridge measurements. The use of a narrow band amplifier in bridge measurements gives some protection against interfering signals of other frequencies when extremely sensitive measurements are being made.

7.2   Input and output impedance

An amplifier may be represented by the box symbol shown in Fig. 7.4. It has one input and one output, the input signal being applied between the input connection and a common connection, while the amplifier output is available between the output connection and the same common connection. Note that in this diagrammatic representation, no attempt is made to show the battery or power supply discussed in the previous section. We are concerned at this stage merely with producing a model which will enable us to consider the signal conditions in the system.

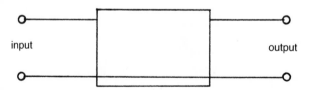

input   output

Fig. 7.4 Amplifier as a box.

The impedance between the input connections, which is presented by the amplifier to the signal source providing the input signal, is called the amplifier *input impedance* Z_1. In general this impedance must be represented by resistance and either inductance or capacitance, but over a limited frequency range about the middle of the amplifier operating range may be represented as an input resistance R_i. Figure 7.5 shows an amplifier of input resistance R_i when driven by a signal source of voltage v_s and source resistance R_s.

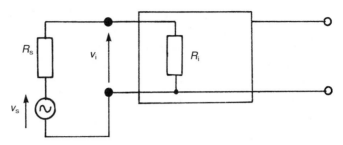

Fig. 7.5 Input impedance.

Only if the amplifier input resistance is infinite will the full source voltage v_s be applied to the amplifier input. For any non-infinite value of input resistance the amplifier input voltage is given by

$$v_i = \frac{R_i}{R_i + R_s} \times v_s \tag{7.3}$$

and the source resistance should be made as low as possible when compared with the source resistance, in order to prevent a considerable reduction in overall voltage gain.

The amplifier output may be considered as an output voltage source, together with an *output impedance Z_o*. Again the output impedance may be taken to be resistive, and of value R_o over a limited mid-range of frequencies. Figure 7.6 shows the amplifier connected to a load resistor R_L. For maximum voltage gain, the output resistance R_o should be small when compared to the load resistor R_L.

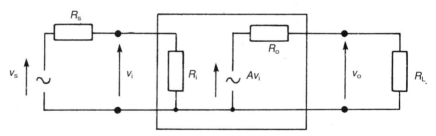

Fig. 7.6 Amplifier with load.

Example An amplifier of open-circuit voltage gain $A_v = 100$ (i.e. with $R_L = \infty$) has an input resistance of $2000\,\Omega$ and an output resistance of $20\,\Omega$. The amplifier is driven by a signal source of e.m.f. $10.0\,\text{mV}$ and source resistance $600\,\Omega$. The load connected to the amplifier output is a $100\,\Omega$ resistor. Determine (a) the overall voltage gain, and (b) the power gain.

(a) Figure 7.7 shows the circuit arrangement. The amplifier input voltage is given by

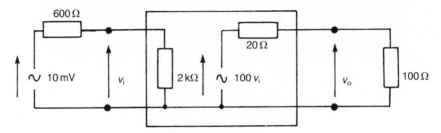

Fig. 7.7 Example.

$$v_i = \frac{2000}{2000 + 600} \times 10\,\text{mV} = 7.69\,\text{mV}$$

The open-circuit output voltage is

$$Av_i = 100 \times 7.69 = 0.769\,\text{V}$$

The output voltage when driving a $100\,\Omega$ load is

$$v_o = \frac{R_L}{R_L + R_o} \times Av_i$$

$$= \frac{100}{120} \times 0.769 = 0.641\,\text{V}$$

Before determining the value of the overall voltage gain, some thought must be given to the meaning of this quantity. The open-circuit e.m.f. of the signal source is known to be 10.0 mV. By using the amplifier described in the question we have achieved an output voltage of 0.641 V, i.e. an overall voltage gain of

$$\frac{0.641}{10.0} \times 1000 = 64.1$$

However, if the amplifier voltage gain was measured using voltmeters connected across the output (to determine v_o) and across the input (to determine v_i) then the gain would be found to be

$$\text{amplifier voltage gain} = \frac{v_o}{v_i} = \frac{0.641}{7.69} \times 1000 = 83.4$$

Obviously the exact meaning of *voltage gain* must be considered carefully. Note also that if the amplifier input resistance is very large when compared with the source resistance R_s then the two voltage gains would be the same.

(b) Defining the power gain as

$$\text{power gain } A_p = \frac{\text{power output to amplifier load}}{\text{power input to amplifier}}$$

then we obtain the following

$$\text{power output to load} = \frac{v_o^2}{R_L} = \frac{0.641^2}{100}W = 4.11\,\text{mW}$$

$$\text{power input to amplifier} = \frac{v_i^2}{R_i} = \frac{7.69^2}{10^3} \times \frac{1}{2000}W$$

$$= 0.0296\,\mu\text{W}$$

$$\text{power gain} = \frac{4.11}{10^3} \times \frac{10^6}{0.0296} = 138\,800$$

7.2.1 *Power transfer*

A point which has to be considered is whether an amplifier is accepting from the source all the available signal power. Consider a signal source of e.m.f. v_s and source resistance R_s connected directly to a load R. This is shown in Fig. 7.8.

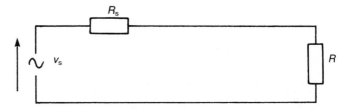

Fig. 7.8 Signal source delivering power to a load.

The current in the circuit is $i = \dfrac{v_s}{R_s + R}$

power in the load resistor $= i^2 R$

$$\text{or } P = \frac{v_s^2}{(R_s + R)^2} \times R$$

For a given source, in which the source e.m.f. and the source resistance are fixed, the load resistor R may be varied to obtain maximum load power.

We can determine the maximum load power as follows:

$$\frac{\text{d}P}{\text{d}R} = v_s^2 \frac{(R_s + R)^2 - R \times 2(R_s + R)}{(R_s + R)^2}$$

Equating $\dfrac{\text{d}P}{\text{d}R}$ to zero, maximum load power will be obtained when

$$(R_s + R)^2 - 2R(R_s + R) = 0$$

i.e. when $R = R_s$ (7.4)

This is a general result which is covered by the *maximum power transfer theorem* which states: 'For maximum power transfer from a source to a load, the load resistance should equal the source resistance.'

The maximum power obtainable from the source is therefore

$$P_{max} = \frac{v_s^2 R_s}{(R_s + R_s)^2} = \frac{v_s^2}{4R_s} \tag{7.5}$$

Example   For the previous example, Fig. 7.7, what could be the maximum power available from the source?

Returning to the source in the example, the maximum power available is

$$P_{max} = \frac{v_s^2}{4R_s} = \frac{0.01^2}{4 \times 600} = 0.0417 \,\mu W$$

As we determined previously, the actual power input to the amplifier is only $0.0296\,\mu W$. It is clear that the conditions which exist in an amplifier chosen to give maximum voltage gain are different from those needed to give maximum power gain.

7.2.2   *Determination of input and output resistance*

The input resistance of an amplifier may be measured directly by applying a known input signal voltage and measuring the signal current passing into the amplifier input, as shown in Fig. 7.9(a). The measurement must, however, be taken with conditions in the amplifier exactly as in normal use. Thus the input signal voltage will necessarily be quite small, and difficulty would, in general, be found in making the measurements, especially in the case of the input current, which would be very small indeed unless the input resistance is very low. Measurements of signal voltages and currents are, perhaps, best undertaken using an oscilloscope, which, with its very high input impedance, is ideally suited to measurements on high impedance circuits. It must always be borne in mind that if a measuring instrument, such as a voltmeter, abstracts an appreciable current from the circuit to which it is connected, then the voltage and current values in the circuit will be altered and an incorrect reading obtained. In amplifier circuits, where resistors may have values of many thousands of ohms, only high impedance measuring circuits may be used.

A better method for determining the input resistance is illustrated in Fig. 7.9(b). The amplifier is fed from a signal source whose source resistance is negligible in comparison with the amplifier input resistance R_i. The calibrated variable series resistor R_x is set at zero. The amplifier input voltage v_i will, in this condition, equal the source e.m.f. v_s. The amplifier output voltage v_o can be monitored using either a voltmeter or an oscilloscope. The series resistor R_x is now increased, until the amplifier output voltage is reduced to one half of its initial value. The amplifier input voltage, v_i, must also have been reduced to

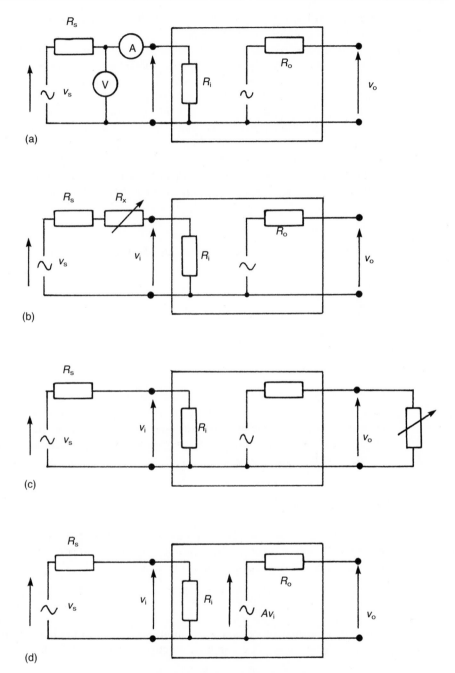

Fig. 7.9 Measurement of input and output resistances.

one half of its initial value. Examination of the input circuit shows that, if R_s is negligible, this occurs when $R_x = R_i$.

The output resistance may also be determined using a calibrated variable resistor. Figure 7.9(c) shows the amplifier loaded by the variable resistor. With the amplifier driven normally, and with R_x open circuit ($R_x = \infty$), the output voltage v_o is measured. The series resistor R_x is then reduced until the output voltage v_o is also reduced to one half of its open circuit value, which occurs when $R_x = R_o$. Notice that an absolute determination of the accurate value of the output voltage is not necessary; it is only necessary to know when the voltage has been reduced to one half of its value. The output resistance could be determined by direct measurements at the amplifier output terminals.

Referring to Fig. 7.9(d), the open-circuit output voltage (i.e. with no load connected) is

$$V_{oc} = Av_i$$

The output current with the output terminals short-circuited is

$$i_{sc} = \frac{Av_i}{R_o}$$

The output resistance can then be calculated from

$$\text{output resistance} = \frac{V_{oc}}{i_{sc}} = \frac{Av_i}{\dfrac{Av_i}{R_o}} = R_o \tag{7.6}$$

Obviously care would be necessary in practice in taking the short-circuit current measurement. In many cases such a measurement would not be possible while still maintaining the normal operating conditions in the amplifier. However, if not always practically a feasible method, analytically this method is often very useful.

7.3   Logarithmic gain units

Instead of expressing the gain of an amplifier as a simple ratio of the output power to the input power, advantages are found, in practice, by the use of a logarithmic unit. Thus power gain is expressed as

$$\text{power gain} = \lg \frac{P_o}{P_I} \text{ bels} \tag{7.7}$$

where P_o = output power and P_I = input power. Since the bel is rather large for many applications, it is common practice to express the gain as

$$\text{power gain} = 10 \lg \frac{P_o}{P_I} \text{ decibels (dB)} \tag{7.8}$$

The decibel unit may also be applied to voltage or current ratios. Consider two equal resistors, R_1 and R_2, each of value R, with applied voltages V_1 and V_2 respectively. In resistor R_1

$$\text{power} = P_1 = \frac{V_1^2}{R}$$

and in resistor R_2

$$\text{power} = P_2 = \frac{V_2^2}{R}$$

The power ratio is thus given by

$$\text{power ratio} = 10 \lg \frac{P_2}{P_1} \, \text{dB}$$

$$= 10 \lg \frac{V_2^2}{R} \times \frac{R}{V_1^2}$$

$$= 20 \lg \frac{V_2}{V_1} \, \text{dB} \tag{7.9}$$

Example An amplifier, of input resistance $1 \, \text{k}\Omega$ is driven by a source of e.m.f. $10 \, \text{mV}$ and negligible source resistance. The amplifier has an open circuit voltage gain of 200 and an output resistance of $1 \, \text{k}\Omega$. The amplifier load resistance is $1 \, \text{k}\Omega$. What are the voltage and power gains?

In Fig. 7.10, the amplifier input voltage is $v_i = 10 \, \text{mV}$. The amplifier output voltage is thus

$$v_o = A v_i \times \frac{R_L}{R_o + R_L} = 200 \times \frac{10}{10^3} \times \frac{1 \times 10^3}{2 \times 10^3} \times 1.0 \, \text{V}$$

$$\text{voltage gain} = A_v = 20 \lg \frac{1}{10^{-3}} = 60 \, \text{dB}$$

$$\text{input power} = \frac{v_L^2}{R_i} = \frac{(10^{-3})^2}{10^3} = 10^{-9} \, \text{W}$$

$$\text{output power} = \frac{v_o^2}{R_L} = \frac{(1)^2}{10^3} = 10^{-3} \, \text{W}$$

$$\text{power gain} = A_p = 10 \lg \frac{10^{-3}}{10^{-9}} = 60 \, \text{dB}$$

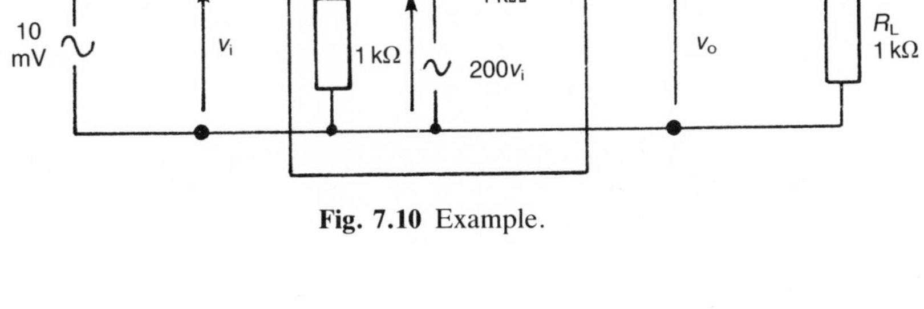

Fig. 7.10 Example.

Thus for equal resistance, the voltage ratio and the power ratio have the same numerical value when the decibel unit is used. This logarithmic system is in widespread use, and is applied generally, even between load resistors of different values, giving power and voltage (or current) gains which are not equal. Notice also that where a gain is less than unity, the gain expressed in decibels is a negative quantity.

7.3.1 Amplifiers in cascade

Consider two amplifiers, of gains A_1 and A_2, connected in cascade (Fig. 7.11). For an input voltage to amplifier 1 of v_{i1} V, the output voltage is

$$v_{o1} = A_1 v_{i1}$$

The overall gain is thus the product of the two individual gains

i.e. $\quad \dfrac{v_{o2}}{v_{i1}} = A_1 A_2$

Expressing this in decibels,

$$\text{gain} = 20 \lg A_1 A_2$$
$$\qquad\qquad = 20 \lg A_1 + 20 \lg A_2 \qquad\qquad (7.10)$$

In decibel form, therefore, the overall gain is given by adding together the individual gains. We have, however, made one assumption, that is, we have assumed that in connecting the two amplifiers in cascade, their individual gains are not altered. This is not always true in practice.

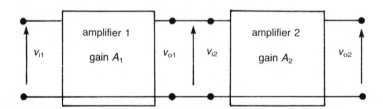

Fig. 7.11 Cascaded amplifiers.

Example   Figure 7.12 shows two amplifiers in cascade, each with a voltage gain of 10, an output resistance of $1\,k\Omega$, and an input resistance of $3\,k\Omega$. The cascaded pair is driven from a signal source of source resistance $1\,k\Omega$, and is loaded with a $3\,k\Omega$ load. Determine the overall gain.

The open circuit gain of each individual amplifier is given as 10, i.e.

$$\text{gain} = 20 \lg 10 = 20\,\text{dB}$$

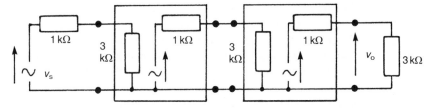

Fig. 7.12 Example.

At each coupling, however, between the source and the input of amplifier 1, between the two amplifiers, and also between amplifier 2 and its load, a voltage attenuation will occur, of ratio given by

$$\text{attenuation} = \frac{3 \times 10^3}{3 \times 10^3 + 1 \times 10^3} = 0.75$$

i.e.

$$\text{attenuation} = 20 \lg 0.75 = -2.5 \, \text{dB}$$

The overall gain is the sum of each individual gain, as follows:

(1) A gain of $-2.5 \, \text{dB}$ due to the load effect of amplifier 1 on the source.
(2) A gain of $20 \, \text{dB}$ in amplifier 1.
(3) A gain of $-2.5 \, \text{dB}$ due to the loading effect of amplifier 2 on amplifier 1.
(4) A gain of $20 \, \text{dB}$ in amplifier 2.
(5) A gain of $-2.5 \, \text{dB}$ due to the loading effect of the load upon amplifier 2.

The overall gain is thus

$$-2.5 + 20 - 2.5 + 20 - 2.5 = 32.5 \, \text{dB}$$

It is instructive to obtain this result directly using the gain ratios. Thus we have

(1) A gain of 0.75 due to the loading effect of amplifier 1 on the source.
(2) A gain of 10 in amplifier 1.
(3) A gain of 0.75 due to the loading effect of amplifier 2 on amplifier 1.
(4) A gain of 10 in amplifier 2.
(5) A gain of 0.75 due to the loading effect of the load upon amplifier 2.

The overall gain is thus

$$0.75 \times 10 \times 0.75 \times 10 \times 0.75 = 42.18$$

or, in decibel form

$$\text{gain} = 20 \lg 42.18 = 32.5 \, \text{dB}$$

7.4 Amplifier bandwidth

The bandwidths of various types of amplifier were considered briefly in Section 7.1.1. The upper and lower cut-off frequencies, i.e. the frequencies at which the gain has reduced by 3 dB, are fixed either by the characteristics of the active device used to make the amplifier or, alternatively, by frequency dependent components in the amplifier circuit. Every active amplifying device, e.g. the transistor, has an upper limit to its operating frequency, often decided by the time of transit of charge through the device. However, many amplifiers have upper cut-off frequencies far lower than the cut-off frequency of the active device. In fact, unavoidable stray capacitance between the signal leads and earth has the effect of reducing the upper cut-off frequency. At the lower end of the frequency spectrum the devices will respond easily to frequencies down to 0 Hz. However, any slow change in device characteristics, due to ageing or to temperature effects, may cause a corresponding change in the d.c. potential at the amplifier output terminal, which would be indistinguishable from a d.c. (0 Hz) output signal.

To remove these drift effects, a low frequency cut-off characteristic is sometimes included, usually by the inclusion of capacitors in series with one or more input connections.

7.4.1 The effect of stray capacitance

Let us consider a signal source of e.m.f. v_s and source resistance R_s feeding a load resistor R_L, as shown in Fig. 7.13. Across the load resistor is a capacitor C to represent the circuit stray capacitance.

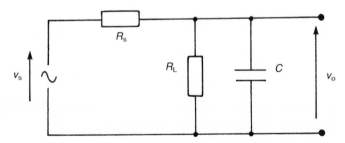

Fig. 7.13 Effect of stray capacitance.

By applying Thévenin's theorem, the circuit may be redrawn as in Fig. 7.14. Here the source resistance and the load resistance have been included together as

$$R = \frac{R_s R_L}{R_s + R_L} \tag{7.11}$$

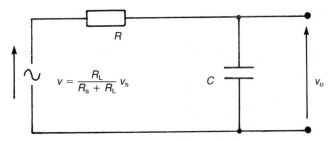

Fig. 7.14 Thévenin equivalent circuit.

while the generator now has the value

$$v = \frac{R_L}{R_s + R_L} \times v_s \tag{7.12}$$

The circuit is thus equivalent to a single resistor–capacitor network. If the capacitor C is negligibly small, then the circuit response is aperiodic, i.e. the circuit output is the same at all frequencies.

If, however, the capacitor is too large to neglect, then

$$\frac{v_o}{v} = \frac{\dfrac{1}{j\omega C}}{R + \dfrac{1}{j\omega C}} = \frac{1}{1 + j\omega CR}$$

By defining an angular frequency ω_o as

$$\omega_o = \frac{1}{CR}$$

then

$$\frac{v_o}{v} = \frac{1}{1 + j\dfrac{\omega}{\omega_o}} \tag{7.13}$$

To study the magnitude and phase of the output voltage, the expression should be rationalized and converted into the polar form

$$\frac{v_o}{v} = \frac{1}{1 + \left(\dfrac{\omega}{\omega_o}\right)^2} \times \left(1 - j\dfrac{\omega}{\omega_o}\right)$$

$$= \frac{\sqrt{[1 + (\omega/\omega_o)^2]}}{1 + (\omega/\omega_o)^2} \Big/ \tan^{-1} - \frac{\omega}{\omega_o}$$

$$= \frac{1}{\sqrt{[1 + (\omega/\omega_o)^2]}} \Big/ \tan^{-1} - \frac{\omega}{\omega_o} \tag{7.14}$$

The form of the response is conveniently examined as follows:

(1) when $\omega \ll \omega_o$. If ω is small in comparison with ω_o, then the term $(\omega/\omega_o)^2$ may be neglected in comparison with 1.
The expression therefore becomes

$$\frac{v_o}{v} = \frac{1}{\sqrt{1}} \bigg/ \tan^{-1} - \frac{\omega}{\omega_o} = 1 \bigg/ \tan^{-1} - \frac{\omega}{\omega_o}$$

While this approximation holds, therefore, the output voltage is constant in magnitude, and in decibels

$$20 \lg \frac{v_o}{v} = 20 \lg 1 = 0\,\text{dB}$$

The output voltage lags the input voltage by a very small angle

$$\phi = \tan^{-1} \frac{\omega}{\omega_o}$$

(2) when $\omega = \omega_o$. At this particular frequency, the function becomes

$$\frac{v_o}{v} = \frac{1}{\sqrt{(1 + 1)}} \bigg/ \tan^{-1} -1 = \frac{1}{\sqrt{2}} \bigg/ -45°$$

The magnitude of the output voltage has fallen to

$$v_o = \frac{1}{\sqrt{2}} v$$

i.e. the gain has reduced by

$$20 \lg \sqrt{2} = 3\,\text{dB}$$

The output voltage now lags the input voltage by 45°. This special frequency when $\omega = \omega_o = \dfrac{1}{CR}$ is therefore the upper cut-off frequency.

(3) when $\omega \gg \omega_o$. At frequencies, when ω is large in comparison with the cut-off frequency ω_o, then $(\omega/\omega_o)^2$ is large in comparison with unity. The expression becomes

$$\frac{v_o}{v} = \frac{1}{\sqrt{(\omega/\omega_o)^2}} \bigg/ \tan^{-1} - \frac{\omega}{\omega_o}$$

$$= \frac{\omega_o}{\omega} \bigg/ \tan^{-1} - \frac{\omega}{\omega_o}$$

The gain is now inversely proportional to the frequency; by doubling the frequency, therefore, the gain is halved, i.e. reduced by

$$20 \lg 2 = 6\,\text{dB}$$

If $\omega \gg \omega_o$ therefore, the gain reduces by 6 dB per octave (20 dB per decade). The output voltage now lags the input voltage by an angle greater than 45°, and which would increase to 90° at infinite frequency.

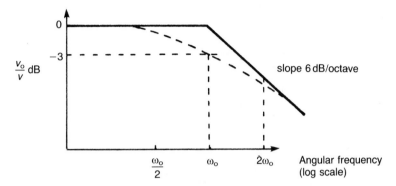

Fig. 7.15 Magnitude response.

The magnitude response can be plotted in an approximate form as shown in Fig. 7.15. If plotted as gain in decibels against frequency to a logarithmic scale, the response can be approximated to two straight lines corresponding to the two approximations,

$$\omega \ll \omega_o \quad \text{and} \quad \omega \gg \omega_o$$

For $\omega \ll \omega_o$, the response approximates to a straight line

$$\frac{v_o}{v} = 0 \, \text{dB}$$

For $\omega \gg \omega_o$, the response approximates to a straight line, reducing from 0 dB at $\omega = \omega_o$ at a constant slope of 6 dB per octave.

The true response is, of course, asymptotic to both of these lines, and must pass through the -3 dB point at $\omega = \omega_o$. The true response can be sketched in with reasonable accuracy. This form of response plot, with gain in decibels plotted against frequency to logarithmic scale, is known as a Bode plot. Note that with a logarithmic frequency scale, equal lengths along the frequency axis do not correspond to equal frequency steps, but rather to equal ratios. Arising directly out of this form of plot, the cut-off frequency ω_o is also known as the *corner frequency*.

We have now considered the effect of one shunt capacitor across a load driven by a signal source. Stray capacitance will, of course, occur at all points throughout the amplifier circuit and must be considered at each point. Its effect will be more serious, i.e. the cut-off frequency will be lower, the higher are the values of the load and source resistors.

7.4.2   The effect of series capacitance

Figure 7.16 shows a voltage source of source resistance R_s and e.m.f. v_s coupled to its load R_L via a series capacitor C. The output voltage may be written

$$v_o = \frac{R_L}{R_s + R_L + \dfrac{1}{j\omega C}} \times v_s$$

$$= \frac{\dfrac{R_L}{R_s + R_L}}{1 + \dfrac{1}{j\omega C(R_s + R_L)}} \times v_s \qquad (7.15)$$

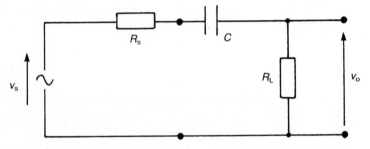

Fig. 7.16 Effect of series capacitance.

At frequencies at which the reactance of C is negligible

$$v_o = \frac{R_L}{R_s + R_L} \times v_s = v$$

Substituting this in Equation (7.15)

$$v_o = \frac{v}{1 + \dfrac{1}{j\omega C(R_s + R_L)}}$$

or

$$\frac{v_o}{v} = \frac{1}{1 + \dfrac{\omega_o}{j\omega}} = \frac{1}{1 - j\dfrac{\omega_o}{\omega}} \qquad (7.16)$$

where ω_o is defined as $\dfrac{1}{C(R_s + R_L)}$

Again, to study the magnitude and phase, this expression must be rationalized and converted into the polar form.

$$\frac{v_o}{v} = \frac{1}{1 - j\frac{\omega_o}{\omega}} = \frac{1 + j\frac{\omega_o}{\omega}}{1 + \left(\frac{\omega_o}{\omega}\right)^2}$$

or

$$\frac{v_o}{v} = \frac{\sqrt{[1 + (\omega_o/\omega)^2]}}{1 + (\omega_o/\omega)^2} \Big/ \tan^{-1}\frac{\omega_o}{\omega}$$

$$= \frac{1}{\sqrt{[1 + (\omega_o/\omega)^2]}} \Big/ \tan^{-1}\frac{\omega_o}{\omega} \qquad (7.17)$$

We can now examine the form of the response.

(1) when $\omega \gg \omega_o$. If $\omega \gg \omega_o$, then the term $(\omega_o/\omega)^2$ may be neglected when compared with unity. The expression therefore becomes

$$\frac{v_o}{v} = \frac{1}{\sqrt{1}} \Big/ \tan^{-1}\frac{\omega_o}{\omega} = 1 \Big/ \tan^{-1}\frac{\omega_o}{\omega}$$

While this approximation holds, the output voltage is constant in magnitude, and in decibels

$$20\lg\frac{v_o}{v} = 20\lg 1 = 0\,dB$$

The output voltage leads the input voltage by a very small angle

$$\phi = \tan^{-1}\frac{\omega_o}{\omega}$$

(2) when $\omega = \omega_o$. At this particular frequency, the expression becomes

$$\frac{v_o}{v} = \frac{1}{\sqrt{(1 + 1)}} \Big/ \tan^{-1}1 = \frac{1}{\sqrt{2}} \Big/ 45°$$

ω_o is, therefore, the lower cut-off or corner frequency. The gain is 3 dB down in comparison with that at mid range frequencies, and the phase angle has increased to 45°.

(3) when $\omega \ll \omega_o$. At frequencies when $(\omega_o/\omega)^2$ is large in comparison to unity, the expression may be written as

$$\frac{v_o}{v} = \frac{\omega}{\omega_o} \Big/ \tan^{-1}\frac{\omega_o}{\omega}$$

The voltage ratio is now directly proportional to frequency; doubling the frequency will also double the output voltage giving an increase of

$$20\lg 2 = 6\,dB$$

While this approximation holds, therefore, the output increases with frequency at the rate of 6 dB per octave (20 dB per decade). The output voltage now leads the input voltage by an angle greater than 45°, the phase angle approaching 90° as the frequency reduces to 0 Hz.

The magnitude response can, as in the previous case, be shown as a Bode plot, approximated as two straight lines, corresponding to the two approximations

$$\omega \ll \omega_o \quad \text{and} \quad \omega \gg \omega_o$$

The response plot is shown in Fig. 7.17.

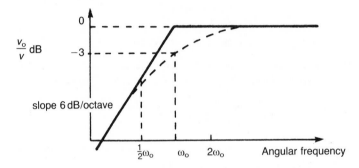

Fig. 7.17 Magnitude response.

Notice that in the response plots of Figs. 7.15 and 7.17, in each case 0 dB corresponds to the response in the mid frequency range, i.e. to the response at frequencies at which the effect of the capacitive reactance is negligible. This is quite an arbitrary choice, and arises because the ordinate of the graph is v_o/v expressed in decibels. When the ordinate is taken as v_o/v and expressed in decibels, then the voltage ratio in the mid frequency range would have been

$$\frac{v_o}{v_s} = 20 \lg \frac{R_L}{R_s + R_L} \, \text{dB}$$

and the complete response would have been lowered by this amount. To assist in sketching on to the Bode plot the final response, the following points should be noted:

(1) At the cut-off frequency, the final response is 3 dB below the mid range value,
(2) At $\omega = 2\omega_o$, the response is

$$20 \lg \frac{1}{\sqrt{[1 + (\omega_o/\omega)^2]}} = 20 \lg \frac{1}{\sqrt{1.25}} \simeq -1 \, \text{dB}$$

In fact, for both the series and the shunt capacitor responses, there is a difference of very nearly 1 dB between the actual response and the approximated response at both $\omega = 2\omega_o$ and $\omega = 0.5\omega_o$.

Example An amplifier has an open circuit voltage gain of 160, an input resistance of $20\,k\Omega$ and an output resistance of $2\,\Omega$. It drives directly a resistive load of $15\,\Omega$. The amplifier is driven from a source of resistance $5\,k\Omega$ via a series coupling capacitor of $0.2\,\mu F$. The amplifier input capacitance (including stray capacitance) is $500\,pF$. Determine the lower and upper cut-off frequencies and sketch the system response.

The amplifier is shown in Fig. 7.18. In the mid frequency range, when the effect of the capacitors is negligible, then

$$\frac{v_i}{v_s} = \frac{20 \times 20^3}{20 \times 10^3 + 5 \times 10^3} = \frac{20}{25}$$

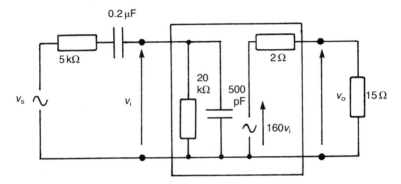

Fig. 7.18 Example.

i.e. due to the input and source resistance

$$\text{voltage ratio} = 20\lg\frac{20}{25} = -1.94\,dB$$

The amplifier open-circuit voltage gain is 160 or

$$20\lg 160 = 44.08\,dB$$

Due to the output and load resistances, there is a gain reduction given by

$$20\lg\frac{15}{2 + 15} = -1.09\,dB$$

The total mid frequency gain is thus

$$-1.94 + 44.08 - 1.09 = 41.05\,dB$$

The lower cut-off frequency is given by

$$\omega = \frac{1}{C_c(R_s + R_I)}$$

$$= \frac{1}{\dfrac{0.2}{10^6}(20 \times 10^3 + 5 \times 10^3)} = 200\,rad/s$$

i.e.

$$f = \frac{200}{2\pi} = 31.8\,\text{Hz}$$

The upper cut-off frequency is given by

$$\omega = \frac{1}{C_s R}$$

where R is the value of R_s and R_l in parallel

$$R = \frac{5 \times 20}{5 + 20}\,\text{k}\Omega = 4\,\text{k}\Omega$$

$$\omega = \frac{1}{\dfrac{500}{10^{12}} \times 4 \times 10^3} = \frac{10^6}{2}\,\text{rad/s}$$

i.e.

$$f = \frac{10^6}{2} \times \frac{1}{2\pi} = 79.6 \times 10^3\,\text{Hz}$$

$$= 79.6\,\text{kHz}$$

The amplifier response is shown in Fig. 7.19.

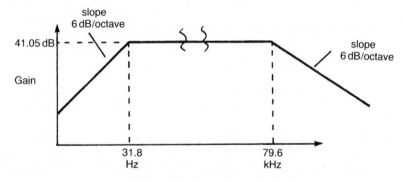

Fig. 7.19 Example.

7.4.3 *Amplifier frequency response*

An amplifier frequency response can, in general, be divided into three sections. These are

(1) The mid frequency range, in which the gain is substantially constant, and effects due to shunt and series capacitors are negligible.

(2) The high frequency range, in which the shunt capacitance causes the gain to reduce as the frequency increases.
(3) The low frequency range, in which series capacitance causes the gain to reduce as the frequency reduces.

The frequencies which form the boundaries between the ranges are, as we have seen, determined by the circuit values, and it is not possible to give any general figures. Thus a frequency which would, for one particular amplifier system, be considered a low frequency, may quite easily fall in the high frequency section of the response of a different amplifier.

7.5   The application of feedback to amplifiers

In our work so far, we have considered the use of amplifiers which have fixed characteristics, i.e. as shown in Fig. 7.6, the amplifier is specified by its input resistance R_i, its output resistance R_o and its open circuit gain A. In practice, these parameters are not constant and would, in fact, vary with changes in ambient temperature, with changes in power supply voltages or perhaps with age. These changes may be minimized by applying negative feedback to the amplifier. In a more general way, feedback, either positive or negative, can be used to modify the characteristics of an amplifier.

The term *feedback* is used when part of the output signal is fed back into the input. It is *negative feedback* if the fed back signal opposes the input signal, i.e. subtracts from it, and *positive feedback* if it adds to it.

7.5.1   Effect of amplifier gain

Figure 7.20 shows a basic amplifier of gain A, to which negative feedback has been applied. In order to simplify the analysis, assume that the amplifier input resistance is sufficiently large, in comparison with the source resistance, that loading effects may be neglected. The feedback has been applied by connecting a potential divider across the amplifier output and tapping off a fraction β of the output voltage, this fraction being connected in series with the input signal in such polarity as to oppose the input signal. The arrows in Fig. 7.20 show the effective directions of the signal voltages. A further assumption is that the loading effect of the feedback potential divider upon the output is negligible, i.e. the resistance of the potential divider is high in comparison with the amplifier output resistance.

Since the feedback voltage opposes the input signal, the amplifier input voltage is

$$v_i = v - \beta v_o \qquad (7.18)$$

The amplifier output voltage is then

$$v_o = A v_i$$

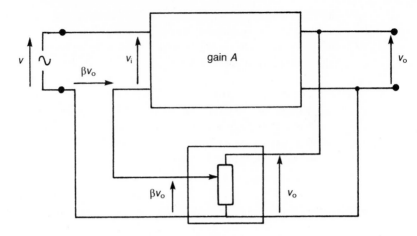

Fig. 7.20 Amplifier with feedback.

or

$$v_o = Av - \beta A v_o \tag{7.19}$$

The overall gain of the system is, therefore

$$A_f = \frac{v_o}{v} = \frac{A}{1 + \beta A} \tag{7.20}$$

The feedback, since it opposes the original input signal, is termed negative feedback. Its effect is to reduce the gain from that of the basic amplifier, i.e. by the factor $1 + \beta A$.

If the original gain A was reasonably large, so that the factor βA is large compared to unity, then the overall gain becomes

$$A_f = \frac{v_o}{v} = \frac{A}{\beta A} = \frac{1}{\beta} \tag{7.21}$$

This is, of course, independent of the gain of the original amplifier and, in fact, the overall system is decided only by the feedback fraction β. Therefore, any changes in the gain of the amplifier, due to any cause, will not affect the gain of the overall system, provided that the product βA always is large compared with unity.

The sensitivity of the system to changes in the amplifier gain may be determined as follows:

The overall gain with feedback is

$$A_f = \frac{v_o}{v} = \frac{A}{1 + \beta A} \tag{7.22}$$

Differentiating with respect to A

$$\frac{\mathrm{d}A_f}{\mathrm{d}A} = \frac{(1 + \beta A) - \beta A}{(1 + \beta A)^2} = \frac{1}{(1 + \beta A)^2} \tag{7.23}$$

Combining Equations (7.22) and (7.23),

$$\frac{\mathrm{d}A_f}{A_f} = \frac{1}{1 + \beta A} \times \frac{\mathrm{d}A}{A}$$

Thus the effect of a fractional change $\mathrm{d}A/A$ in the gain of the amplifier is a fractional change of

$$\frac{\dfrac{\mathrm{d}A}{A}}{1 + \beta A} \tag{7.24}$$

in the overall system.

One of the main causes of change in the gain of an amplifier is the reactive effects of series capacitance, and of shunt stray capacitance, as explained in Section 7.4. The effects of these reactances upon the gain of the system will be reduced by negative feedback, giving a resultant increase in the bandwidth of the amplifier. Summarizing the effects of negative feedback on the system gain:

(1) The overall gain of the system is reduced by the factor $(1 + \beta A)$.
(2) The sensitivity of the system gain to changes in the amplifier gain is reduced by the factor $(1 + \beta A)$.

Example An amplifier, of gain 200, has negative feedback applied by feeding back one quarter of the output voltage in series with its input. What is the effect on the gain of the feedback being reduced by 20%?

$$\text{Amplifier gain, with feedback} = \frac{A}{1 + \beta A}$$

$$= \frac{200}{1 + \frac{1}{4} \times 200} = 3.92$$

If the amplifier gain should reduce by 20% to 160

$$\text{gain, with feedback} = \frac{160}{1 + \frac{1}{4} \times 160} = 3.90$$

The overall gain has, therefore, changed by about 0.5% for a 20% reduction in gain in the amplifier.

7.5.2 *Effect on the input resistance*

The input resistance of any system is defined as the resistance R_i given by

$$R_i = \frac{v}{i}$$

where v is the input signal voltage applied to the system and i is the resulting signal current which flows into the system input.

Figure 7.21 shows the input of the amplifier. With no applied feedback, i.e. $\beta v_o = 0$, then

$$v = v_i$$

and the input resistance is

$$R_i = \frac{v_i}{i} = \frac{v}{i} \qquad (7.25)$$

With applied negative feedback, however, the actual amplifier input v_i is much less than the system input v because

$$v_i = v - \beta v_o$$

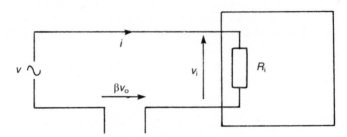

Fig. 7.21 Amplifier input circuit.

For a given value of input voltage v, therefore, the signal current flowing into the amplifier is less with feedback than without. The effect of applying negative feedback in this manner· is thus to raise the input resistance of the system.

Since

$$v_o = A v_i$$

then

$$v_i = v - \beta A v_i$$

or

$$v = v_i(1 + \beta A)$$

The input resistance, with applied feedback, is therefore

$$R_{i_f} = \frac{v}{i} = \frac{v_i(1 + \beta A)}{i}$$

or

$$R_{i_f} = R_i(1 + \beta A) \qquad (7.26)$$

The effect of negative feedback applied in this manner is to increase the input resistance by the factor $(1 + \beta A)$.

7.5.3   *Effect on the output resistance*

In Section 7.2 it was shown that an amplifier output resistance is given by

$$R_\mathrm{o} = \frac{\text{open-circuit voltage}}{\text{short-circuit current}}$$

The output circuit of the amplifier is shown in Fig. 7.22. The assumption is still made that the loading effect of the feedback potentiometer on the amplifier output is negligible, i.e. the resistance of the potentiometer is large in comparison with R_o.

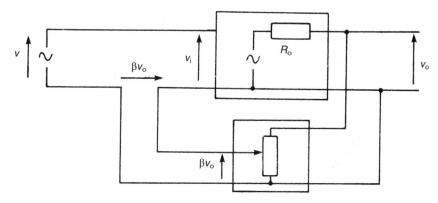

Fig. 7.22 Amplifier output circuit.

Without feedback, therefore, (i.e. with $\beta v_\mathrm{o} = 0$)

open-circuit output voltage $v_\mathrm{oc} = Av_\mathrm{i} = Av$

and

short-circuit output current $i_\mathrm{sc} = \dfrac{Av_\mathrm{i}}{R_\mathrm{o}} = \dfrac{Av}{R_\mathrm{o}}$

$$\text{output resistance} = \frac{v_\mathrm{oc}}{i_\mathrm{sc}} = \frac{Av}{\dfrac{Av}{R_\mathrm{o}}} = R_\mathrm{o} \qquad (7.27)$$

With negative feedback applied,

open-circuit output voltage $v_\mathrm{oc} = Av_\mathrm{i}$

$$= A(v - \beta v_\mathrm{oc})$$

$$= \frac{Av}{1 + \beta A}$$

With the output short-circuited, in order to determine the short-circuit current, the output voltage is obviously zero. There is, therefore, no feedback voltage, and the feedback is removed from the system. Thus the short-circuit output current is

$$i_{sc} = \frac{Av_i}{R_o} = \frac{Av}{R_o}$$

The output resistance with feedback is, therefore, given by

$$R_{o_f} = \frac{v_{oc}}{i_{sc}} = \frac{Av}{1 + \beta A} \times \frac{R_o}{Av} = \frac{R_o}{1 + \beta A} \tag{7.28}$$

The effect of negative feedback, applied in this manner, is thus to reduce the output resistance of the amplifier by the factor $(1 + \beta A)$.

7.5.4   *Effect on signals*

Ideally an amplifier should produce at its output an exact, amplified replica of the signal applied to its input. However, in all practical amplifiers the output will contain, in addition to the amplified input signal, components at other frequencies. These unwanted output components may be considered as internally generated signals. They arise from many sources.

In all circuits, for example, there exist signal components which are caused by the random motions of electric charges in the circuit. At low current levels, the particle nature of an electric current becomes more obvious and there is a continuous random variation in the magnitude from instant to instant. Because the magnitude of this variation is dependent upon temperature interfering signals, due to this effect, are described as *thermal noise*. The noise energy is uniformly distributed over the whole of the frequency spectrum, and so produces an output at all frequencies within the amplifier frequency range. Similar wideband noise signals are also produced in the active devices used in the amplifier. If the amplifier is an audio amplifier, designed to drive a loudspeaker, the noise signal would manifest itself as an audible background hiss.

A further source of interference is the 50 Hz (or 60 Hz) main power system. This is very often used to produce the d.c. power supply for the amplifier circuit, and can affect the signal circuits by means of e.m.f.s, electrostatically induced voltages in the higher impedance sections of the circuit, or electromagnetically induced currents in circuit loops. This interference would manifest itself in the amplifier output as a 50 Hz (or 100 Hz, depending upon the type of power supply) signal, which would give a deep hum note in the output loudspeaker.

These interfering signals form a lower limit to the allowable input signal to the amplifier, below which the signal to noise ratio is inadequate. Internally generated noise of this sort normally causes difficulties only in the input stage of an amplifier, because after amplification in the input stage, the signal level is sufficiently large for noise generated in the second amplifier stage to be negligible.

The amplitudes of the noise signals described are normally independent of the amplitude of the input signal, since the noise effects are properties of the

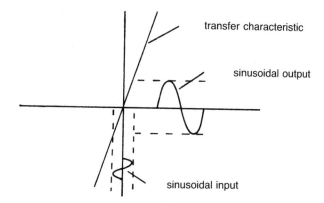

Fig. 7.23 Input/output characteristic.

amplifier circuit itself. There is, however, a further class of internally generated signals whose amplitude is a function of the amplitude of the signal itself. Figure 7.23 shows the input/output characteristic (or transfer characteristic) of an ideal amplifier. Because of the linear form of the characteristic, a sinusoidal input signal results in an amplified sinusoidal output signal. However, any departure from perfect linearity in the transfer characteristic will produce a non-sinusoidal output, as is shown in exaggerated form in Fig. 7.24.

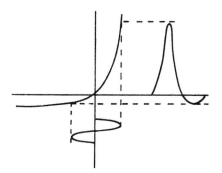

Fig. 7.24 Effect of non-linear characteristic.

The non-sinusoidal output can be shown (by Fourier analysis) to consist of a sinusoidal signal component at the input signal frequency, together with other harmonically related signals. These internally generated signals are normally described as *distortion*. Because most practical amplifier stage transfer characteristics are reasonably linear for small input signals, distortion occurs mainly in the amplifier stages where the signal amplitude is large, i.e. in the amplifier output stage. There is, therefore, the fundamental difference between internally generated distortion signals and internally generated wideband noise signals in that they are produced in different parts of the amplifier circuit.

Consider now the effect of applying negative feedback to an amplifier which produces some internally generated noise or distortion signal. Let the noise output from the amplifier without feedback be n. Let the noise output from the amplifier with feedback be n_f. With feedback, a fraction βn_f of the output noise is fed back into the input circuit and is then amplified A times. The amplifier noise output is thus

$$n_f = n - \beta A n_f$$

since the applied feedback does not alter the amount of noise produced in the amplifier. Thus

$$n_f = \frac{n}{1 + \beta A} \tag{7.29}$$

The output noise component is thus reduced by the same factor $(1 + \beta A)$. It must be carefully considered, however, whether this reduction in noise output provides a useful increase in the signal to noise ratio. In reducing the generated noise by the factor $(1 + \beta A)$, by the addition of negative feedback, we must also have produced a decrease in overall voltage gain by the same factor. The signal to noise ratio must, therefore, be the same with feedback as without. If, however, we can insert in the amplifier an extra amplifying stage which will increase the gain of the system without increasing the noise output, then the signal to noise ratio would be improved.

Consider first, noise entering the system at the amplifier input, and including the thermal noise and transistor noise associated with the input amplifying stage. The application of negative feedback will cause a reduction in both the noise output and the signal output by the factor $(1 + \beta A)$. If an extra amplifying stage is included in the amplifier to return the gain to its original value, this will affect equally the noise signal and the input signal. No improvement will be obtained in the signal to noise ratio.

Consider now, noise entering the system anywhere else except at the input stage, and including, of course, the distortion signals generated in the later stages of the amplifier. Negative feedback will, as before, reduce the output of noise, and of the signal, by $(1 + \beta A)$ but, in this case, the inclusion of an extra amplifying stage in the system at any point earlier in the signal path than the noise is generated will return the gain to its original value without affecting the noise output. The overall signal to noise ratio will, therefore, be improved.

The following general conclusions may, therefore, be drawn:

(1) The signal to noise ratio cannot be improved by feedback if the noise is entering the amplifier at the input or in the input stage.
(2) The signal to noise ratio can be improved by feedback if an additional amplifying stage, free from noise, can be included in the system prior to the point where the noise signal is being generated.

7.5.5 Positive feedback and amplifier stability

In the previous work in this section, we have shown that the effect of negative feedback upon the gain of the amplifier of Fig. 7.20 is to give a gain, with feedback equal to

$$A_f = \frac{A}{1 + \beta A}$$

In Fig. 7.20, the feedback voltage βA is produced by means of a potential divider and is fed back in series with the input voltage in such a phase as to oppose the input voltage. If the phase of the feedback signal is reversed, then the overall gain will become

$$A_f = \frac{A}{1 - \beta A} \tag{7.30}$$

The overall gain of the system will, therefore, be increased by the feedback, which is now termed *positive feedback* since the fed back signal and the input signal are now additive.

An interesting condition arises if the feedback factor is such as to make $\beta A = 1$. Then

$$A_f = \frac{A}{1 - \beta A} = \infty$$

In fact, the feedback voltage is equal in amplitude and phase to the original input signal and can, therefore, replace it. The amplifier thus produces its own input signal and is, therefore, an oscillator. This is inherently an unstable situation and the output signal will increase until it is limited in amplitude by non-linearity in the oscillator circuit. The output will not in general be sinusoidal, and may take the form of a distorted sine wave, or even of a square or pulse waveform. All oscillating systems thus require positive feedback, and in specially designed oscillators, for use as signal sources, the magnitude of the oscillation is carefully controlled to give the desired output waveform.

In practical amplifying systems, negative feedback is very often applied in order to produce the circuit gain and characteristics required. In considering the effects of negative feedback, it has been assumed that the feedback voltage is always in antiphase with the input signal voltage. However, we have also considered the effects of series and shunt capacitances, one effect of which is to cause the phase change in the amplifier to vary. The maximum phase change possible with one reactive element was shown to be 90° and this will occur at either zero or infinite frequency, depending upon the connection. However, practical amplifiers may contain several reactive components and the phase change in the amplifier may be such that the feedback, designed to be negative, may at some frequency within the range of the amplifier become positive. The amplifier system will then oscillate at this frequency if the feedback magnitude is sufficiently great to make $\beta A = 1$. Phase or gain correcting networks are sometimes necessary to prevent instability of this sort.

Example An amplifier of gain 20 has 3.75% of its output voltage fed back in series with the input as positive feedback. What is the gain of the system?

Gain with positive feedback is

$$A_f = \frac{20}{1 - \dfrac{3.75}{100} \times 20}$$

$$= \frac{20}{1 - 0.75}$$

$$= 80$$

Problems

1. An amplifier has an open-circuit voltage gain of 1000, an input resistance of 15 kΩ and an output resistance of 20 Ω. It has an input from a signal source of internal resistance 5 kΩ and feeds a resistive load of 180 Ω. What is the output signal voltage when the input signal is 300 μV?

2. An amplifier has an open-circuit voltage gain of 100, an input resistance of 100 kΩ and an output resistance of 10 Ω. It has an input from a signal source of resistance 10 kΩ and feeds a resistive load of 100 Ω. What is the overall voltage gain?

3. An amplifier has a voltage gain of 40 dB when it has a resistive load of 500 Ω. When there is an input signal of 20 mV, what is the magnitude of the output signal voltage?

4. An amplifier has an open-circuit voltage gain of 60 dB and an output resistance of 1 kΩ. What is the overall voltage gain, in dB, when there is a load resistance of 3 kΩ?

5. A two stage amplifier consists of two identical stages, each having an open-circuit gain of 100, an input resistance of 100 kΩ and an output resistance of 100 Ω. What is the overall voltage gain when there is a load of 1 kΩ and the input source has a resistance of 10 kΩ?

6. An amplifier, of open-circuit voltage gain 400, is correctly matched to its source and load resistances, which are equal. Calculate, both as a ratio and in decibels
   (a) the voltage gain between the load and the amplifier input,
   (b) the voltage gain between the load and the source e.m.f.,
   (c) the overall power gain.

7. An amplifier A of open-circuit voltage gain 20 dB, input resistance 5 kΩ and output resistance 1 kΩ is driven by a signal source of e.m.f. 10 mV and source resistance 1 kΩ. The amplifier output drives a second amplifier B of open-circuit voltage gain 25 dB, input resistance 3 kΩ and output resistance 8 Ω. The load for amplifier B is a 16 Ω resistor. Calculate
   (a) the signal voltage at the input of amplifier A,
   (b) the signal voltage at the input of amplifier B,
   (c) the load voltage,

(d) the overall voltage gain as a ratio and in decibels between the load and the source e.m.f.

8. In the system of Problem 7 an impedance matching transformer is introduced between the two amplifiers. Determine the required transformer ratio and repeat the calculations of Problem 7.

9. (a) A low frequency cut-off is introduced into the amplifier of Problem 7, by including a 1 µF capacitor in series with the output of amplifier A. Determine the resulting cut-off frequency. At what frequency has the gain reduced by 5%?
    (b) It is also desired to limit the upper frequency response by including a capacitor in parallel with the output of amplifier A. What capacitor is required to establish an upper cut-off frequency of 8 kHz?

10. An amplifier has an open-circuit voltage gain of 10 000, an input resistance of 1 kΩ and a very low output resistance. Voltage negative feedback is applied by feeding back one fiftieth of its output voltage in series with its input. Calculate the resulting voltage gain and input resistance.

11. The amplifier of Problem 10 reduces in gain to 8000 due to a faulty component. What is its effective gain with feedback?

12. An amplifier is required with a gain of 20, guaranteed to 0.5%. If production tolerances may produce a gain variation of 15%, what must be the value of β and of the amplifier gain without feedback for this specification to be met?

13. An amplifier with a negative feedback fraction of 0.005 has a gain of 100. What is the gain of the amplifier in the absence of feedback?

14. An amplifier has a gain of 150. What feedback is required if the overall gain is to become 10?

15. An amplifier is required with an overall voltage gain of 100, the gain not varying by more than 1%. Negative feedback is to be used with a basic amplifier for which the gain might vary by 20%. What must be the value of the voltage gain for this basic amplifier and the feedback factor?

Chapter 8

Transistor Amplifiers

8.1 Small signal common emitter amplifier

Figure 8.1(a) shows a single transistor with a collector load resistor, in the common emitter connection. In Fig. 8.1(b) typical common emitter characteristics are shown, with the load line for the collector resistor. The characteristics are typical of a transistor which would be used as a low level signal amplifier.

The d.c. conditions in the circuit are established by driving the transistor base with a steady direct current I_B; for the characteristic shown a current $I_B = 40\,\mu A$ is chosen. The circuit operating point is given by the intersection of the characteristic for a base current of $40\,\mu A$ with the collector resistor load line, labelled B in the diagram. Under these static conditions, the collector voltage and current have the values 5 V and 0.5 mA respectively. An alternating signal current may be applied to the circuit by adding to the steady base current, a sinusoidal current, a convenient value for the circuit being $20\,\mu A$ peak. This will result in the base current varying between $20\,\mu A$ and $60\,\mu A$ in a sinusoidal manner. The circuit operating point will, therefore, move along the load line and will oscillate between the extreme limits marked A and C, these being the intersections of the load line with the $20\,\mu A$ and $60\,\mu A$ characteristics. The collector current swings from 0.3 mA to 0.7 mA, rising and falling in phase with the base current. The signal current gain of the circuit is thus

$$A_I = \frac{\text{change in load current}}{\text{change in input current}}$$

$$= \frac{7 - 3}{0.06 - 0.02}$$

$$= 100$$

The collector voltage falls as the collector current rises, due to the increased voltage drop across the load resistor. The signal voltage at the collector is, therefore, $180°$ out of phase with the input signal current. The signal voltage swing is between 3 and 7 V, giving a peak sinusoidal output voltage of 2 V. The input signal voltage cannot be determined from these characteristics; however, since it will be the voltage change required to change the current in the base–emitter diode in a sinusoidal manner with a peak value of $20\,\mu A$ about a mean value of $40\,\mu A$, an estimated voltage swing of about 0.2 V is reasonable. The circuit voltage gain may then be estimated as

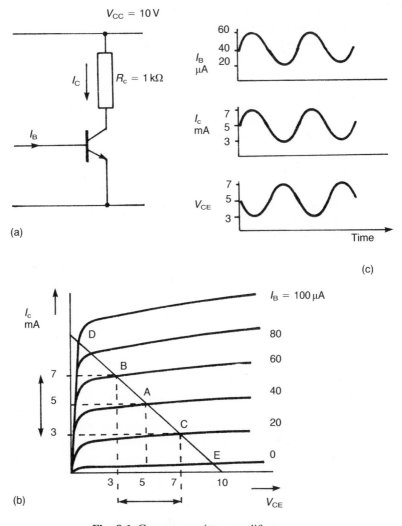

Fig. 8.1 Common emitter amplifier.

$$A_V = \frac{\text{output voltage change}}{\text{input voltage change}}$$

$$= \frac{7 - 3}{0.2}$$

$$= 20$$

The following points should be noted. In estimating the voltage and current values from the characteristics, equal swings in the values about the mean values have been assumed. Because, however, the characteristics are not exactly parallel, and not equally spaced, the circuit output voltage will not be exactly sinusoidal for a sinusoidal input current. The departure from a true sinusoidal signal will be greater for larger input signals. In particular, if the input current swing is sufficient to take the operating point to the points D or E, i.e. to bottom the transistor (point D), or to cut off the transistor (point E), then the collector voltage waveform, for a sinusoidal input current, will start to flatten at its peaks. It follows from this argument that, if such an amplifier stage is to be designed to allow the maximum sinusoidal output voltage possible, then the initial operating point must be positioned in the centre of the operating characteristics. The steady collector voltage will normally, therefore, be equal to $\frac{1}{2}V_{cc}$. The input signal to the circuit was specified as a sinusoidal base current. The impedance presented to the input signal generator is the forward biased base–emitter diode, which is, of course, a non-linear impedance. For a sinusoidal current, therefore, the input voltage must be non-sinusoidal. In fact, the output voltage changes are reasonably linearly related to the input base current; the driving circuits should, therefore, be designed to produce a sinusoidal current.

8.1.1 Equivalent circuits

It is also possible to derive for the transistor an equivalent circuit which may be used to give an estimation of the performance of the circuit under signal conditions. In fact there are many such small signal equivalent circuits, varying in complexity. It is, however, possible to obtain a reasonable estimate of circuit performance using only a very simple equivalent circuit, if its use is not extended beyond relatively low frequency operation.

Figure 8.2 shows the simple equivalent circuit for a transistor connected in common emitter. The values of the parameters will obviously vary as the d.c. operating conditions are varied. The equivalent circuit is termed a *small signal* equivalent circuit because the parameters are assumed to remain constant in value. It is one of the main functions of the d.c. design of a transistor circuit to set the conditions in the transistor so that the a.c. parameters have the optimum values for the purpose of the design. Further, the d.c. design should be such that the a.c. parameters remain as constant as possible, with changes in external conditions, such as temperature, etc.

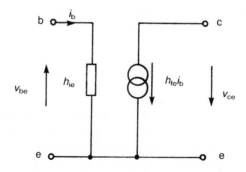

Fig. 8.2 Simple equivalent circuit.

The two parameters of the equivalent circuit are the resistor h_{ie}, which represents the impedance presented by the transistor to the source generator, and the current gain between the output and the input h_{fe}. The value of h_{ie} is determined by the resistance of the forward biased base–emitter junction, together with the effective resistance of the narrow base region. A typical value for a transistor, similar to that represented by Fig. 8.1, is about $1000\,\Omega$. The current gain h_{fe} will be of the order of 100. Figure 8.3 shows the equivalent circuit with a resistive load R_C of $1\,\mathrm{k}\Omega$, as in Fig. 8.1.

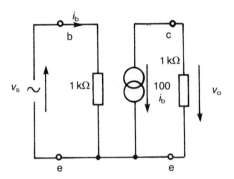

Fig. 8.3 Typical equivalent circuit values.

For an input current of i_b, the load resistor current is $100i_b$, giving the current gain as

$$A_I = 100$$

For an input voltage of v_s, the input current i_b is given by

$$i_b = \frac{v_s}{1000}$$

The output voltage is then

$$v_o = h_{fe}i_bR_C$$

$$= 100 \times \frac{v_s}{1000} \times 10\,000$$

The voltage gain is therefore

$$A_v = \frac{v_o}{v_s} = 100$$

In general, for this simple equivalent circuit we have for the transistor equivalent circuit (Fig. 8.2) with a load R_L a current gain A_I of

$$A_I = h_{fe} \tag{8.1}$$

The transistor input resistance R_i is v_{be}/I_b. But $v_{be} = h_{ie}i_b$. Hence

$$R_i = h_{ie} \tag{8.2}$$

Note that the above is the transistor input resistance and does not take account of any internal resistance that the signal source might have. The current through the load is $h_{fe}i_b$ and so the voltage across it is $h_{fe}i_bR_L$. The input voltage $v_s = h_{ie}i_b$. The voltage gain A_V is thus

$$A_V = \frac{h_{fe}R_L}{h_{ie}} \tag{8.3}$$

The term *mutual conductance* g_m is often used for

$$g_m = \frac{\text{current gain}}{\text{input resistance}} = \frac{h_{fe}}{h_{ie}} \tag{8.4}$$

Thus

$$A_v = g_m R_L \tag{8.5}$$

The power gain A_p for the transistor is

$$A_p = \frac{i_L^2 R_L}{i_b^2 h_{ie}} = \frac{(h_{fe}i_b)^2 R_L}{i_b^2 h_{ie}} = A_V A_I \tag{8.6}$$

Example   Using the equivalent circuit given in Fig. 8.2, determine (a) the current gain and (b) the voltage gain, when a source of internal resistance $2\,k\Omega$ supplies a signal of $50\,mV$ to a transistor with $h_{ie} = 1\,k\Omega$ and $h_{fe} = 50$ and with a load of resistance $4\,k\Omega$.

(a) The current gain is $h_{fe} = 50$.

(b) For the input circuit there is a total resistance of $(2 + 1) = 3\,k\Omega$ and hence the current i_b is $(50 \times 10^{-3})/3 \times 10^3 = 16.7\,\mu A$. For the output circuit there is a current generator of $h_{fe}i_b = 50 \times 16.7 = 835\,\mu A$. This is the current through the load. Hence the voltage across the load is $4 \times 10^3 \times 835 \times 10^{-6} = 3.34\,V$. The voltage amplification is thus $3.34/0.050 = 66.8$. Note that the voltage gain of the transistor, without any consideration of the source resistance, would have been $h_{fe}R_L/h_{ie} = 200$.

8.1.2   Single stage transistor amplifier

Figure 8.4 shows two versions of the single transistor amplifier, together with their equivalent small signal circuits. In Fig. 8.4(a) the steady base bias current is provided via a resistor R_1 from the collector supply voltage V_{CC}. This method is not good, being susceptible to changes in the operating point conditions due to small variations in the transistor current gain or in the leakage currents, both of which are temperature sensitive. In the equivalent circuit, note that both the positive and negative voltage supply lines are

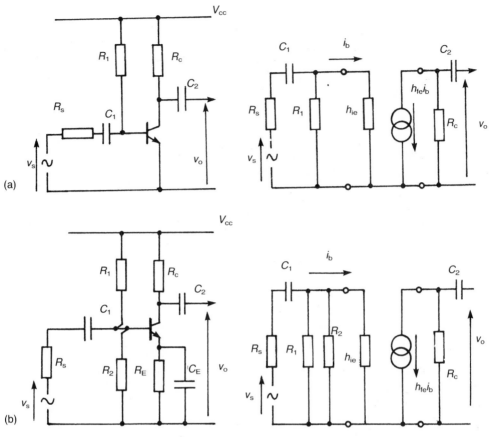

Fig. 8.4 Single transistor amplifier.

effectively the same point. Although they are separated by a direct voltage V_{CC}, they are, in fact, shorted together to a.c. signals via the very low a.c. impedance of the power supply. The bias resistor R_1 is effectively in parallel with the transistor input resistance h_{ie}.

A better circuit is shown in Fig. 8.4(b) in which the bias current is provided by fixing the d.c. potential of the transistor base by the potentiometer R_1 and R_2; the base and emitter currents are then determined by including a resistor R_E in series with the emitter. In order to remove the effect of R_E at signal frequencies, it is short-circuited by a capacitor C_E. As long as C_E is sufficiently low in reactance, R_E and C_E need not appear on the signal equivalent circuit. The advantage of this circuit is that it is self compensating for small changes in d.c. conditions. Thus, if the emitter current tends to rise, due perhaps to a temperature effect, then the emitter voltage will also rise. If the base voltage is adequately fixed by the potentiometer $R_1 R_2$, this will effectively reduce the base–emitter voltage, so reducing the emitter current again. Resistors R_1 and R_2 should, therefore, be as low in value as possible, consistent with not reducing the gain of the amplifier too much. In both circuits, the capacitors C_1 and C_2 prevent the external circuits from upsetting the d.c. conditions established in the transistor.

To illustrate the use of the equivalent circuits, consider the circuit in Fig. 8.4(a). The collector current i_c is the fraction of the current $h_{fe}i_b$ passing through the load resistor R_L. Thus

$$i_c = \frac{h_{fe}i_b R_c}{R_c + R_L}$$

The current gain A_I is, when the source impedance is assumed to be zero,

$$A_I = \frac{i_c}{i_b} = \frac{h_{fe}R_c}{R_c + R_L} \tag{8.7}$$

The above is often referred to as the stage amplification. When the source has a resistance R_s, then since $i_b = i_s R_1/(R_1 + h_{ie})$ the gain is reduced to

$$A_{Is} = \frac{i_c}{i_s} = A_I \frac{R_1}{R_1 + h_{ie}} \tag{8.8}$$

Example   For the transistor amplifier shown in Fig. 8.4(a), the components have the values of $R_1 = 80\,k\Omega$, $R_C = 10\,k\Omega$, $R_s = 5\,k\Omega$ and the load is $10\,k\Omega$. The transistor parameters are $h_{ie} = 2\,k\Omega$ and $h_{fe} = 50$. The reactances of the capacitors can be neglected at the frequencies used. What is the power dissipated across the load when the input signal v_s is 0.3 V?

The equivalent circuit is as shown in Fig. 8.4(a). Considering the input circuit, the total resistance is

$$R_s + \frac{R_1 h_{ie}}{R_1 + h_{ie}} = 5 + \frac{80 \times 2}{80 + 2} = 6.95\,k\Omega$$

Thus the current i_s from the source is 0.3/6.95 = 0.0432 mA and thus the base current i_b is

$$i_b = \frac{R_1 i_s}{h_{ie} + R_1} = \frac{80 \times 0.0432}{2 + 80} = 0.0421\,\text{mA}$$

Considering the output circuit, the current generator delivers a current of $h_{fe}i$ = 50 × 0.0421 = 2.105 mA. The current i_L through the load is thus

$$i_L = \frac{R_C \times 2.105}{R_C + R_L} = \frac{10 \times 2.105}{10 + 10} = 1.053\,\text{mA}$$

The power dissipated in the load is

$$P_L = i_L^2 R_L = (1.053 \times 10^{-3})^2 \times 10 \times 10^3 = 0.0111\,\text{W}$$

8.2   Emitter follower circuit

The emitter follower, shown schematically in Fig. 8.5, together with its small signal equivalent circuit, is a widely used amplifier circuit. It also provides an interesting example of the use of an equivalent circuit in the determination of the properties of transistor circuits.

Because the circuit does not have a collector load resistor, the collector is connected directly to earth in the equivalent circuit. The circuit output voltage is taken from across the emitter resistor. Two basic equations may be written for the circuit

$$v_i = ih_{ie} + v_o \tag{8.9}$$

and

$$v_o = i(1 + h_{fe})R_E \tag{8.10}$$

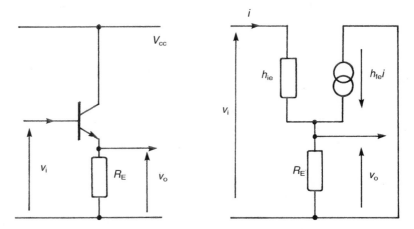

Fig. 8.5 Emitter follower.

If the current i is eliminated from these two equations, the circuit voltage gain is shown to be

$$A_v = \frac{R_E(1 + h_{fe})}{R_E(1 + h_{fe}) + h_{ie}} \tag{8.11}$$

With typical values of $h_{fe} = 100$, $h_{ie} = 1\,k\Omega$ and using $R_E = 1\,k\Omega$

$$A_v = 0.99$$

The circuit input resistance may also be determined, from the equations, to be

$$R_i = \frac{v_i}{i} = h_{ie} + R_E(1 + h_{fe}) \tag{8.12}$$

giving a value of

$$R_i = 102\,k\Omega$$

The output impedance of any circuit may be estimated by determining the ratio

$$R_o = \frac{\text{open-circuit voltage}}{\text{short-circuit current}} \qquad (8.13)$$

where the open-circuit voltage is the output voltage with no load connected across the output terminals and the short-circuit current is the current obtained in a short circuit placed across the output terminals. This method may actually be used in the laboratory if the d.c. conditions are not altered by the short circuit. The use of an a.c. short circuit (a capacitor) will normally allow the measurements to be made. The method is, however, very useful analytically.

Referring to Fig. 8.5, the open-circuit output voltage, with no load connected across v_o, is

$$v_{oc} = A_v v_i \qquad (8.14)$$

The current in a short circuit placed across the output is

$$i_{sc} = \frac{v_i}{h_{ie}}(1 + h_{fe}) \qquad (8.15)$$

The output resistance is then given by

$$R_o = \frac{v_{oc}}{i_{sc}}$$

$$= A_v \frac{h}{1 + h_{fe}} \qquad (8.16)$$

Substituting for A_v, from Equation (8.11) we obtain

$$R_o = \frac{h_{ie}}{1 + h_{fe} + (h_{ie}/R_E)} \qquad (8.17)$$

Typical values give

$$R_o = 9.8\,\Omega$$

The emitter follower circuit has approximately unity gain, zero phase shift, a very high input resistance and a very low output resistance. It finds great use as an impedance matching device, for example in instrument probes, or to couple effectively the signal from a high impedance device to a transmission line. Bias current for the base of the emitter follower is normally obtained via a resistor connected from the base to the voltage supply V_{CC}. If a very high input resistance is required, however, it must be borne in mind that this resistor is effectively in parallel with the circuit input, so giving a lower input resistance.

Problems

1. A transistor has the parameters $h_{ie} = 1\,k\Omega$ and $h_{fe} = 50$. Using the simplified equivalent circuit of Fig. 8.2, what is (a) the input resistance, (b) the current gain, (c) the voltage gain and (d) the power gain, when there is a load of $4\,k\Omega$?

2. Using the equivalent circuit for a transistor shown in Fig. 8.2, determine the voltage amplification for a transistor circuit with $h_{fe} = 50$ and $h_{ie} = 800\,\Omega$, when there is an input of 50 mV from a source of internal resistance $2\,k\Omega$ and the transistor has a load of $5\,k\Omega$.

3. A transistor amplifier has a circuit of the form shown in Fig. 8.4(a), with $R_s = 2.2\,k\Omega$, $R_1 = 10\,k\Omega$, $R_c = 10\,k\Omega$, $h_{ie} = 1\,k\Omega$, $h_{fe} = 50$ and a load of $10\,k\Omega$. The reactances of the capacitors may be neglected. What is (a) the base current, (b) the power gain of the amplifier stage if the output load dissipates a power of 10 mW?

4. A transistor amplifier has a circuit of the form shown in Fig. 8.4(a), with $R_s = 2\,k\Omega$, $R_1 = 100\,k\Omega$, $R_c = 10\,k\Omega$, $h_{ie} = 1.2\,k\Omega$, $h_{fe} = 50$ and a load of $5\,k\Omega$. The reactances of the capacitors may be neglected. What is the power dissipated across the load if the input signal is 20 mV?

5. A transistor amplifier stage has a circuit of the form shown in Fig. 8.4(b), with $R_1 = 82\,k\Omega$, $R_2 = 10\,k\Omega$, $R_c = 4.7\,k\Omega$, $R_E = 1\,k\Omega$, $h_{ie} = 700\,\Omega$, $h_{fe} = 250$ and a load resistance of $4.7\,k\Omega$. The reactances of the capacitors may be neglected. What is (a) the input resistance, (b) the current gain?

6. Using the equivalent circuit for the transistor given in Fig. 6.19, determine an improved equivalent circuit for the amplifier circuit in Fig. 8.4(b) and hence determine the current and voltage gains for the stage when $R_1 = 20\,k\Omega$, $R_2 = 10\,k\Omega$, $R_c = 2\,k\Omega$, $R_E = 1.6\,k\Omega$, $h_{ie} = 1\,k\Omega$, $h_{fe} = 50$, $h_{oe} = 5 \times 10^{-5}\,S$. Assume that the reactances of the capacitors can be ignored and that $h_{re}i_c$ is negligible.

7. A transistor has the parameters $h_{ie} = 1.0\,k\Omega$, $h_{fe} = 50$, $h_{oe} = 20\,\mu S$ and h_{re} is insignificant. The input signal is supplied by a source of 50 mV and internal resistance $4\,k\Omega$. The load between the collector and emitter is $5\,k\Omega$. What is (a) the transistor current gain, (b) the transistor voltage gain and (c) the power supplied to the load?

8. A transistor, connected in common emitter, has a $2\,k\Omega$ collector load and is biased by a single $100\,k\Omega$ resistor connected from its base to the collector power supply. The transistor parameters are $h_{ie} = 1.25\,k\Omega$, $h_{fe} = 150$. Determine the amplifier mid frequency gain. If the signal source is connected to the transistor base via a $1\,\mu F$ capacitor, determine the cut-off frequency.

9. A transistor is connected in common emitter in the circuit of Fig. 8.4(b). The collector load $R_c = 2.2\,k\Omega$, $R_2 = 12\,k\Omega$ and $R_1 = 100\,k\Omega$. The emitter capacitor C_E is of negligible reactance. The transistor parameters are $h_{fe} = 150$, $h_{ie} = 1.25\,k\Omega$. The signal source is an e.m.f. of 10 mV with a source resistance of $600\,\Omega$. What is the signal output voltage? The frequency is such that the capacitor C_1 and C_2 are of negligible reactance.

10. In the amplifier of Problem 9, it is desired that the lower cut-off frequency is 30 Hz. What should be the value of C_1?

Chapter 9
Operational Amplifiers

9.1 The operational amplifier

The operational amplifier, or *op amp*, is an amplifier which has a high input resistance, a low output resistance, a very large voltage gain and responds only to the difference between two input voltages. Operational amplifiers were first designed for use with analogue computers in order to carry out mathematical operations, hence the term operational amplifier. Nowadays the amplifier finds widespread use in many areas of electronics, having become a major building block in electronic systems.

The modern operational amplifier is an integrated circuit. An integrated circuit is a complete circuit fabricated on a small chip of silicon. One of the most popular operational amplifier versions is the 741 type. The chip is encapsulated in plastic with dual in-line rows of pins for making connections to the device. Figure 9.1 shows the form of the encapsulated chip and the pin connections.

The operational amplifier has two input connections, the inverting and the non-inverting inputs, and produces an output which is proportional to the difference between these two inputs. Thus for a voltage v_1 input to the inverting input and v_2 to the non-inverting input, the output v_o is

$$v_o = A(v_2 - v_1) \tag{9.1}$$

where A is the gain. Figure 9.2 shows the circuit symbol and the basic form of the characteristic. The gain is very large, typically in the range 10 000 to 200 000. The input resistance is in the range 100 kΩ to 10 MΩ and the output resistance between about 4 Ω and 75 Ω. A reasonable approximation to such characteristics, the ideal operational amplifier, is an infinite voltage gain, an infinite input impedance and zero output impedance.

Small imbalances in the internal circuitry of the operational amplifier mean that even with both the input terminals at the same potential there can be an output voltage. This can be cancelled by the use of external circuitry to provide a bias voltage or by connecting an external potentiomer between the amplifier *offset null* pins and adjusting it to give zero output when there is a zero differential input.

9.2 The inverting amplifier

Figure 9.3 shows the operational amplifier connected as an inverting amplifier. A feedback path from the amplifier output to the input is provided via resistor

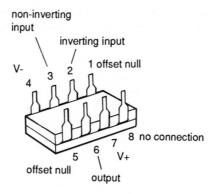

Fig. 9.1 The 741 integrated circuit.

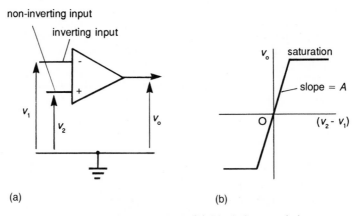

(a)   (b)

Fig. 9.2 (a) Circuit symbol, (b) ideal characteristic.

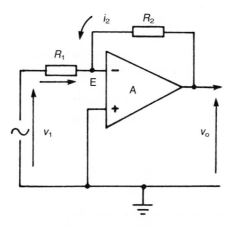

Fig. 9.3 Inverting amplifier.

R_2. Because the amplifier input resistance is very high, the current flowing into the input terminal is very nearly zero. Thus for node E we must have

$$i_1 = -i_2$$

The output voltage is typically limited to about 10 V (the saturation voltage) and thus with a gain of, say, 100 000 then the input voltage will be 0.000 01 V. This is the voltage at point E. Thus E is effectively at earth potential and is referred to as the *virtual earth*. Hence the potential difference across R_1 is effectively v_1 and across R_2 is v_o. Thus $i_1 = v_1/R_1$ and $i_2 = v_o/R_2$. Hence

$$\frac{v_1}{R_1} = -\frac{v_o}{R_2}$$

and thus the overall circuit voltage gain is

$$\text{overall gain} = \frac{v_o}{v_1} = -\frac{R_2}{R_1} \tag{9.2}$$

The overall circuit gain is thus determined by the external resistances R_1 and R_2 and is independent of the actual gain of the amplifier. The input impedance of this circuit is v_1/i_1 and hence is R_1.

Figure 9.4 shows the equivalent circuit of the inverting amplifier when used with a source having a voltage v_s and internal resistance R_s and a load of resistance R_L.

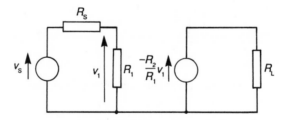

Fig. 9.4 Equivalent circuit.

Example Determine the values of the resistors needed to obtain an inverting operational amplifier circuit with an input resistance of 1 kΩ and a gain of −20.

With the circuit shown in Fig. 9.3, then the input resistance is R_1 and thus $R_1 = 1\,\text{k}\Omega$. For a gain of −20 we must have, using Equation (9.2), $R_2/R_1 = 20$ and so $R_2 = 20\,\text{k}\Omega$.

9.2.1 The summing amplifier

The summing amplifier is a circuit derived from the inverting voltage amplifier. Figure 9.5 shows the circuit. The current i into the virtual earth point is the

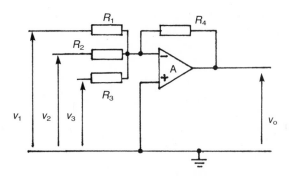

Fig. 9.5 The summing amplifier.

sum of the currents through the three resistors R_1, R_2 and R_3. Thus

$$i = \frac{v_1}{R_1} + \frac{v_2}{R_2} + \frac{v_3}{R_3}$$

The feedback current through R_4 is v_o/R_4. Thus since the sum of i and the feedback current is effectively zero

$$\frac{v_o}{R_4} = -\left(\frac{v_1}{R_1} + \frac{v_2}{R_2} + \frac{v_3}{R_3}\right) \tag{9.3}$$

If $R_1 = R_2 = R_3 = R$, then

$$v_o = -\frac{R_4}{R}(v_1 + v_2 + v_3) \tag{9.4}$$

9.2.2   *The integrating amplifier*

If a capacitor is used for the feedback element with the inverting operational amplifier, as in Fig. 9.6, then we have for the current through the input resistance v_1/R_1 and for the current through the capacitor $C\,dv_o/dt$. Thus since at the virtual earth point the sum of these currents is effectively zero,

$$\frac{v_1}{R_1} = -C\frac{dv_o}{dt}$$

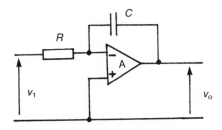

Fig. 9.6 Integrating circuit.

and so the output voltage v_o is

$$\int_{v_o(0)}^{v_o(t)} dv_o = -\frac{1}{CR_1} \int_0^t v_1 dt$$

$$v_o(t) = -\frac{1}{CR_1} \int_0^t v_1 dt + v_o(0) \tag{9.5}$$

The output voltage is thus proportional to the integral with time of the input voltage v_1, plus the initial voltage condition of the output $v_o(0)$.

A differentiator circuit can be produced if the resistance is the feedback element and the capacitor the input element.

Example   What is the output voltage signal when the input voltage to an integrator operational amplifier circuit is $4 \sin 500t$ V, the input resistance R is $10\,k\Omega$ and the feedback capacitance is $50\,nF$?

Using Equation (9.5) and taking $v_o(0)$ as 0, then

$$v_o(t) = -\frac{1}{CR_1} \int v_1 dt$$

$$= -\frac{1}{50 \times 10^{-9} \times 10 \times 10^3} \int 4 \sin 500t \, dt$$

$$= 16 \cos 500t \text{ V}$$

9.3   The non-inverting amplifier

Figure 9.7 shows the operational amplifier connected as a non-inverting amplifier. Due to the very high input resistance of the operational amplifier the input current to it at node X is negligible. Thus we can consider the output voltage v_o to be across $R_2 + R_1$ and so the voltage at X due to feedback is $v_o R_1/(R_2 + R_1)$, the arrangement being a potential divider. The operational amplifier amplifies the difference between the voltages at its two inputs, this difference being $v_1 - v_X$. Thus

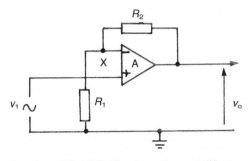

Fig. 9.7 Non-inverting amplifier.

$$v_o = A(v_1 - v_X)$$

where A is the gain of the operational amplifier. Hence

$$v_o = A\left(v_1 - \frac{v_o R_1}{R_2 + R_1}\right)$$

and so we can write

$$\frac{v_o}{v_1} = \frac{1}{[(1/A) + R_1/(R_2 + R_1)]}$$

Because A is very large the $1/A$ term can be neglected and so the overall gain is

$$\text{overall gain} = \frac{R_2 + R_1}{R_1} = \frac{R_2}{R_1} + 1 \tag{9.6}$$

The input impedance is v_1/I_1. Since the input current has to flow through the input terminal of the operational amplifier and it has such a high impedance that any input current is effectively zero, then $V_1/I_1 \approx V_1/0 \approx \infty$. Such a circuit has thus a very high input impedance. The output impedance is very low, effectively zero. It is effectively $(R_1 + R_2)$ in parallel with the output impedance R_o of the operational amplifier and so is the fraction $(R_1 + R_2)/(R_1 + R_2 + R_o)$ of R_o. Since R_o is already low, this arrangement makes it even lower.

Figure 9.8 shows the equivalent circuit of the non-inverting amplifier when connected to a source of voltage v_s with internal impedance R_s. The input impedance of the amplifier is taken as infinity, hence the open circuit. The output impedance is taken to be zero.

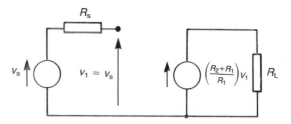

Fig. 9.8 Equivalent circuit.

Example   Design a non-inverting amplifier to provide a voltage gain of $+100$.

Using Equation (9.6), then we must have $R_2/R_1 = 100 - 1 = 99$. Suitable values of these resistances might thus be $R_1 = 10\,\text{k}\Omega$ and $R_2 = 990\,\text{k}\Omega$.

9.3.1   The voltage follower

Figure 9.9 shows a special form of non-inverting amplifier. The input resistance is the same as the feedback resistance, both being virtually zero. This gives the condition $R_2/R_1 = 0$ and so the gain is 1, using Equation (9.6). The output voltage thus *follows* the variations of the input voltage. The system has a high input impedance and a virtually zero output impedance. The circuit is thus used to buffer a high resistance source and minimize loading effects.

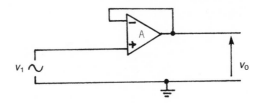

Fig. 9.9 Voltage follower.

9.4   Differential amplifier

Figure 9.10 shows the basic circuit of a differential amplifier. The operational amplifier amplifies the difference in voltage between its two input terminals.

Suppose that the input to the +terminal is zero and v_1 to the −terminal. Since E is the virtual earth, then the voltage across R_1 is v_1 and the voltage across R_2 is v_o. Thus, since virtually no current flows into the terminal of the operational amplifier we must have $v_1/R_1 = -v_o/R_2$ and so the voltage gain is

$$\frac{v_o}{v_1} = -\frac{R_2}{R_1} \tag{9.7}$$

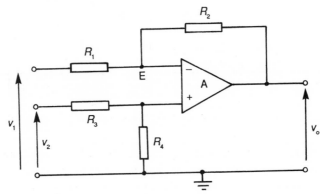

Fig. 9.10 Differential amplifier.

Now suppose that the input to the −terminal is zero and v_2 to the +terminal. The resistors R_3 and R_4 act as a voltage divider for the voltage v_2

and so the voltage at the + input terminal is $v_2R_4/(R_3 + R_4)$. There will also be a voltage at the −terminal, since R_1 and R_2 act as a potential divider for the voltage v_0. The voltage is thus $v_0R_2/(R_1 + R_2)$. The difference in voltage between the two terminals is amplified to give the output voltage v_0. Thus

$$v_o = A\left(\frac{v_2R_4}{R_3 + R_4} - \frac{v_oR_2}{R_1 + R_2}\right)$$

Thus the voltage gain is

$$\frac{v_o}{v_2} = \frac{R_4/(R_3 + R_4)}{(1/A) + R_2/(R_1 + R_2)}$$

Because A is so large $(1/A)$ can be neglected. Hence

$$\frac{v_o}{v_2} = \frac{R_4}{R_3 + R_4}\left(1 + \frac{R_2}{R_1}\right) \tag{9.8}$$

The output voltage when both inputs v_1 and v_2 are present is the sum of the inputs given by Equations (9.7) and (9.8). Thus

$$v_o = \frac{R_4}{R_3 + R_4}\left(1 + \frac{R_2}{R_1}\right)v_2 - \frac{R_2}{R_1}v_1 \tag{9.9}$$

If $R_1 = R_3$ and $R_2 = R_4$, then Equation (9.9) reduces to

$$v_o = \frac{R_2}{R_1}(v_2 - v_1) \tag{9.10}$$

The output is thus proportional to the difference in the two inputs.

Example What is the output from a differential amplifier with inputs of $+0.6\,$V and $-1.5\,$V if $R_1 = R_3 = 100\,\text{k}\Omega$ and $R_2 = R_4 = 200\,\text{k}\Omega$?

Using Equation (9.10),

$$v_o = \frac{R_2}{R_1}(v_2 - v_1) = \frac{200}{100}(0.6 + 1.5) = 4.2\,\text{V}$$

9.4.1 *Common mode rejection ratio*

In an ideal system the output from the differential amplifier depends only on the difference between the signals applied to its two inputs and is independent of the size of the inputs. In practice this is not the case, the output is affected by what is termed the *common mode voltage*, this being the average input voltage. This is because the gain for the non-inverting input is not precisely the same as the gain for the inverting input. Thus for an input of v_1 to the non-inverting input with a gain A_1 and for an input of v_2 to the inverting input with a gain A_2, the output v_0 is

$$v_o = A_1 v_1 - A_2 v_2$$

This can be written as

$$v_o = (\tfrac{1}{2}A_1 + \tfrac{1}{2}A_1)v_1 - (\tfrac{1}{2}A_2 + \tfrac{1}{2}A_2)v_2$$
$$= \tfrac{1}{2}(A_1 + A_2)(v_1 - v_2) + (A_1 - A_2)\tfrac{1}{2}(v_1 + v_2)$$

The first term in the above equation is the differential mode for the system. The differential mode gain is thus $\tfrac{1}{2}(A_1 + A_2)$, i.e. the average gain. The second term arises because of the difference between the two gains with $\tfrac{1}{2}(v_1 + v_2)$ being the average input voltage, the so-termed common mode voltage. Thus $(A_1 - A_2)$ is termed the common mode gain. A figure of merit used for a differential amplifier is the *common mode rejection ratio* (CMRR). This is defined as

$$\text{CMRR} = \frac{\text{differential mode gain}}{\text{common mode gain}} \qquad (9.11)$$

The CMRR should be large so that output errors due to common mode signals are minimized.

Example A differential amplifier is required which will produce an output for a 100 mV difference signal input. However, the signal is superimposed on a common mode voltage of 4 V. What differential mode gain and CMRR is required if the output signal due to the differential input is to be 10 V and is to be affected by the common mode signal by less than 1%?

The output for the differential signal of 10 V with a 100 mV input means a differential gain of 10/0.1 = 100. For the common mode signal of 4 V the required output is to be 1% of 10 V, i.e. 0.1 V. Thus the common mode gain has to be 0.1/4 = 0.025. The CMRR is thus 100/0.025 = 4000. It is common practice to express the CMRR in decibels. Thus $20\lg 4000 = 72\,\text{dB}$.

Problems

1. Design an inverting operational amplifier circuit to have an input impedance of $2\,\text{k}\Omega$ and a gain of -100.

2. A summing inverting operational amplifier has two inputs. What is the output voltage when the inputs are $+0.5\,\text{V}$ and $+1.2\,\text{V}$ through the respective input resistors of $200\,\text{k}\Omega$ and $100\,\text{k}\Omega$ and the feedback resistor is $200\,\text{k}\Omega$?

3. A summing inverting operational amplifier has two inputs, one having a resistance of $5\,\text{k}\Omega$ and an input voltage of $4\,\text{V}$ and the other a resistance of $8\,\text{k}\Omega$ and an input voltage of $4\sin 1000t\,\text{V}$. If the feedback resistance is $10\,\text{k}\Omega$, what is the output voltage?

4. Show that the input voltage v_1 – output voltage v_o relationship for an inverting operational amplifier having a resistor R as the feedback element and a capacitor C as the input element is

$$v_o = -CR\frac{dv_1}{dt}$$

5. What are the values of the circuit components needed for the input and feedback elements with an inverting operational amplifier if it is to carry out the operation

$$v_o t = -10 \int v_1 dt + v_o(0)$$

and have an input resistance of $10\,k\Omega$?

6. What is the output voltage signal when the input voltage to an integrator operational amplifier circuit is $10\sin 1000t\,V$ when the input resistance R is $100\,k\Omega$ and the feedback capacitance is 50 nF?

7. Design a non-inverting operational amplifier to have a voltage gain of $+20$.

8. A differential amplifier has $R_1 = R_2 = R_3 = R_4$ and two simultaneous inputs of $2\sin 500t$ and $4\sin 500t$. What is the output voltage?

9. A differential amplifier is required for use with a transducer which gives a differential input of $40\,mV$. Superimposed on it is a common mode voltage of $2\,V$. What differential mode gain and CMRR is required if the output signal due to the differential input is to be $1\,V$ and is to be affected by the common mode signal by less than 0.5%?

10. An operational amplifier is specified as having an inverting gain of $150\,000$ and a non-inverting gain of $149\,950$. What is the CMRR?

Chapter 10
Oscillators

10.1 Criteria for oscillation

An oscillator is an electronic circuit which is used to produce an alternating signal of a particular frequency and waveform. This chapter is concerned with those oscillators which generate sinusoidal signals.

An oscillator circuit is essentially an amplifier with positive feedback from its output to input and which is capable of sustaining an output without the need for an external signal. Thus suppose we have an amplifier with a voltage gain A_v and a feedback loop feeding back the fraction β of the output. In the absence of the feedback the output v_o will be related to the input v_i by

$$v_o = A_v v_i$$

Now with the feedback connected the fed back signal will be

$$\text{feedback signal} = \beta v_o = \beta A_v v_i$$

If we can make this feedback signal equal to v_i then the circuit will not need any external input signal. This will be the case if

$$\beta A_v = 1 \tag{10.1}$$

The quantity βA_v is known as the *loop gain* and the equation is known as the *Barkhausen criterion*.

In general, both the amplifier gain and the feedback fraction are complex and thus, in polar form, we can write Equation (10.1) as

$$|\beta|\underline{/\phi} \times |A_v|\underline{/\theta} = 1\underline{/0°}$$

$$|\beta|\,|A_v|\underline{/\phi + \theta} = 1\underline{/0°} \tag{10.2}$$

where $|\beta|$ is the magnitude of the feedback fraction and ϕ its phase angle, and $|A_v|$ is the magnitude of the gain and θ its phase angle. Thus the conditions for oscillation are that

$$|\beta|\,|A_v| = 1 \tag{10.3}$$

and

$$\underline{/\phi} + \theta = 0° \tag{10.4}$$

i.e. the loop gain magnitude must be unity and the phase shift of the loop gain must be 0°, or a multiple of 360°.

The above are the conditions for oscillation. If an oscillator is required which oscillates at some particular frequency, then the conditions must be only met for that frequency and no other.

Example   An oscillator consists of an amplifier with a gain of -5000 and a frequency dependent feedback loop with a feedback fraction $1/[Q + j(\omega C - 1/\omega L)]$. What will the feedback fraction need to be if there are to be oscillations at the angular frequency ω?

Using Equation (10.1) then we must have $\beta A_v = 1$, i.e.

$$\frac{-5000}{Q + j(\omega C - 1/\omega L)} = 1$$

For the phase condition to be met we must have the j term equal to zero. Thus

$$\omega = \frac{1}{\sqrt{(LC)}}$$

We must also have

$$Q = -5000$$

Thus the feedback fraction is $1/[-5000 + j(\omega C - 1/\omega L)]$ and at the oscillation frequency is $-1/5000$.

10.2   LC oscillators

The general form of an LC oscillator is shown in Fig. 10.1. It consists of an amplifier, in the figure it is shown as an operational amplifier but a similar circuit is possible with a transistor, with a feedback loop. The amplifier system has a gain of $-A_v$, an infinite input impedance and an output resistance of R_o. The load Z_L across the output terminals is that of jX_3 in parallel with $(jX_1 + jX_2)$, i.e. $jX_3(jX_1 + jX_2)/(jX_1 + jX_2 + jX_3)$. The output from the amplifier is thus dropped across R_o in series with this load Z_L and thus the overall gain is given by

$$\text{overall gain} = \frac{-A_v Z_L}{R_o + Z_L} = \frac{A_v(X_1 + X_2)/j(X_1 + X_2 + X_3)}{R_o - (X_1 + X_2)/j(X_1 + X_2 + X_3)}$$

$$= \frac{A_v(X_1 + X_2)}{jR_o(X_1 + X_2 + X_3) - (X_1 + X_2)}$$

The feedback fraction is $X_2/(X_1 + X_2)$, the three reactances acting as a potential divider across the output voltage v_o. Thus the loop gain is given by

$$\text{loop gain} = \frac{A_v X_2}{jR_o(X_1 + X_2 + X_3) - (X_1 + X_2)}$$

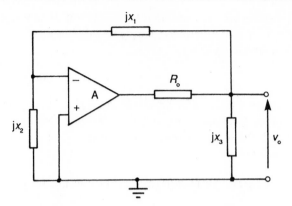

Fig. 10.1 General form of LC oscillator.

For the phase condition to be met for oscillation then the j term in the denominator must be zero, i.e.

$$X_1 + X_2 + X_3 = 0 \tag{10.5}$$

For the loop gain to have a magnitude of 1 at the frequency specified by the above condition, then we must have

$$1 = \frac{-A_v X_2}{X_1 + X_2} = \frac{A_v X_2}{X_3} \tag{10.6}$$

10.2.1 Hartley oscillator

This form of LC oscillator has a tuned LC circuit with a tapped coil for inductive feedback (Fig. 10.2). The impedances are $X_1 = 1/j\omega C_1$, $X_2 = j\omega L_1$ and $X_3 = j\omega L_2$. Equation (10.5) thus gives

$$\frac{1}{j\omega C_1} + j\omega L_1 + j\omega L_2 = 0$$

Hence

$$\omega = \frac{1}{\sqrt{[C_1(L_1 + L_2)]}} \tag{10.7}$$

The gain required for the operational amplifier circuit is given by Equation (10.6) as

$$A_v = \frac{X_3}{X_2} = \frac{L_2}{L_1} \tag{10.8}$$

Example A Hartley oscillator has a capacitor of 0.01 μF. What will be the frequency of oscillation if the inductor has an inductance of 12 mH and is tapped at its mid point?

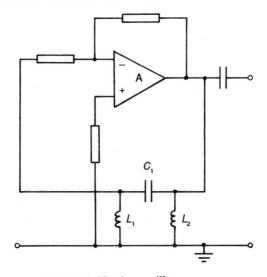

Fig. 10.2 Hartley oscillator.

With this arrangement $L_1 = L_2 = 6\,\text{mH}$. Hence, using Equation (10.7),

$$f = \frac{1}{2\pi\sqrt{[C_1(L_1 + L_2)]}} = \frac{1}{2\pi\sqrt{[0.01 \times 10^{-6} \times 0.012]}} = 14.5\,\text{kHz}$$

10.2.2 The Colpitts oscillator

This is similar to the Hartley oscillator but uses capacitive feedback instead of inductive feedback (Fig. 10.3). The impedances are $X_1 = j\omega_1 L_1$, $X_2 = 1/j\omega C_1$ and $X_3 = 1/j\omega C_2$. Thus Equation (10.5) gives

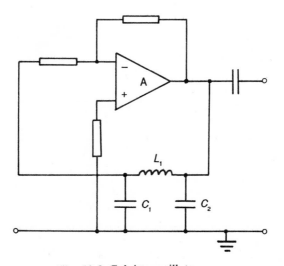

Fig. 10.3 Colpitts oscillator.

$$j\omega L_1 + \frac{1}{j\omega C_1} + \frac{1}{j\omega C_2} = 0$$

Hence

$$\omega = \sqrt{\left[\frac{1}{L_1}\left(\frac{1}{C_1} + \frac{1}{C_2}\right)\right]} \tag{10.9}$$

The gain required for the operational amplifier circuit is given by Equation (10.6) as

$$A_v = \frac{X_3}{X_2} = \frac{C_1}{C_2} \tag{10.10}$$

10.3   RC oscillators

The capacitance and inductance values required for the capacitor and inductor in the LC oscillator to obtain oscillation at audio frequencies are high and so RC oscillators tend to be used at such frequencies. Figure 10.4 shows an RC oscillator involving an operational amplifier. An RC ladder network (Fig. 10.4(b)) is in the feedback loop. The three capacitors are of equal value, as are the three resistors.

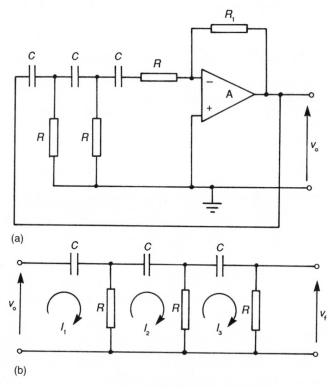

Fig. 10.4 RC oscillator.

Applying Kirchoff's voltage law, using mesh analysis (see Chapter 2), to the ladder network gives the following three equations for the three meshes:

$$v_o - (1/j\omega C)i_1 - R(i_1 - i_2) = 0$$

$$-R(i_2 - i_1) - (1/j\omega C)i_2 - R(i_2 - i_3) = 0$$

$$-R(i_3 - i_2) - (1/j\omega C)i_3 - Ri_3 = 0$$

From these three simultaneous equations we can obtain an equation relating i_3 and v_o. Since the feedback voltage $v_f = -Ri_3$, then the equation obtained can be written as

$$\frac{v_f}{v_o} = \frac{1}{-1 + 5(1/\omega CR)^2 + j[6(1/\omega CR) - (1/\omega CR)^3]} \tag{10.11}$$

This is the feedback fraction β. Since the operational amplifier circuit has a gain of $-R_1/R$ then for the phase condition for oscillation to be satisfied we must have the j term in the denominator equal to zero, i.e.

$$6(1/\omega CR) = (1/\omega CR)^3$$

$$\omega = \frac{1}{\sqrt{6}CR} \tag{10.12}$$

At this frequency the loop gain magnitudes must be equal to 1. Hence

$$A_v \frac{1}{-1 + 5(1/\omega CR)^2} = 1$$

and so using the value for ω given by Equation (10.12),

$$A_v = -1 + 5 \times 6 = 29 \tag{10.13}$$

This is the minimum gain the amplifier circuit must have if oscillations are to occur.

Example   Design an RC operational amplifier oscillator which oscillates at 200 Hz.

Using Equation (10.12) then

$$f = \frac{1}{2\pi\sqrt{6}RC} = 200$$

If we take C to be, say, $0.1\,\mu F$ then $R = 3.25\,k\Omega$. For the gain of the amplifier circuit to be 29 then the feedback resistor R_1 must be $29 \times 3.25 = 94.25\,k\Omega$.

10.3.1   The Wien bridge oscillator

The Wien bridge oscillator (Fig. 10.5) is another form of RC oscillator. The fed back voltage is the fraction $Z_4/(Z_3 + Z_4)$ of the output voltage, the arrangement acting as a voltage divider. Thus

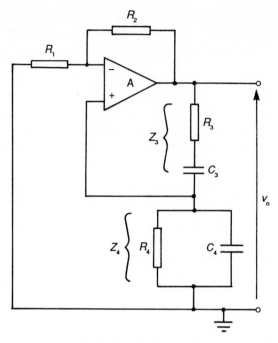

Fig. 10.5 Wien bridge oscillator.

$$\beta = \frac{R_4(1/j\omega C_4)/[R_4 + (1/j\omega C_4)]}{R_3 + (1/j\omega C_3) + R_4(1/j\omega C_4)/[R_4 + (1/j\omega C_4)]}$$

$$= \frac{\omega R_4 C_3}{\omega C_4 R_4 + \omega C_3 R_3 + \omega C_3 R_4 + j(\omega^2 R_3 C_3 R_4 C_4 - 1)}$$

To satisfy the phase condition for oscillation, then the j term should be zero. Thus

$$\omega = \frac{1}{\sqrt{(R_3 C_3 R_4 R_4)}}$$

Generally we have $R_3 = R_4$ and $C_3 = C_4$. Hence $\omega = 1/RC$. For the magnitudes of the loop gain to be 1 we must have, at this frequency,

$$A_v \frac{\omega R_4 C_3}{\omega C_4 R_4 + \omega C_3 R_3 + \omega C_3 R_4} = 1$$

Hence, with $R_3 = R_4$ and $C_3 = C_4$, we have $A_v = 3$.

10.4 Crystal oscillator

The stability of oscillators is improved by using a quartz crystal in the feedback path. Quartz is a piezoelectric material, i.e. when it is squeezed an electrical potential is developed across it and conversely when a voltage is applied across it the crystal becomes stretched or compressed. An alternating voltage applied across such a crystal causes it to vibrate. The resonant frequency of this vibration depends on the crystal dimensions and the resonance curve is very narrow, i.e. the crystal has a high Q factor (in the range 1000 to 100 000). When the crystal is used as the tuning load in a feedback circuit then it is effectively an LC circuit with a resonance frequency determined by its mechanical resonance curve. This frequency is much less affected by temperature changes than conventional LC circuits and so the frequency stability of the oscillator is considerably improved.

Problems

1. An amplifier has a gain of $-A/j\omega$ and a feedback fraction of $5000/(300 + j\omega)^2$. At what frequency will the circuit oscillate and what will be the required value of A?

2. An amplifier has a gain of -2×10^5 and a feedback fraction of $A/[2000 + j(10^{-6}\omega - 20/\omega)]$. At what frequency will the circuit oscillate and what will be the required value of A?

3. A Hartley oscillator has a capacitor of 5 pF. What will be the frequency of oscillation if the inductor has an inductance of 10 mH and is tapped at is mid point?

4. A Colpitts oscillator has capacitors of 20 nF and 10 nF with an inductance of 10 mH. What will be the frequency of oscillation?

5. A Colpitts oscillator has capacitors of 4.7 nF and 0.1 nF and a variable inductor. What range of inductance values is needed if the frequency of the oscillator is to vary between 40 kHz and 1 MHz?

6. Design an RC operational amplifier oscillator for oscillation at 2 kHz. The capacitors are each 5 nF.

7. Design a Wien bridge oscillator to oscillate at 5 kHz. The capacitors are each 5 nF.

8. A Wien bridge oscillator uses resistors of 10 kΩ and capacitors of 10 nF. What will be the frequency of oscillation and the required voltage gain of the operational amplifier?

Chapter 11

Power Supplies

11.1 Rectifying equipment

The majority of the direct current supplies required for electronic and control equipment is provided by means of rectifying equipment from the national alternating current mains supply. The term power supply in this context is used to refer to the complete circuitry which performs the conversion from a.c. to d.c., including the mains transformer which is normally used to isolate the d.c. supply from the a.c. mains, and which will usually also be required to alter the alternating line voltage. The term also includes any further circuitry which may be included to improve the operation of the supply, including ripple filters, voltage or current regulators or overvoltage or overcurrent protection. The direct current requirements vary over a wide range, from, say, 5 V at a current of 20 or 30 mA for small electronic circuits to 80 or 100 V at a current of several amperes for high power audio equipment. Welding equipment or motor controllers may have current requirements of several hundred amperes. The rectifying devices used in the majority of power supplies are silicon diodes, but thyristors are used to a large extent in controlled rectifier applications.

11.2 The half wave rectifier

The circuit of a half wave rectifier supplying a resistive load R is shown in Fig. 11.1, together with the load voltage and current waveforms. For a sinusoidal alternating voltage supply, the load voltage and current waveforms are half sinusoidal in shape, corresponding in time with the half cycle of the supply voltage which foward biases the diode.

The peak voltage across the load resistor will be slightly smaller than the peak value of the applied voltage, because of the voltage drop in the diode (of the order of 1 V). If, however, the diode voltage drop is neglected, the voltage across the load resistor may be expressed as

$$V_L = V_m \sin \omega t \qquad 0 < \omega t < \pi$$

$$V_L = 0 \qquad \pi < \omega t < 2\pi \qquad (11.1)$$

A moving coil d.c. voltmeter connected across the load resistor will read the average value of the load voltage. The d.c. voltage is, therefore,

$$V_{DC} = \frac{1}{2\pi}\int_o^\pi V_m \sin \omega t \, d\omega t + \frac{1}{2\pi}\int_\pi^{2\pi} 0 \, d\omega t$$

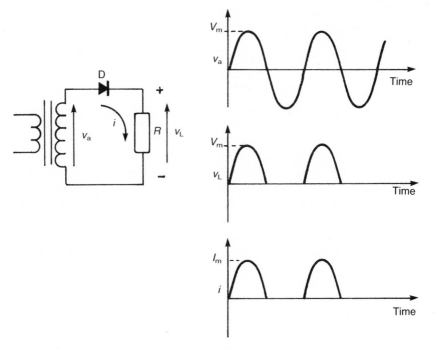

Fig. 11.1 Half wave rectifier.

$$= \frac{1}{\pi}V_m$$

$$= 0.318V_m \tag{11.2}$$

The direct current is

$$I_{DC} = \frac{V_{DC}}{R} = \frac{1}{\pi}\frac{V_m}{R} \tag{11.3}$$

For a sinusoidal supply voltage, the root mean square (rms) value of the voltage, V, is

$$V = \frac{V_m}{\sqrt{2}} \tag{11.4}$$

The direct load voltage may, therefore, be written

$$V_{DC} = \frac{1}{\pi}V_m$$

$$= \frac{\sqrt{2}V}{\pi} = 0.45\,V \tag{11.5}$$

During the half cycle of the a.c. supply in which the diode is reverse biased, no current flows in the circuit. All the supply voltage is applied across the reverse biased diode. The diode must, therefore, be of sufficiently high reverse voltage rating to withstand safely the peak supply voltage. The *peak inverse voltage* (PIV) is defined as the maximum voltage appearing across the diode during the non-conducting period. For the half wave rectifier, therefore,

$$\text{PIV rating of rectifier} = V_m \tag{11.6}$$

The load waveforms show that whereas an ideal output would consist of a constant magnitude, steady voltage, or current, the output from a half wave rectifier consists of half sinusoidal pulses, and that, in fact, for 50% of the period, the voltage and current are actually zero. It can be considered that the output involves, in addition to the d.c. component of magnitude $0.318V_m$, a *ripple* consisting of alternating components at frequencies which are harmonically related to the input supply frequency. Only the d.c. output is of use and, although in some applications the presence of such ripples may be tolerated, often ripple filtering circuits must be included to reduce the ripple magnitude to a low level. As an example, in a battery charging application, the efficiency of the circuit does not depend upon a smooth charging current, but depends rather upon the total amount of charge passed into the battery. A pulsating charging current is quite acceptable. In audio equipment, however, the presence of excessive ripple components in the power supply voltage will result in signals at the ripple frequencies appearing in the system output. In the case of the half wave rectifier, the largest ripple component occurs at the power supply frequency; this would result in audio equipment in a very low frequency (50 Hz) hum in the loudspeaker.

11.3 The full wave rectifier

In the full wave rectifier circuit of Fig. 11.2, a centre tapped transformer is used to provide effectively two secondary windings, which, together with separate rectifier diodes, act as two half wave rectifiers conducting during alternate half cycles of the a.c. supply and supplying a common load resistor. In the figure, when A is positive with respect to B, B is also positive with respect to C. Diode D_2 is, therefore, reverse biased but diode D_1 is conducting, the current flowing in the direction shown through the load resistor R. During the next half cycle of the a.c. supply, diode D_1 is reverse biased, but diode D_2 conducts, and its current flows in the same direction through the load resistor.

As in the case of the half wave rectifier, the load voltage and current waveforms consist of half sinusoids, but as one occurs for each half cycle of the supply, the direct voltage and current values are double those of the half wave circuit. The load voltage may be expressed as

$$V_L = V_m \sin \omega t \qquad 0 < \omega t < \pi$$

$$V_L = -V_m \sin \omega t \qquad \pi < \omega t < 2\pi \tag{11.7}$$

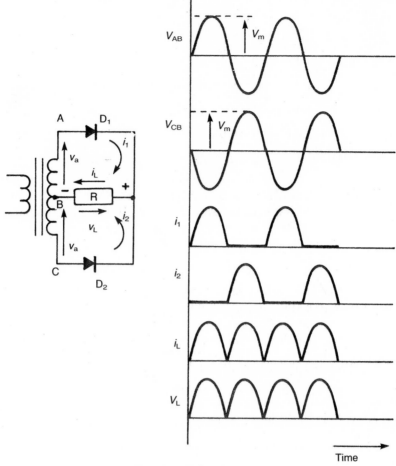

Fig. 11.2 Full wave rectifier.

The average value of the load voltage is then

$$V_{DC} = \frac{1}{2\pi}\int_{o}^{\pi} V_m \sin \omega t \; d\omega t + \frac{1}{2\pi}\int_{\pi}^{2\pi} - V_m \sin \omega t \; d\omega t$$

$$= \frac{2}{\pi}V_m$$

$$= 0.636V_m \tag{11.8}$$

The direct current is

$$I_{DC} = \frac{V_{DC}}{R} = \frac{2\,V_m}{\pi\,R} \tag{11.9}$$

In terms of the rms value of the supply voltage V,

$$V_{DC} = \frac{2}{\pi}\sqrt{2}\,V$$

$$= 0.90\,V \qquad\qquad (11.10)$$

It is important to note that each half of the transformer secondary winding is required to produce an alternating voltage of

$$v_a = V_m \sin \omega t$$

In other words, the alternating voltage between points A and C in the figure is twice this value.

We have seen that the peak value of the load voltage is equal to the peak value of the supply voltage, and both occur at the same time. The reverse voltage across the non-conducting rectifier diode is equal to the load voltage plus the voltage across half of the secondary winding. Because both of these voltages have peak values of V_m, the maximum reverse voltage is equal to $2V_m$. For the full wave rectifier, therefore,

$$PIV = 2V_m \qquad\qquad (11.11)$$

The load current and voltage waveforms, shown in Fig. 11.2, consist again of repeated half sinusoids, both falling instantaneously to zero twice per cycle of the supply. The waveform may be shown, as in the half wave case, to consist of the direct value, together with harmonics of the supply frequency. In the full wave case, however, the lowest frequency component present is a component at twice the supply frequency. This makes the reduction of the magnitude of the ripple components using filtering techniques slightly easier than is the case with a half wave rectifier.

11.3.1   The bridge rectifier

A disadvantage of the circuit of Fig. 11.2 is the need for the centre tapped transformer, which is inefficiently utilized, since only half of the transformer is conducting at any given time. An alternative full wave rectifier circuit is shown in Fig. 11.3 which does not require a centre tapped transformer, but does require four diode rectifiers instead of two. This is the bridge rectifier circuit. The transformer conducts during both half cycles of the supply voltage, using two rectifiers in each cycle. Thus, when A is positive with respect to B, diodes D_1 and D_3 conduct, giving a half sinusoidal current pulse in the load R. During the succeeding half cycle, when B is positive with respect to A, diodes D_2 and D_4 conduct giving again a half sinusoidal current pulse in the load R, in the same direction as in the preceding half cycle. The load voltage and current waveforms are similar, therefore, to those of Fig. 11.2. The reverse biased diodes are subjected to the sum of the voltage across the load resistor R, and the secondary voltage of the transformer, the maximum reverse voltage thus being $2V_m$. However, since this voltage is shared between the two diodes, the peak inverse voltage of either diode is given by

$$\text{peak inverse voltage} = V_m \qquad\qquad (11.12)$$

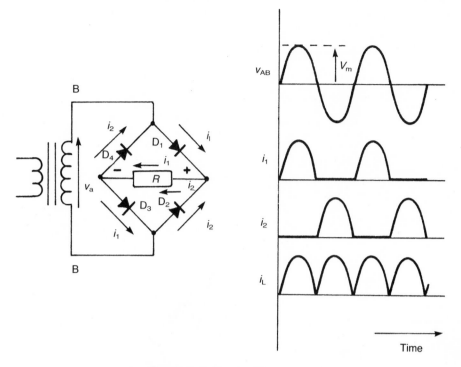

Fig. 11.3 Bridge rectifier.

In the forward conduction half cycle also, two diodes are in series together with the load R. The load voltage will, therefore, be less than the transformer voltage by two diode forward voltage drops. However, if the forward voltage drop is neglected, the direct output voltage is the same as that of the previous circuit, i.e.

$$V_{DC} = \frac{2}{\pi}V_m$$

$$= 0.636V_m \tag{11.13}$$

11.4 Polyphase rectifiers

The full wave rectifier of Fig. 11.2 may be considered as a two phase half wave rectifier circuit, each phase of the two phase supply conducting for one half of each cycle of the a.c. supply. The voltage produced by a single phase half wave rectifier has been shown to be

$$V_{DC} = 0.318V_m$$

while that of a two phase half wave rectifier is

$$V_{DC} = 0.636V_m$$

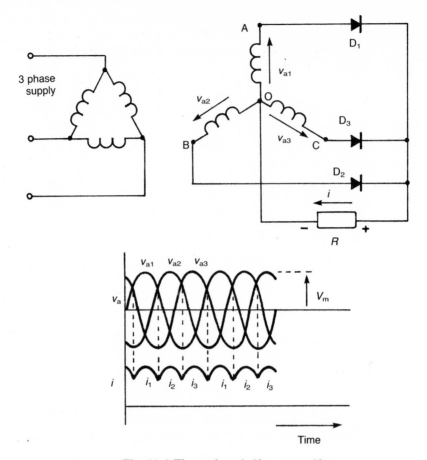

Fig. 11.4 Three phase half wave rectifier.

A further gain in rectified voltage is obtained if the number of phases is increased.

Figure 11.4 shows a three phase half wave rectifier fed from the star connected secondary windings of a three phase transformer. The figure also shows the voltage waveforms of the three phases of the transformer secondary, which are each sinusoidal and of peak value V_m V. The three voltages are separated in phase by $2\pi/3$ radians (120°).

Consider the instant that the phase voltage v_{a1} is at its peak value, with A positive with respect to the star point O. Both of the other two phase voltages are such that B and C are negative with respect to the star point O. Diode D_1 is, therefore, conducting while diodes D_2 and D_3 are both non-conducting. Phase voltage v_{a2} is, however, starting to rise positively while v_{a1} is starting to reduce. At the instant voltage v_{a2} exceeds voltage v_{a1}, diode D_2 commences to conduct raising the load voltage and reverse biasing D_1. Each diode, therefore, conducts for the fraction of the supply cycle that its phase voltage is in the forward biasing direction, and is greater in magnitude than both the other phase voltages. Each diode, therefore, conducts for one third of the supply cycle, giving a load current and voltage waveform as shown in Fig. 11.4. It is easily seen that one advantage of increasing the number of rectifier phases is to decrease the magnitude of the ripple component of the rectified voltage. The lowest frequency ripple component is also increased to three times the frequency of the supply.

It can be seen from Fig. 11.4 that the voltage (or current) waveform consists of repeated identical segments of a sinusoidal curve, each segment being $2\pi/3$ radians in width. The equation to the waveform is best written in terms of a cosine function to give

$$v_L = V_m \cos \omega t \qquad -\frac{\pi}{3} < \omega t < +\frac{\pi}{3}$$

$$v_L = V_m \cos \left(\omega t - \frac{2\pi}{3} \right) \qquad \frac{\pi}{3} < \omega t < \pi$$

$$v_L = V_m \cos \left(\omega t - \frac{4\pi}{3} \right) \qquad \pi < \omega t < \frac{5\pi}{3} \qquad (11.14)$$

Bearing in mind that the direct value of the rectified voltage is given by the average value of the voltage, and also that the average value of three identical consecutive segments is the same as the average of one of the segments, then

$$V_{DC} = \frac{3}{2\pi} \int_{-\pi/3}^{+\pi/3} V_m \cos \omega t \, d\omega t$$

$$= 0.827 V_m \qquad (11.15)$$

In terms of the rms value of the supply voltage V, where

$$V = \frac{V_m}{\sqrt{2}}$$

then

$$V_{DC} = 0.827 \times \sqrt{2} \, V$$

$$= 1.17 \, V \qquad (11.16)$$

In practice, because of the inductance of the circuit, due mainly to the transformer windings, the switch from one diode to the next diode does not take place instantaneously and, in fact, a short overlap occurs during which time both diodes may conduct together. This tends to reduce slightly the value of the rectified output voltage.

In power rectifier systems, the use of polyphase rectifiers is quite common and rectifiers are built with a large number of phases. In general, with an m phase rectifier using m diodes, each diode will conduct for $2\pi/m$ radians per cycle. Neglecting losses the rectified output voltage may be written in general as

$$V_{DC} = \frac{m}{2\pi} \int_{-\pi/m}^{+\pi/m} V_m \cos \omega t \, d\omega t$$

$$V_{DC} = V_m \frac{\sin \dfrac{\pi}{m}}{\dfrac{\pi}{m}} \tag{11.17}$$

Figure 11.5 shows a graph of the direct rectified output voltage magnitude against m, the number of phases.

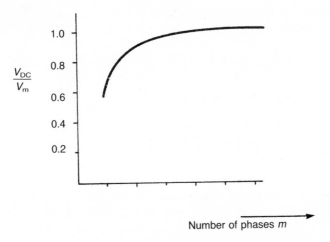

Fig. 11.5 Polyphase rectification.

11.5 Ripple reduction

For many applications, especially those in which the power supply is required for electronic equipment, the ripple produced by the rectifier circuits described in the previous sections is too large in magnitude. Additional ripple reducing circuits must be included.

11.5.1 The reservoir capacitor

Figure 11.6 shows the half wave rectifier circuit described in Section 11.2, with the addition of a capacitor in parallel with the load. The effect of the added capacitor is to modify the output voltage waveform to that shown in the figure.

During the period AB, the transformer supplies current, via the diode, to supply the load and also to charge the capacitor. Because of the short charging time constant, the capacitor voltage (which is also the load voltage) follows almost exactly the waveform of the supply voltage. After its peak value, the supply voltage starts to reduce. If the capacitor is sufficiently large, its voltage does not fall as quickly as that of the supply. The diode, therefore, becomes reverse biased, and non-conducting. All the load current is now supplied from

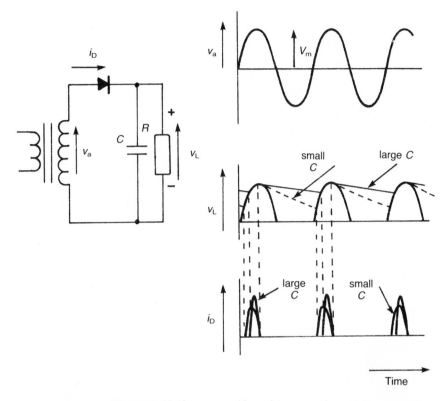

Fig. 11.6 Half wave rectifier with reservoir capacitor.

the capacitor; the capacitor voltage falls exponentially with a time constant determined by the value of the capacitor and by the effective resistance of the load. The diode remains non-conducting for the remainder of the supply cycle, and also during the next cycle, until the transformer voltage rises sufficiently to exceed the capacitor voltage and hence to forward bias the diode again. The action of the capacitor is very similar to that of a reservoir, in that it is charged by the supply, while the supply voltage is high; when the supply voltage is low, or negative, the current required by the load is supplied from the stored charge on the capacitor. This is the origin of the name used, the capacitor being termed a *reservoir capacitor*.

An important point to note is that, whereas without the capacitor the diode conducts for a complete half cycle, with a reservoir capacitor the diode conducts for only a fraction of the half cycle. Further, the conducting fraction of the half cycle is reduced if the value of the reservoir capacitor is increased. The capacitor acts, however, only as a reservoir of charge; all the charge flowing into the load must be supplied during the conducting period, via the diode. Increasing the capacitor value, therefore, shortens the period for which the diode conducts, but also increases the magnitude of the current pulse which flows. Care should be taken that the reservoir capacitor chosen for a particular design is not so large that the diode peak current rating is exceeded.

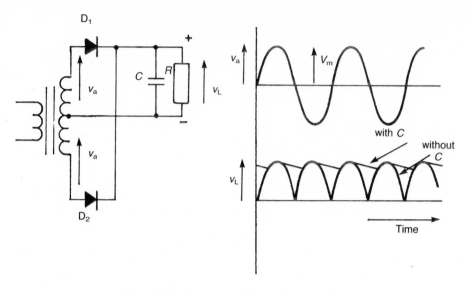

Fig. 11.7 Full wave rectifier with reservoir capacitor.

Figure 11.7 shows a full wave rectifier with an added reservoir capacitor, together with the rectified voltage waveform. It is interesting to note that if the load current taken from a power supply, which includes a reservoir capacitor, is reduced to zero, then the capacitor will charge up to the peak voltage of the supply. No further current will then flow in the rectifier diode. In practice this effect may occur if the power supply is disconnected from the circuit which it is supplying. Even if the a.c. supply is then switched off, the capacitor will retain the d.c. voltage across its terminals. Because the diode is reverse biased, the capacitor cannot discharge via the transformer secondary. If the reservoir capacitor is a good quality component, with high insulation resistance, the stored charge may be maintained for long periods. This dangerous condition is prevented in good designs by including, in parallel with the capacitor, a leakage resistor of sufficiently high value not to load the power supply excessively, but sufficiently low to discharge the capacitor within a reasonable time if the power supply is switched off after the load has been disconnected.

A measure of the effectiveness of a rectification system is given by the *ripple factor* which is defined as

$$R = \text{ripple factor} = \frac{\text{rms value of ripple components}}{\text{average (d.c.) component}}$$

An estimation of the peak to peak ripple amplitude may be made by the approximation illustrated in Fig. 11.8. For reasonably large values of reservoir capacitor, the direct output voltage will not fall excessively between successive supply voltage peaks. In the figure the exponential capacitor discharge curve has been approximated by a linear discharge curve which has been extended in duration until the time of the next supply voltage peak. The ripple voltage

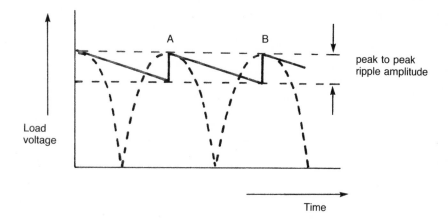

Fig. 11.8 Ripple.

waveform has thus been approximated to a triangular waveform. During the discharge period AB, the load current i_L is supplied by the capacitor. We may write

$$i_L = C\frac{dv_L}{dt}$$

whence

$$\frac{dv_L}{dt} = \frac{1}{C}i_L \tag{11.18}$$

The use of a reservoir capacitor also has an effect on the *peak inverse voltage* of a diode used in a half wave circuit. This was shown to be equal to V_m when used without a capacitor. The peak inverse voltage of the non-conducting diode is equal to the sum of the peak supply voltage and the peak direct voltage. As the reservoir capacitor maintains the direct voltage, during the non-conducting half cycle, the peak inverse voltage is increased to be

$$\text{peak inverse voltage} = 2V_m \tag{11.19}$$

for the half wave circuit. The peak inverse voltage for the full wave circuit is unchanged by the addition of a reservoir capacitor.

Example   Estimate the peak to peak ripple voltage for (a) a half wave rectifier, (b) a full wave rectifier designed to produce from a 50 Hz supply a current of 2 A when used with a reservoir capacitor of 5000 μF.

(a) Using the approximation, for a half wave rectifier, the duration of the discharge will be equal to the time of one full cycle of the a.c. supply

$$\text{discharge time} = \frac{1}{50} = 0.02\,\text{s}$$

The rate of change of load voltage will be, using Equation (11.18),

$$\frac{dv_L}{dt} = \frac{1}{C} i_L$$

$$= \frac{10^6}{5000} \times 2 = 400 \text{ V/s}$$

Fall in voltage during the discharge time is then

$$\delta V = \text{discharge time} \times \frac{dv_L}{dt}$$

$$= 0.02 \times 400$$

$$= 8 \text{ V}$$

Peak to peak ripple voltage $= \delta V = 8$ V

(b) For the full wave rectifier, the rate of fall of voltage will be the same as for the half wave rectifier.

$$\frac{dv_L}{dt} = 400 \text{ V/s}$$

The discharge time will, however, be equal to one half cycle of the a.c. supply.

$$\text{discharge time} = \frac{1}{2} \times \frac{1}{50} = 0.01 \text{ s}$$

Peak to peak ripple voltage is thus

$$\delta V = 0.01 \times 400$$

$$= 4 \text{ V}$$

This example shows that, for a given ripple voltage, a smaller reservoir capacitor may be used with a full wave rectifier than with a half wave rectifier.

When using the approximation of Fig. 11.8 the ripple waveform is a triangular wave. The root mean square value of a triangular wave of peak to peak magnitude V_p is given by

$$\text{rms value} = \frac{1}{\sqrt{3}} V_P$$

Example   The full wave rectifier of the previous example produces an average direct output voltage of 40 V. Estimate the ripple factor. The peak to peak ripple voltage is 4 V.

The rms value of a triangular waveform is

$$\text{rms value} = \frac{1}{\sqrt{3}} \times \text{peak to peak value}$$

$$= \frac{1}{\sqrt{3}} \times 4 = 2.30 \, \text{V}$$

$$\text{ripple factor} = R = \frac{2.30}{40} = 0.0577$$

The ripple factor is often expressed as a percentage, i.e.

$$\text{ripple factor } R = 5.77\%$$

11.5.2 Ripple filters

For applications where the use of a reservoir capacitor does not reduce the ripple magnitude to a sufficiently low value, further ripple filtering circuits may be added. A filter in its simplest form is in effect a potentiometer whose voltage output is a fraction of its input voltage, the fraction being designed to be different for signals of different frequencies. Figure 11.9(a) shows a potentiometer, consisting of two impedances Z_A and Z_B. The potentiometer output voltage is given by

$$v_o = \frac{Z_B}{Z_A + Z_B} v_i \tag{11.20}$$

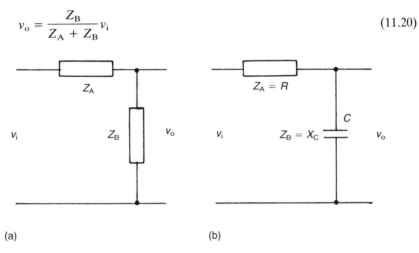

(a)   (b)

Fig. 11.9 Ripple filter circuit.

Two conditions should be considered.

(1) $Z_B \gg Z_A$. If Z_B is large in comparison to Z_A then

$$\frac{v_o}{v_i} = \frac{Z_B}{Z_A + Z_B} \simeq \frac{Z_B}{Z_B} \simeq 1 \tag{11.21}$$

(2) $Z_A \gg Z_B$. If Z_A is large in comparison to Z_B then

$$\frac{v_o}{v_i} = \frac{Z_B}{Z_A + Z_B} \simeq \frac{Z_B}{Z_A} \simeq 0 \tag{11.22}$$

A *resistor–capacitor ripple filter* is shown in Fig. 11.9(b). Considering the input voltage v_i to consist of two components, a direct voltage and an a.c. voltage, the effect of the filter on each component may be considered separately.

Direct voltage component. To the direct component, the impedance of the capacitor C is very high indeed, being virtually infinite. The value of the resistor R is kept as low as possible. Thus the direct output voltage is

$$v_{oDC} = \frac{\infty}{R + \infty} \times v_{iDC} \simeq v_{iDC}$$

The filter has little effect on the direct component.

Alternating voltage component. To the alternating component, the impedance of the capacitor C is very low, a large value capacitor being chosen. Thus the alternating output voltage is

$$v_{oAC} = \frac{X_C}{\sqrt{(R^2 + X_C^2)}} \times v_{iAC}$$

$$\simeq 0 \text{ if } X_C \ll R$$

where X_C is the reactance of the capacitor C and is equal to

$$X_C = \frac{1}{2\pi f C}$$

at a frequency of f Hz.

Figure 11.10 shows a half wave rectifier circuit, with a reservoir capacitor C_R and a resistor–capacitor ripple filter, R_F–C_F. As explained, the filter is required to reduce the ripple component to as low a value as possible. Thus C_F is made as large as is convenient, while R_F should also be large. However, in order that the value of the direct voltage is not reduced by the flow of the load current through R_F, its value must be kept reasonably low. The value of R_F is thus a compromise between those two conflicting requirements. One solution to this problem is to replace the resistor R_F by an inductor L_F. The inductor, which would normally be wound upon a magnetic core to allow a large

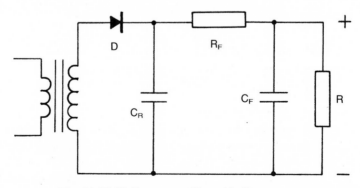

Fig. 11.10 Half wave rectifier with filter.

inductance value, would have a large impedance at the alternating ripple frequency while having a low resistance to the direct current. The use of an inductor, especially for high current power supplies, is relatively expensive and the use of a resistor is more common.

11.6 Voltage multiplying rectifiers

Various circuits are used in which capacitors, charged via diode rectifiers, are connected in series in order to produce higher output voltages. Figure 11.11 shows one such circuit which operates as a voltage doubling rectifier.

When A is positive with respect to B, capacitor C_1 charges, with the polarity shown via the diode D_1. During the next half cycle, when B is positive with respect to A, capacitor C_2 charges with the polarity shown via the diode D_2. The load voltage, being taken from across the two capacitors connected in series, is the sum of the two voltages. With the load R disconnected, each capacitor would charge to the peak value of the voltage v_a; the direct output voltage would then be equal to $2v_a$.

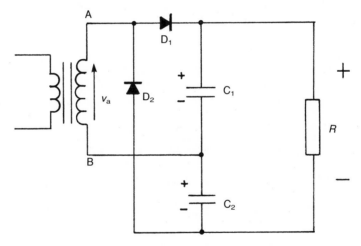

Fig. 11.11 Voltage multiplying rectifier.

11.7 Power supply regulation

The main requirement of a power supply is that it shall produce its full load output current with as little change in output voltage as possible. The rectifier circuits discussed so far in this chapter all produce an output voltage which is related to the voltage of the a.c. supply to the transformer primary winding. Any variation in the supply voltage will, therefore, cause a variation in the direct output voltage. Further, due to the impedance of the transformer windings, to the voltage drop across a conducting non-ideal diode and to the mode of operation of the reservoir capacitor, the direct output voltage is also reduced as the current drawn from a power supply is increased.

The regulation of a power supply may be defined generally as the amount that the output voltage (or current if it is designed as a constant current supply) changes when an external parameter changes. Thus *load regulation* may be defined as

$$\text{load regulation} =$$

$$\frac{\text{no load output voltage} - \text{full load output voltage}}{\text{full load output voltage}} \times 100\% \qquad (11.23)$$

Regulation may be similarly specified in terms of variation of the a.c. supply voltage or of ambient temperature changes. The effect of variations of load current upon the load voltage may also be specified in terms of an apparent internal resistance of the power supply. As the output load current taken from the supply is increased, the output voltage falls, and the power supply behaves as if it had an internal resistance. Suppose that a change of output current δI_L produces a change of output voltage δV_L. Then the internal resistance of the supply is given by

$$R_S = -\frac{\delta V_L}{\delta I_L} \qquad (11.24)$$

Although a load voltage/load current characteristic for a rectifier circuit would not, in general, exhibit a linear fall in voltage, the concept of an internal resistance is very useful in comparing different power supplies.

Example   A 1 ampere power supply using a half wave diode rectifier circuit with a reservoir capacitor has a *no load* voltage of 35 V and a full load voltage of 25 V. Determine the regulation and its effective internal resistance.

$$\text{regulation} = \frac{35 - 25}{25} \times 100 = 40\%$$

$$\text{effective internal resistance } R_1 = \frac{\delta V_L}{\delta I_L}$$

$$= \frac{35 - 25}{1}$$

$$= 10 \, \Omega$$

11.7.1   *Regulated power supplies*

For applications in which the regulation of a normal half or full wave rectifier power supply is not adequate (for TTL logic devices, for example, the direct supply voltage must be maintained within the range 5 V ± 0.25 V), a regulator

series regulator

shunt regulator

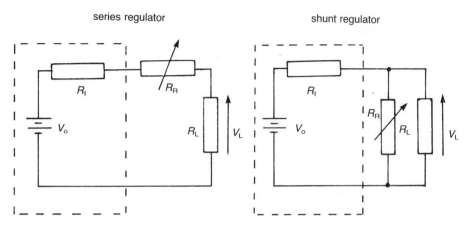

Fig. 11.12 Regulator circuits.

circuit may be added to reduce the effective internal resistance of the supply.

Two basic alternative regulator circuits are in common use and are illustrated in Fig. 11.12. The power supply in the figure is represented by a battery of voltage equal to the no load voltage of the supply, together with a resistor representing its internal resistance. In the series regulator circuit, a variable resistor R_R is included in series with the load. The value of the variable resistor is continually adjusted so that, at all values of load current, the output voltage is the same. In the shunt regulator circuit, a variable resistor R_R is included in parallel with the load, its value being continually adjusted to maintain the total current taken from the supply, and hence the load voltage constant.

A very common, simple regulator circuit makes use of a Zener diode as a shunt regulator, as illustrated in Fig. 11.13. The Zener diode, when reverse biased, has an almost constant voltage over a wide range of reverse currents. In the circuit shown the output load voltage V_L is equal to the Zener voltage and the no load voltage of the power supply V_o is dropped across the series resistance of the circuit, $R_I + R_S$, the resistor R_S being included to prevent the

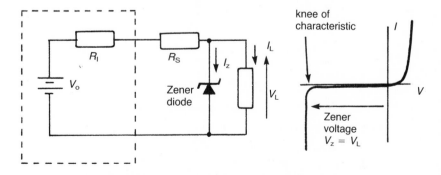

Fig. 11.13 Regulator circuit with a Zener diode.

power dissipation in the diode exceeding its maximum rating, if the load R_L is disconnected. If the load current I_L is increased, the Zener current I_Z will decrease by an equivalent amount to maintain the total current and hence the voltage drop across the series resistors constant. The stabilizing effect of the Zener diode ceases when the load current I_L is increased and the Zener current reduces to a very low value. The Zener diode operating point then moves round the knee of the characteristic. For more accurate stabilization of the output voltage, regulating devices are included which incorporate an amplifier to increase the sensitivity of the device.

Figure 11.14 shows a series regulator and a shunt regulator, both using a transistor as the regulating device. In the shunt regulator circuit, if the load voltage increases, the transistor collector potential will rise positively with respect to the power supply negative line. A Zener diode regulator circuit is used to hold the potential of the transistor base constant with respect to the positive line; as the load voltage increases, therefore, the transistor base potential will rise positively with respect to its emitter. This, in turn, increases the current flowing via the transistor, increasing the voltage drop across the series resistors R_1 and R_S, so counteracting the original rise in the load voltage. In the series regulator circuit, the difference between the power supply voltage and the load voltage is dropped across the series regulator transistor T. An increase of load current tends to cause the load voltage to fall. However, because the base of the transistor is fixed in potential by the Zener diode regulator circuit, as the load voltage starts to fall, the base emitter voltage of the transistor is increased, so increasing the transistor current and reducing the voltage drop across the transistor. This compensates for the original fall in load voltage.

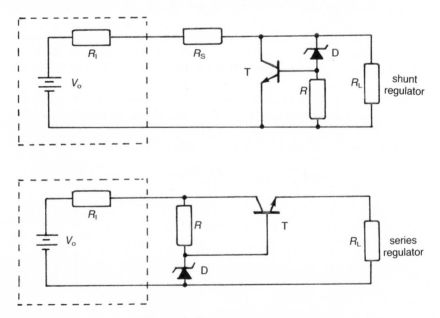

Fig. 11.14 Regulators using transistors.

The series regulator arrangement is more common, and the power dissipated in the transistor is normally smaller for a given power output. The disadvantage of the series connection is that, if the output of the regulated supply is short-circuited, the full power supply voltage is applied directly across the regulator transistor, which will probably be destroyed. Series regulator circuits are, therefore, normally used in conjunction with protection circuits, which prevent the application of excessive load currents.

11.8   Controlled power rectification

For higher power systems, rectified a.c. mains supplies are often constructed utilizing the thyristor as the rectifying element. The controlled conduction properties of the thyristor allow control of the power in the load.

A very simple method of obtaining controlled triggering of the thyristor is illustrated in Fig. 11.15. The circuit shown is a half wave rectifier circuit in which the thyristor conducts during the half cycle in which its anode is positive with respect to its cathode. The thyristor differs from a simple diode rectifier, however, in that it does not conduct until a certain critical anode–cathode voltage is reached, as described in Section 6.10.1. The critical voltage may be varied, being lowered by increasing the gate current, which is provided via the resistor R. Increasing the value of R will thus raise the critical anode–cathode voltage. Variation of the point of conduction can be achieved over the first quarter cycle of the a.c. mains supply, i.e. up to the instant of the peak voltage of the supply; a further decrease of gate current will then prevent the rectifier conducting altogether. Using this simple circuit, a reduction of the power to the load of up to 50% of that obtained using a normal half wave rectifier may be achieved. The diode is included in the circuit between the gate and cathode of the thyristor to prevent the full voltage of the supply being applied in

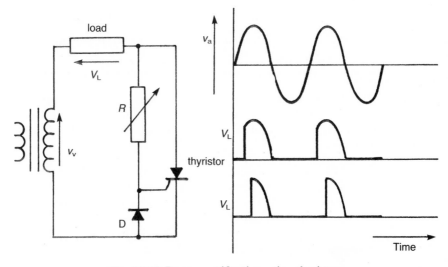

Fig. 11.15 Power rectification using thyristors.

reverse direction between the cathode and gate during the non-conducting half cycle. Conduction continues in the thyristor until the anode–cathode voltage is reduced almost to zero at the end of the conducting half cycle.

Problems

1. A half wave rectifier is fed from a transformer with a secondary voltage of 35 V rms. Neglecting the voltage drop across the rectifier diode, determine (a) the average voltage output, (b) the rms voltage output, (c) the peak inverse voltage.

2. If a reservoir capacitor is added to the rectifier of Problem 1, determine (a) the peak voltage across the capacitor, (b) the peak inverse voltage, (c) the mean output voltage on no load, (d) the rms output voltage on no load.

3. A d.c. output of average value 300 V is required from a half wave rectifier operating on a 230 V rms sinusoidal supply via a transformer. Neglecting rectifier voltage drop, determine (a) the transformer ratio, (b) the peak inverse voltage.

4. Repeat Problem 3 for a full wave rectifier.

5. A half wave rectifier with a 16 μF reservoir capacitor is required to supply an average current of 40 mA at an average voltage of 250 V. The mains supply is 240 V 50 Hz. Determine the required transformer ratio.

6. A full wave rectifier, operating from a 50 Hz supply, has a peak output of 400 V. It is loaded with a resistor of 8 kΩ and is shunted by a reservoir capacitor. Calculate the capacitance for a ripple of 20 V peak to peak.

7. A half wave rectifier, operating from a 50 Hz supply, has a peak output voltage of 20 V. It is loaded with a resistor of 200 Ω and is shunted by a reservoir capacitor of 1250 μF. What is the peak value of the ripple voltage?

8. A bridge rectifier circuit is supplied with 240 V rms and the output is across a load of 100 Ω. What is the mean load current if the diodes can be assumed to have negligible forward resistances and infinite reverse resistances?

9. A 50 Hz power supply consists of a transformer giving an output of 50 V rms to a bridge rectifier. The output current is 2 A and a reservoir capacitor of 2500 μF is used. Calculate the peak output voltage and estimate the peak to peak ripple.

10. A stabilized voltage source consists of a 5.6 V Zener diode which is connected via a 220 Ω limiting resistor R_s across a d.c. supply of voltage 10 V ± 10%. (a) What is the maximum load current which can be taken from the stabilized source if the Zener diode current must not fall below 1 mA? (b) If the maximum permissible power dissipation of the diode is 250 mW, is there a minimum value of load current which must be taken from the voltage source?

Chapter 12
Transformers

12.1 Transformers

The transformer is the electrical equivalent to the mechanical gear box. The gear box transmits mechanical power at different speed and torque levels whilst the transformer *transforms* electrical power from one voltage and current level to another. Both devices work with very little power loss, the efficiency of electrical transformers being of the order of 98%.

The alternating current transmission and distribution systems have developed largely because of the transformer. Electrical power is generated at 11 000 V, transformed up to 33 000 or 132 000 V and transmitted to the centres of power demand, where it is transformed down to a voltage level of 415 or 240 V and distributed for industrial or domestic use.

12.1.1 Transformer action

Consider the simplest form of single phase transformer shown in Fig. 12.1 in which there are two coils wound on a ferrous core. Usually the coil which is connected to the supply is called the *primary* and that which is connected to the load the *secondary*. Transformers are completely reversible, i.e. power can flow in either direction and transformation of voltage and current can be up or down.

Let a sinusoidal alternating voltage V_1 be applied to the primary winding. A current I_1 will flow in the primary winding and a flux ϕ will be produced in the core. If it is assumed that the flux links both windings perfectly, i.e. there is no flux *lost*, then by Faraday's Law, Equation (1.27),

$$E_1 = N_1 \frac{d\phi}{dt}$$

and

$$E_2 = N_2 \frac{d\phi}{dt}$$

Therefore

$$\frac{E_1}{E_2} = \frac{N_1}{N_2} \tag{12.1}$$

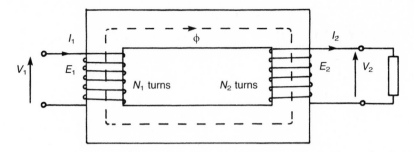

Fig. 12.1 Single-phase transformer.

The ratio N_1 to N_2 is termed the *transformation ratio* of the transformer. The difference between the applied and induced primary voltages is the primary winding impedance voltage drop I_1Z_1 which for most practical transformers is very small, so that $V_1 \approx E_1$. This is also the case in the secondary winding and $V_2 \approx E_2$. Therefore

$$\frac{V_1}{V_2} \approx \frac{N_1}{N_2} \tag{12.2}$$

When a load is connected to the transformer a current I_2 will flow in the secondary winding producing a flux which, by Lenz's law, tends to oppose the main flux ϕ. If the main flux were reduced then E_1 would be reduced and the primary current I_1 would increase according to the equation

$$\mathbf{V_1} = \mathbf{E_1} + \mathbf{I_1}Z_1 \tag{12.3}$$

The increased primary current produces a flux which opposes the flux produced by the secondary current and so tends to maintain the main flux ϕ. The steady state condition is such that there is an ampere–turn balance between the primary and secondary windings, and in practice the main flux in the transformer is only marginally reduced as the secondary current is increased from no load to full load. For ampere–turn balance

$$I_1N_1 = I_2N_2$$

$$\frac{I_1}{I_2} = \frac{N_2}{N_1} \tag{12.4}$$

12.2   Equivalent circuit

When dealing with electrical devices such as transformers and machines, an attempt is usually made to devise a simple equivalent circuit consisting of R, L and C to facilitate analytical solutions. The quantities to be modelled in a transformer are

(1) The transformation ratio of the transformer.
(2) The copper losses in the windings.

(3) Useful and leakage fluxes.
(4) Iron losses in the core.

12.2.1 The ideal transformer

The ideal transformer, shown in Fig. 12.2, consists of two coils which simply model the transformation ratio of the transformer. The coils possess no resistance or inductance and thus the ideal transformer can be characterized by the simple transformation equations,

$$\frac{V_1}{V_2} = \frac{N_1}{N_2} = \frac{I_2}{I_1} \qquad (12.5)$$

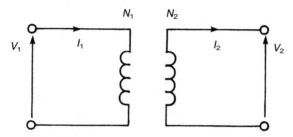

Fig. 12.2 Ideal transformer equivalent circuit.

12.2.2 Copper losses

Both the primary and secondary coils of an actual transformer possess resistance so that when current flows there is a *copper loss* (I^2R loss) associated with both coils. These winding resistances can be represented in the equivalent circuit by resistances included in series with the ideal transformer.

12.2.3 Useful and leakage fluxes

The useful or main flux is that flux which links both coils, as shown in Fig. 12.3. Considering sinusoidal operation of the transformer and assuming that the instantaneous value of the useful flux may be written as

$$\phi = \phi_m \sin \omega t$$

then the induced e.m.f. is

$$e_1 = N_1 \frac{d\phi}{dt} = \omega N_1 \phi_m \cos \omega t$$

Thus the flux is proportional to the e.m.f. and lags it by 90°. The current producing this useful flux is called the *magnetizing current* and is designated I_m. It is in phase with the flux and therefore lags the induced voltage E_1 by 90°.

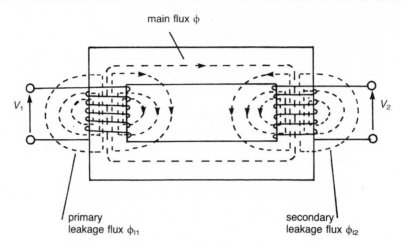

main flux ϕ

V_1

V_2

primary
leakage flux ϕ_{l1}

secondary
leakage flux ϕ_{l2}

Fig. 12.3 Leakage flux.

The effect of this magnetizing current can be included in the equivalent circuit by connecting an inductance, called the *magnetizing inductance*, in parallel with the primary coil of the ideal transformer. An inductance is used because the current in an inductance lags the voltage across it by 90° and it is connected in parallel because the flux is directly proportional to the e.m.f.

The *leakage flux* is that flux which links only one winding and produces no magnetic coupling between the coils. When the transformer is operating on load the primary and secondary currents produce fluxes which are mutually opposing, causing leakage flux paths as shown in Fig. 12.3. The leakage flux path is largely in air and thus the relationship between the leakage fluxes and the current producing them is linear.

ϕ_{l1} is proportional to and in phase with I_1
ϕ_{l2} is proportional to and in phase with I_2

The induced e.m.f.s produced by the leakage fluxes are proportional to and leading them by 90°, similar again to the conditions existing in an inductor. The effect of the leakage fluxes can therefore be included in the equivalent circuit by including an inductance in series with each coil of the ideal transformer.

12.2.4 Iron losses

It was shown in Sections 1.6.2 and 1.6.4 that iron cores which are subjected to fluxes which vary with time exhibit hysteresis and eddy current losses. Therefore, when a sinusoidal voltage is applied to the primary winding, hysteresis and eddy current losses are present in the transformer core. These losses are termed the *iron* or *core losses* of the transformer. For a particular transformer, the iron losses are constant for a fixed value of supply voltage, therefore they can be represented in the equivalent circuit by a resistance connected in parallel with the primary coil of the ideal transformer.

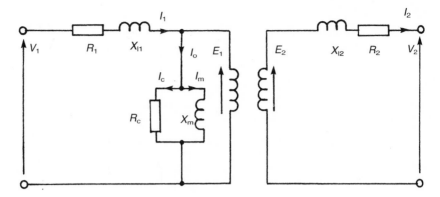

Fig. 12.4 Equivalent circuit.

The equivalent circuit of a practical transformer is obtained by combining the effects outlined above and is shown in Fig. 12.4. R_1 and R_2 are the primary and secondary winding resistances, X_{l1} and X_{l2} are the primary and secondary leakage reactances, X_m is the magnetizing reactance, R_c is the resistance to represent core losses, I_o is the no load current, I_m is the magnetizing current and I_c is the core loss current.

12.2.5 Referred values

The ideal transformer can be eliminated from the equivalent circuit if the secondary parameters are *referred* to the primary side. The secondary current I_2 is replaced by the referred current I_2' flowing in the primary side of the equivalent circuit, where

$$I_2' = \frac{N_2}{N_1} I_2 \tag{12.6}$$

The secondary voltage V_2 is replaced by the referred voltage V_2', where

$$V_2' = \frac{N_1}{N_2} V_2 \tag{12.7}$$

The secondary winding resistance R_2 is replaced by the referred secondary resistance R_2' connected in the primary side of the equivalent circuit. The value of R_2' is such that the power dissipated in R_2' when the referred current I_2' flows through it is equal to the power dissipated in R_2 when the current I_2 flows through it.

$$(I_2')^2 R_2' = I_2^2 R_2$$

$$R_2' = \left[\frac{N_1}{N_2}\right]^2 R_2 \tag{12.8}$$

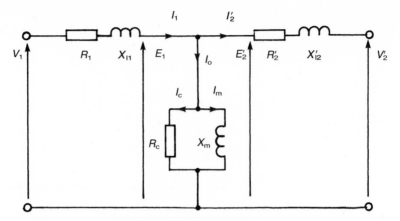

Fig. 12.5 Equivalent circuit.

Similarly X_{12} is replaced by the referred value X'_{12} where

$$X'_{12} = \left[\frac{N_1}{N_2}\right]^2 X_{12} \tag{12.9}$$

The resulting equivalent circuit is shown in Fig. 12.5.

12.2.6 Approximate equivalent circuit

In a practical transformer the no load current is only a small percentage of the full load current. Power transformers have a no load/full load current ratio of about 1/20, therefore, when considering transformers on load (say, above three-quarters full load), the parallel branch of the equivalent circuit is neglected as it has little effect on the overall performance and the approximate equivalent circuit of Fig. 12.6 can be used.

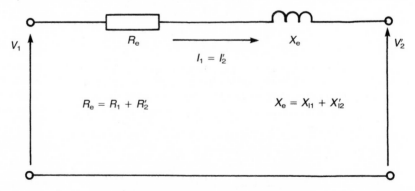

Fig. 12.6 Approximate equivalent circuit.

Example   A single phase transformer with a voltage step-down ratio of $3\,\text{kV}/240\,\text{V}$ has primary and secondary winding resistances of $1.0\,\Omega$ and $0.01\,\Omega$ respectively. The corresponding leakage reactances are $5\,\Omega$ and $0.05\,\Omega$. What is the secondary winding terminal voltage when the load is a resistance of $10\,\Omega$?

Using the approximate equivalent circuit of Fig. 12.6, then the total equivalent resistance is

$$R_\text{e} = R_1 + R_2' = 1.0 + \left(\frac{3000}{240}\right)^2 0.01 = 2.56\,\Omega$$

The total equivalent reactance is

$$X_\text{e} = X_{l1} + X_{l2}' = 5 + \left(\frac{3000}{240}\right)^2 0.05 = 12.81\,\Omega$$

The total circuit impedance is thus $(2.56 + \text{j}12.81) + 10 = 12.56 + \text{j}12.81\,\Omega$, or in polar notation $17.94\underline{/45.6°}\,\Omega$. Hence the circuit current is $3000/17.94\underline{/45.6°} = 167.3\underline{/-45.6°}$ A. Thus the potential difference across the load, i.e. the secondary winding terminal voltage, is $IZ_\text{L} = 10 \times 167.3\underline{/-45.6°} = 1673\underline{/-45.6°}$ V.

12.2.7   *Phasor diagram of a transformer on load*

The phasor diagram shown in Fig. 12.7 is based on the equivalent circuit shown in Fig. 12.5.

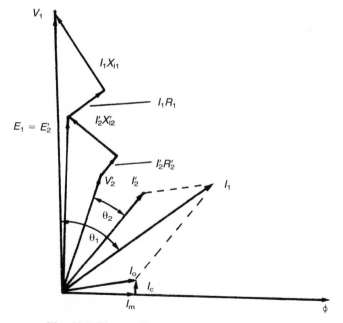

Fig. 12.7 Phasor diagram for transformer on load.

12.3   Transformer tests

In order to determine the values of the parameters in the equivalent circuit, tests are carried out on the transformer. The tests used are the open-circuit test and the short-circuit test. These tests also enable the efficiency and the voltage regulation to be determined.

12.3.1   Short-circuit test

The secondary winding of the transformer is short-circuited and the primary winding is supplied from a variable voltage source with a reduced voltage. An ammeter, voltmeter and wattmeter are connected in the primary circuit and the primary voltage is increased until the rated (full load) current flows in the windings, when readings of voltage current and power are taken. Let these readings be V_{sc}, I_{sc} and P_{sc}, then considering the equivalent circuit of Fig. 12.6 with the secondary short-circuited

$$Z_e = \frac{V_{sc}}{I_{sc}}, \; R_e = \frac{P_{sc}}{I_{sc}^2}$$

$$X_e = \sqrt{(Z_e^2 - R_e^2)} \tag{12.10}$$

12.3.2   Open-circuit test

Full rated voltage is applied to the primary winding and readings of primary voltage current and power are taken with the secondary winding open-circuited. Let these readings be V_{oc}, I_{oc} and P_{oc}, then considering the equivalent circuit shown in Fig. 12.4 and noting that on open-circuit $I_2 = 0$ and also that in a practical transformer R_c and X_m are very much greater than R_1 and X_{11}, R_c and X_m are given by

$$R_c = \frac{V_{oc}}{I_c} \quad \text{and} \quad X_m = \frac{V_{oc}}{I_m} \tag{12.11}$$

where

$$I_c = I_{oc} \cos \theta_{oc} \quad \text{and} \quad I_m = I_{oc} \sin \theta_{oc}$$

and θ_{oc} is the open-circuit power factor angle.
Also

$$\cos \theta_{oc} = \frac{P_{oc}}{V_{oc} I_{oc}}$$

hence

$$I_c = \frac{P_{oc}}{V_{oc}}$$

and, therefore

$$R_c = \frac{V_{oc}^2}{P_{oc}}$$

12.4   Voltage regulation

As in most other voltage sources the terminal voltage of the transformer drops when the load current drawn from it is increased. The difference between the rated no load output voltage and the output voltage on load is termed the regulation, and the percentage regulation of the transformer is defined as

$$\% \text{ regulation} = \frac{\text{voltage on no load} - \text{voltage on load}}{\text{voltage on no load}} \times 100 \qquad (12.12)$$

Using the approximate equivalent circuit shown in Fig. 12.6, the voltage equation is

$$V_2' = V_1 - I_2' Z_e \qquad (12.13)$$

and the phasor diagram for a typical lagging power factor load current is shown in Fig. 12.8.

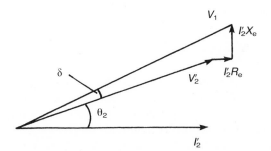

Fig. 12.8 Phasor diagram with lagging power factor.

Now in a practical transformer the voltage drops $I_2' R_e$ and $I_2' X_e$ are small compared to V_2' and V_1, therefore the angle δ is negligibly small and thus

$$V_1 = V_2' + I_2' R_e \cos \theta_2 + I_2' X_e \sin \theta_2 \qquad (12.14)$$

and the percentage regulation is given by

$$\% \text{ regulation} = \frac{V_1 - V_2'}{V_1} \times 100$$

$$\% \text{ regulation} = \frac{100}{V_1}\{I_2' R_e \cos \theta_2 + I_2' X_e \sin \theta_2\} \qquad (12.15)$$

12.5 Efficiency of the transformer

The efficiency of any device is defined as the ratio of the power output to the power input, the difference between the output and input being the inherent power loss in the device, which in the case of the transformer is made up of two components, i.e. copper loss and iron loss. Therefore the efficiency of the transformer is given by

$$\eta = \frac{\text{output}}{\text{input}} = \frac{\text{output}}{\text{output} + \text{copper losses} + F_e \text{ losses}}$$

$$\eta = \frac{V_2 I_2 \cos \theta_2}{V_2 I_2 \cos \theta_2 + I_2^2 R_e + F_e} \tag{12.16}$$

(F_e represent the constant iron losses).

In Equation (12.16) there are two variables, the load current I_2 and the power factor θ_2. Considering operation at a constant power factor, Equation (12.16) may be rewritten as

$$\eta = \frac{V_2 \cos \theta_2}{V_2 \cos \theta_2 + I_2 R_e + \dfrac{F_e}{I_2}} \tag{12.17}$$

differentiating the denominator of Equation (12.17) with respect to I_2 and equating to zero gives

$$\frac{d(\text{den})}{dI_2} = R_e = \frac{F_e}{I_2^2} = 0$$

$$I_2^2 R_e = F_e$$

The second differential shows that this is the minimum condition for the denominator and therefore the maximum efficiency of a transformer occurs when the copper loss is equal to the iron loss. The variation of the efficiency and the losses with the load current is shown in Fig. 12.9(a). If the load current is now assumed constant, then Equation (12.16) becomes

$$\eta = \frac{V_2 I_2}{V_2 I_2 + \dfrac{I_2^2 R_e + F_e}{\cos \theta_2}} \tag{12.18}$$

It is easy to see from Equation (12.18) that as the power factor decreases from unity the efficiency will decrease, as shown in Fig. 12.9(b).

Example A transformer having a rated output of 100 kVA has an efficiency of 98% at full load unity power factor and its maximum efficiency occurs at two-thirds full load unity power factor. Calculate the iron losses and the maximum efficiency.

At maximum efficiency the iron losses equal the copper losses. Also the copper losses vary as the square of the load current. Therefore at two-thirds full load

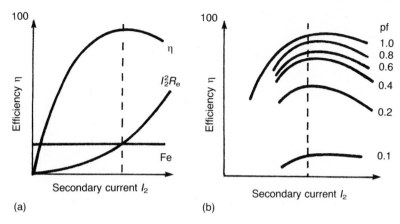

Fig. 12.9 Efficiency graphs.

$(\frac{2}{3})^2 \times$ copper loss at full load = Fe loss

At full load the efficiency is 0.98, i.e.

$$0.98 = \frac{100\,000}{100\,000 + (\text{Fe} + \text{Cu})\text{ loss}}$$

Thus the copper loss at full load = $2000 -$ Fe loss. Therefore

$(\frac{2}{3})^2(2000 - \text{Fe loss}) = \text{Fe loss}$

iron loss = 615.4 W

$$\text{Maximum efficiency} = \frac{\frac{2}{3} \times 100 \times 1000}{\frac{2}{3} \times 100 \times 1000 + 1230.8} \times 100$$

$$= 98.19\%$$

Example   A 5 kVA, 200/400 V, 50 Hz, single phase transformer gave the following test results,

open-circuit test, performed from the low voltage side
   200 V, 0.7 A, 60 W
short-circuit test, performed from the high voltage side
   22 V, 16 A, 120 W

Calculate (a) the equivalent impedance of the transformer referred to the high voltage side, (b) the percentage voltage regulation at full load and 0.9 p.f. lagging, (c) the maximum percentage full load voltage regulation and the power factor at which it occurs, (d) the power factor at which zero regulation occurs, (e) the efficiency at full load and half load and a power factor of 0.8.

(a) $Z_e = \dfrac{22}{16} = 1.375\,\Omega$; $R_e = \dfrac{120}{16^2} = 0.469\,\Omega$

$$X_e = \sqrt{(1.89 - 0.2)} = 1.3 \ \Omega$$

(b) From Equation (12.15)

$$\% \text{ regulation} = \frac{100}{400}(12.5 \times 0.47 \times 0.9 + 12.5 \times 1.3 \times 0.44)$$

$$= 3.1\%$$

(c) $\dfrac{d(\text{reg})}{d\theta} = \dfrac{100}{400} \times 12.5 \ (-R_e \sin \theta_2 + X_e \cos \theta_2)$

Equate to zero for max or min, $\tan \theta_2 = 1.3/0.47$ and $\theta_2 = 70.1°$.

$$\text{max } \% \text{ regulation} = \frac{100}{400} \times 12.5 \ (0.47 \times 0.34 + 1.3 \times 0.94)$$

$$= 4.32\%$$

at a p.f. of 0.34 lagging.

(d) Zero regulation occurs when

$$I_2 R_e \cos \theta_2 + I_2 X_e \sin \theta_2 = 0$$

This can only occur when θ_2 goes negative, and that will only occur for leading currents. In this example zero regulation occurs when

$$\frac{R_e}{X_e} = -\tan \theta_2$$

or

$$\theta_2 = -19.87°$$

and

$$\cos \theta_2 = 0.94 \text{ leading}$$

(e) The iron loss is constant and is the power measured on open circuit. At full load, 0.8 lagging

$$\eta = \frac{5000 \times 0.8}{5000 \times 0.8 + 60 + (12.5)^2 \times 0.47} = 0.9677$$

$$\eta = 96.8\%$$

at $\frac{1}{2}$ full load, 0.8 lagging

$$\eta = \frac{2500 \times 0.8}{2500 \times 0.8 + 60 + (6.25)^2 \times 0.47} = 0.9623$$

$$\eta = 96.2\%$$

12.6 The voltage equation

The relationship between the voltage, flux, turns and frequency can be obtained directly from Faraday's Law. On the primary side,

$$e_1 = N_1 \frac{d\phi}{dt}$$

Let the sinusoidal flux be $\phi_m \sin \omega t$

$$\text{thus } e_1 = N_1 \frac{d(\phi_m \sin \omega t)}{dt}$$

$$e_1 = 2\pi f N_1 \phi_m \cos \omega t$$

The rms value of induced e.m.f. is

$$E_1 = \frac{2\pi f N_1 \phi_m}{\sqrt{2}}$$

$$E_1 = 4.44 f N_1 \phi_m \tag{12.19}$$

Similarly

$$E_2 = 4.44 f N_2 \phi_m \tag{12.20}$$

Example A 3300/250 V, 50 Hz, single phase transformer with 1000 secondary turns is built on a core with an effective cross-sectional area of 100 cm^2. What will be the maximum value of the flux density in the core?

Using Equation (12.20), $E_2 = 4.44 f N_2 \phi_m$, then taking E_2 as 250 V

$$\phi_m = \frac{E_2}{4.44 f N_2} = \frac{250}{4.44 \times 50 \times 1000} = 1.13 \text{ mWb}$$

The maximum flux density $B_m = \phi_m/A = 0.113$ T.

12.7 Auto-transformers

The auto-transformer utilizes only one winding, the secondary voltage and current being obtained from a tapping point along the winding as shown in Fig. 12.10. The auto-transformer shown in Fig. 12.10 is equivalent to a conventional two winding step down transformer; however a considerable saving in copper has been effected since in the two winding transformer the windings were carrying a total current of $I_1 + I_2$ and in the auto-transformer the single winding carries only $I_2 - I_1$.

Auto-transformers find a wide application when a variable voltage is required; the tapping is made into a sliding contact and a secondary voltage of zero, when the contact is at a and to V_1, when the contact is at b, can be obtained. The main disadvantage of the auto-transformer is that the primary and

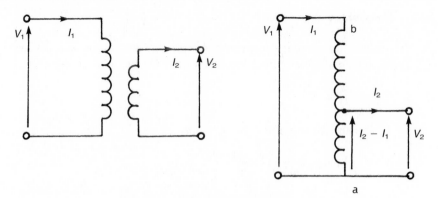

Fig. 12.10 Auto-transformer.

secondary windings are not electrically isolated from each other, a safety feature which is essential on all distribution transformers.

12.8   Three phase transformers

A three phase transformer can be obtained by suitably connecting three single phase units or by using a three-limbed core as shown in Fig. 12.11. The latter is far more common, particularly in distribution transformers, because it is less costly.

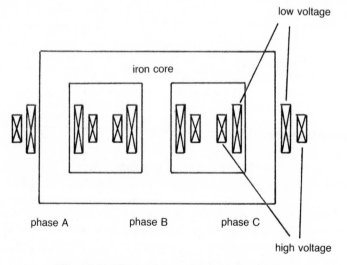

Fig. 12.11 Three-phase core type transformer.

Electrically the three phase transformer is made up of six coils, three primary and three secondary, which may be connected in either star or delta, giving the possibility of twelve different transformer connections. It is important, therefore, to know exactly how a three phase transformer is connected especially when two or more transformers are to operate in parallel. The different transformer connections are arranged in four main groups according to the phase displacement between the primary and secondary line voltages, as shown in the table of Fig. 12.12. For successful parallel operation of three phase transformers it is essential that they belong to the same main group and that their voltage ratios are equal.

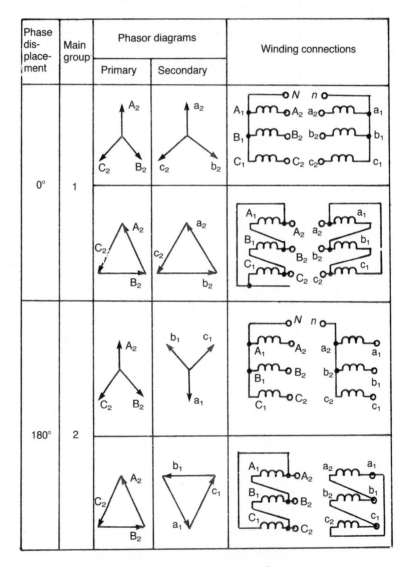

Fig. 12.12 Transformer connections.

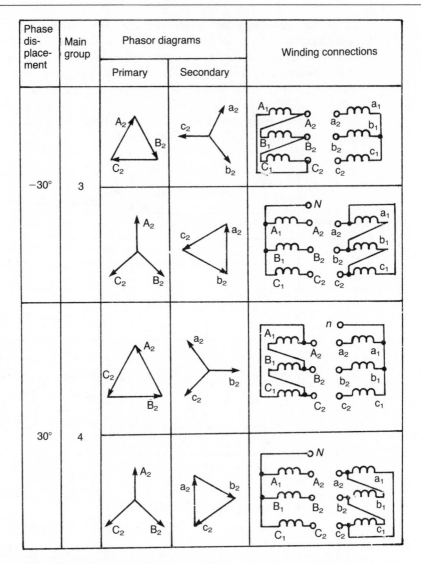

Fig. 12.12 *Continued*

Problems

1. Explain how a transformer can be represented by an equivalent circuit involving a reactance in series with a resistance and the load.

2. A 3200/400 V single phase transformer has a primary resistance of 1.5 Ω and reactance 12 Ω and a secondary resistance of 0.02 Ω and reactance 0.10 Ω. What will be the circuit impedance when these are referred to the primary circuit?

3. A 1440/240 V single phase transformer has a primary resistance of $0.9\,\Omega$ and reactance $5\,\Omega$ and a secondary resistance of $0.03\,\Omega$ and reactance $0.13\,\Omega$. What primary voltage input is required if the secondary current is to be 100 A when the secondary is short-circuited and what will be the power factor?

4. What is the efficiency at half load of a 100 kVA transformer with a power factor of 0.8 when, at full load, the copper loss is 1 kW and the iron loss 1 kW.

5. A 100 kVA transformer has an efficiency of 0.980 at full load with a power factor of 1. What are the iron losses and the maximum efficiency if maximum efficiency occurs at two-thirds full load, power factor 1?

6. The test results on a 5 kVA, 6600/240 V, 50 Hz single phase transformer were

   open-circuit test, on the high-voltage side
   6600 V, 20 W, 0.04 A
   short-circuit test, on the high voltage side
   76 V, 23.1 W, full load current

   Calculate the parameters of the equivalent circuit, assuming $R_1 = R_2'$, and $X_{11} = X_{12}'$.

7. A 110 kVA, 11 000/240 V, 50 Hz, single phase transformer gave the following test results on short circuit. With the secondary short-circuited a voltage of 500 V on the primary produced full load current in the transformer, the power supplied being 3000 W. Calculate (a) the percentage regulation, (b) the secondary terminal voltage, at full load current for power factors of unity, 0.8 lagging and 0.8 leading.

8. A 100 kVA, 10 000/1000 V, 50 Hz transformer gave the following test results. With the low voltage winding open-circuited, the input power to the high voltage side was 1500 W. With the low voltage winding short-circuited, the power supplied to the transformer with full load current flowing was 2680 W.
   Calculate the efficiency of the transformer at (a) full load, unity power factor, (b) 50% full load at 0.8 power factor lagging. At what load does the maximum efficiency occur?

9. A transformer has a maximum efficiency of 0.98 which occurs at 15 kVA and unity power factor. During one day it is loaded as follows:

   12 hours – 4 kW at 0.5 power factor
   8 hours – 12 kW at 0.8 power factor
   4 hours – 9 kW at 0.9 power factor

   Calculate the *all day* efficiency, i.e. the ratio of kWh output k/Wh input.

10. A 3300/250 V, 50 Hz, single phase transformer has a core of effective cross-sectional area 125 cm^2 with a primary winding of 70 turns. What will be the maximum flux density in the core?

11. A welding transformer is required to operate from the 240 V, 50 Hz mains and the output voltage required is 10 V. If the maximum flux in the transformer

core is not to exceed 1 mWb, calculate the number of turns required for the primary and secondary windings. What other factors would be taken into account in the design of such a transformer?

12. A 110 kVA, 11 000/2200 V, two winding transformer may be connected to form an auto-transformer. Calculate the voltage ratios available and the new rating of the auto-transformer.

Chapter 13
Attenuators and Filters

13.1 Attenuators

Attenuators are circuits which reduce the magnitude of an electrical signal to a fraction which ideally is constant irrespective of frequency. If the circuit contains only components such as resistors, inductors and capacitors then it is said to be *passive*, if it contains a device such as an operational amplifier then it is referred to as *active*.

The attenuation produced by a circuit (Fig. 13.1) can be described as the ratio of the output and input powers, i.e. for an input power P_1 and an output power P_2,

$$\text{attenuation} = \frac{P_2}{P_1} \tag{13.1}$$

This ratio is usually expressed in decibels. Thus

$$\text{attenuation in dB} = 10\lg\left(\frac{P_2}{P_1}\right) \tag{13.2}$$

The input power $P_1 = V_1^2/Z_1$ and the output power $P_2 = V_2^2/Z_2$, where Z_1 is the input impedance and Z_2 the output impedance. Thus, for a symmetrical attenuator where $Z_1 = Z_2$ then

$$\text{attenuation in dB} = 10\lg\left(\frac{V_2^2}{V_1^2}\right) = 20\lg\left(\frac{V_2}{V_1}\right) \tag{13.3}$$

The input power can also be written as $P_1 = I_1^2 Z_1$ and the output power $P_2 = I_2^2 Z_0$. Thus for a symmetrical attenuator we can also write

$$\text{attenuation in dB} = 20\lg\left(\frac{I_2}{I_1}\right) \tag{13.4}$$

In the above equations the attenuation has been defined as the ratio of the output to input. Sometimes, however, attenuation is described in terms of the ratio of the input to output. This leads to the same numerical value in dB but the values are positive.

Example A symmetrical attenuator has a current input of 100 mA, what will be the current output if the attenuation is −20 dB?

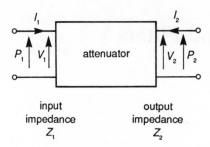

Fig. 13.1 An attenuator.

Using Equation (13.4),

$$\text{attenuation in dB} = 20\lg\left(\frac{I_2}{I_1}\right)$$

$$-20 = 20\lg\left(I_2/100\right)$$

$$I_2 = 10\,\text{mA}$$

Example   The attenuator network shown in Fig. 13.2 has an input from a battery of e.m.f. 1.5 V and negligible internal resistance. Determine the load voltage and the voltage and power attenuation in decibels.

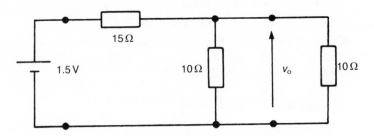

Fig. 13.2 Example.

As the two $10\,\Omega$ are in parallel and so have an effective resistance of $5\,\Omega$, the output voltage v_o across them is

$$v_o = \frac{5}{5 + 15} \times 1.5 = 0.375\,\text{V}$$

Thus the voltage attenuation is $0.375/1.5 = 0.25$, or in dB

$$\text{voltage attenuation} = 20\lg 0.25 = -20\lg\left(1/0.25\right)$$

$$= -12.0\,\text{dB}$$

The input resistance is $15 + 5 = 20\,\Omega$. Thus the input power is $1.5^2/20 = 0.1125\,\text{W}$. The output power is $0.375^2/10 = 0.0141\,\text{W}$. Hence the power attenuation is $0.0141/0.1125 = 0.1253$, or in dB

$$\text{power attenuation} = 10\lg 0.1253 = -10\lg(1/0.1253)$$

$$= -9.0\,\text{dB}$$

13.2 Characteristic impedance

When inserting an attenuator between a source and its load it is often a requirement that the insertion of the attenuator should not change the loading of the source. This means that, whatever the impedance of the source, the input to the attenuator is the same as the input that would have occurred to the load prior to the insertion of the attenuator. The input impedance of the attenuator when connected across the load must thus be the same as that of the load. Such attenuators permit a sequence of identical attenuator networks to be connected together to give a compound attenuation and still present the same input impedance.

The term *characteristic impedance* is used for the impedance of a symmetrical attenuator which presents an input impedance equal to the load impedance. An attenuator is said to be symmetrical if it presents the same impedance regardless of whether the input or output terminals are considered.

13.2.1 *Attenuators in cascade*

Consider a number of symmetrical attenuators, each with the same characteristic impedance Z_0, connected in cascade, i.e. the output from the first attenuator becomes the input to the second and its output becomes the input to the third attenuator, and so on (Fig. 13.3). The final attenuator has a load Z_0. The attenuation of the first attenuator is $20\lg(V_1/V_2)$, that of the second attenuator $20\lg(V_2/V_3)$ and that of the third is $20\lg(V_3/V_4)$. The overall attenuation is $20\lg(V_1/V_4)$. Thus

$$\text{attenuation} = 20\lg\left(\frac{V_1}{V_2} \times \frac{V_2}{V_3} \times \frac{V_3}{V_4}\right)$$

$$= 20\lg(V_1/V_2) + 20\lg(V_2/V_3) + 20\lg(V_3/V_4) \qquad (13.5)$$

Fig. 13.3 Attenuators in cascade.

Thus the total attenuation is the sum of the attenuations, in dB, of the separate attenuators.

Most attenuators are made up of either repeated T or repeated π configurations of impedances. Such configurations may be *symmetrical* or *asymmetrical*.

Example Three identical attenuators are connected in cascade. If each has an attenuation of -10 dB, what is the overall attenuation?

The total attenuation is the sum of the attenuations of the separate attenuator sections and is thus $3 \times (-10) = -30$ dB.

13.2.2 Symmetrical T-section attenuator

Figure 13.4 shows a symmetrical T-section attenuator with a characteristic impedance of Z_0. The impedance looking into the input end of the section is thus Z_0. Then Z_0 must be equal to the impedance due to $\frac{1}{2}Z_1$ in series with a parallel arrangement of Z_2 with $(\frac{1}{2}Z_1 + Z_0)$, i.e.

$$Z_0 = \tfrac{1}{2}Z_1 + \frac{Z_2(\tfrac{1}{2}Z_1 + Z_0)}{Z_0 + \tfrac{1}{2}Z_1 + Z_2}$$

$$Z_0(Z_0 + \tfrac{1}{2}Z_1 + Z_2) = \tfrac{1}{2}Z_1(Z_0 + \tfrac{1}{2}Z_1 + Z_2) + Z_2(\tfrac{1}{2}Z_1 + Z_0)$$

$$Z_0^2 = (\tfrac{1}{2}Z_1)^2 + Z_1 Z_2$$

$$Z_0 = \sqrt{(\tfrac{1}{4}Z_1^2 + Z_1 Z_2)} \tag{13.6}$$

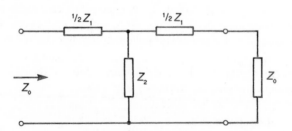

Fig. 13.4 Symmetrical T-section.

Example What is the characteristic impedance of a symmetrical T-section attenuator having a series resistance of $10\,\Omega$ and a shunt resistance of $15\,\Omega$?

$\frac{1}{2}Z_1 = 10\,\Omega$ and $Z_2 = 15\,\Omega$. Thus, using Equation (13.6)

$$Z_0 = \sqrt{(\tfrac{1}{4}Z_1^2 + Z_1 Z_2)} = \sqrt{(\tfrac{1}{4}20^2 + 20 \times 15)} = 20\,\Omega$$

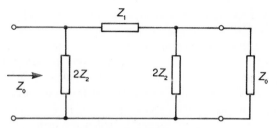

Fig. 13.5 Symmetrical π-section.

13.2.3 *Symmetrical π-section attenuator*

Figure 13.5 shows a symmetrical π-section attenuator with a characteristic impedance Z_0. The impedance looking into the input end of the section is thus Z_0. The load Z_0 is in parallel with $2Z_2$, i.e. an impedance of $Z_0 2Z_2/(Z_0 + 2Z_2)$. This is in series with Z_1, to give an impedance of $Z_1 + Z_0 2Z_2/(Z_0 + 2Z_2)$. This is in parallel with $2Z_2$. Thus, since the total impedance is Z_0,

$$Z_0 = \frac{2Z_2\left(Z_1 + \dfrac{Z_0 2Z_2}{Z_0 + 2Z_2}\right)}{2Z_2 + \left(Z_1 + \dfrac{Z_0 2Z_2}{Z_0 + 2Z_2}\right)}$$

Hence

$$Z_0 = \sqrt{\left(\frac{4Z_1 Z_2^2}{Z_1 + 4Z_2}\right)} \tag{13.7}$$

Example   What is the characteristic impedance of a symmetrical π-attenuator which has a series resistance of $100\,\Omega$ and shunt resistances of $1000\,\Omega$?

$Z_1 = 100\,\Omega$ and $2Z_2 = 1000\,\Omega$. Thus using Equation (13.7)

$$Z_0 = \sqrt{\left(\frac{4Z_1 Z_2^2}{Z_1 + 4Z_2}\right)} = \sqrt{\left(\frac{4 \times 100 \times 500^2}{100 + 4 \times 500}\right)} = 218\,\Omega$$

13.3   Filters

The term filter is used for a network for which the attenuation is frequency dependent, having low attenuation and so passing certain bands of frequencies and having considerable attenuation for other frequencies. The band, or bands, of frequencies for which the attenuation is effectively zero is called the *pass band*. The frequencies at which the attenuation changes from finite to zero are called the *cut-off frequencies*. There are four basic types of filter. Ideally, the *low pass* filter passes signals up to some limiting frequency but not above it, the *high pass* filter passes signals down to some limiting frequency but not below it, the *band pass* filter passes signals over a range of frequencies but not outside it, and the *band stop* filter only passes signals outside a range of frequencies. Figure 13.6 shows these ideal characteristics. In reality the transition in attenuation at the cut-off frequencies is not an abrupt change but gradual, as indicated by the dotted lines in Fig. 13.6. The cut-off frequency is then defined as the frequency at which the gain has dropped by 3 dB (see the discussion of bandwidth in Section 4.5.3).

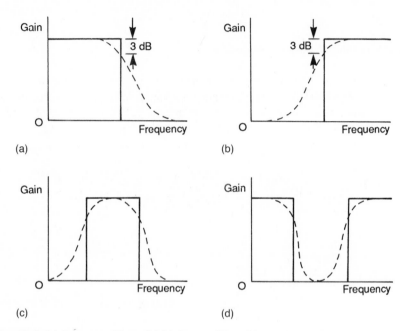

Fig. 13.6 (a) Low pass filter, (b) high pass filter, (c) band pass filter, (d) band stop filter.

13.3.1 CR filters

A CR filter is a potential divider circuit with a capacitor and a resistor in series (Fig. 13.7). An increase in frequency produces a decrease in the reactance of the capacitor and hence an increase in frequency results in a smaller potential drop across the capacitor and consequently a higher potential drop across the resistor. Thus, when the output is taken from across the capacitor the attenuation of the circuit increases, giving a low pass filter, and when the output is taken from across the resistor the attenuation is reduced, giving a high pass filter. This simple type of filter finds widespread use in electronics, e.g. blocking a.c. interference or blocking a d.c. signal.

With a load R_L the total circuit impedance for the circuit in Fig. 13.7(a) is

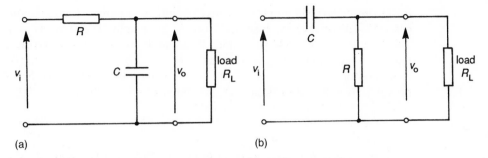

Fig. 13.7 CR filter, (a) low pass, (b) high pass.

$$Z = R + \frac{(1/j\omega C)R_L}{R_L + (1/j\omega C)} = R + \frac{R_L}{1 + j\omega CR_L}$$

The input voltage V_i is thus

$$V_i = ZI_i = I_i\left[R + \frac{R_L}{1 + j\omega CR_L}\right]$$

The output voltage V_o is across the parallel arrangement of C and R_L and is thus

$$V_o = I_i\left[\frac{(1/j\omega C)R_L}{R_L + (1/j\omega C)}\right] = I_i\left[\frac{R_L}{1 + j\omega CR_L}\right]$$

Thus

$$\frac{V_o}{V_i} = \frac{\left[\dfrac{R_L}{1 + j\omega CR_L}\right]}{\left[R + \dfrac{R_L}{1 + j\omega CR_L}\right]} = \frac{R_L}{R + j\omega CR_L R + R_L}$$

With $R_L \gg R$, then

$$\frac{V_o}{V_i} \approx \frac{1}{1 + j\omega CR} \approx \frac{1 - j\omega CR}{1 + \omega^2 C^2 R^2} \qquad (13.8)$$

Thus the voltage gain, i.e. the ratio of the voltage magnitudes, is

$$\frac{|V_o|}{|V_i|} = \sqrt{\left[\frac{1 + \omega^2 C^2 R^2}{(1 + \omega^2 C^2 R^2)^2}\right]} = \frac{1}{\sqrt{(1 + \omega^2 C^2 R^2)}} \qquad (13.9)$$

and the phase difference between V_o and V_i is

$$\phi = \tan^{-1}(-\omega CR) \qquad (13.10)$$

With $\omega CR \ll 1$ then the voltage gain is about 1, and thus there is effectively no attenuation. With ωCR less than 1 then the real term in Equation (13.8) dominates. With $\omega CR \gg 1$ then Equation (13.8) gives $V_o/V_i \approx -j/\omega CR$ and the circuit behaves as a purely reactive circuit. As a purely reactive circuit no power is delivered to the load and thus the attenuation is effectively infinite. The circuit is thus a low pass filter. The transition between the resistive or reactive nature dominating is when $\omega CR = 1$, i.e. $\phi = -45°$. This also is the condition for V_o^2/V_i^2 to be $\frac{1}{2}$, i.e. the power to have dropped by one half, i.e. by 3 dB. Thus the cut-off frequency is defined by

$$\omega_c = \frac{1}{CR} \tag{13.11}$$

For the high pass L-section circuit in Fig. 13.7(b) a similar analysis leads to

$$\frac{V_o}{V_i} \approx \frac{\omega^2 C^2 R^2 + j\omega CR}{1 + \omega^2 C^2 R^2} \tag{13.12}$$

and thus

$$\frac{|V_o|}{|V_i|} = \frac{1}{\sqrt{(1 + 1/\omega^2 C^2 R^2)}} \tag{13.13}$$

$$\phi = \tan^{-1}(1/\omega CR) \tag{13.14}$$

With $\omega CR \ll 1$ then $V_o/V_i \approx j\omega CR$ and the circuit behaves as a purely reactive circuit. With $\omega CR \gg 1$ then $V_o/V_i \approx 1$, and thus there is effectively no attenuation. The transition between the reactive and resistive nature dominating is when $\omega CR = 1$. Thus the cut-off frequency for the high pass filter is defined by

$$\omega_c = \frac{1}{CR} \tag{13.15}$$

Example An RC filter is to be used as a low pass filter with a cut-off frequency of 500 Hz. What capacitance will be required if the resistance is 1 kΩ and what will be the attenuation in dB at 1 kHz?

Using Equation (13.5) $\omega_c = 1/CR$, hence

$$C = \frac{1}{2\pi \times 500 \times 1000} = 0.32\,\mu F$$

Equation (13.13) gives, when $\omega = 2\omega_c = 2/CR$,

$$\frac{|V_o|}{|V_i|} = \frac{1}{\sqrt{(1 + \omega^2 C^2 R^2)}} = \frac{1}{\sqrt{(1 + 4)}}$$

Hence

attenuation $= 20\lg(V_o/V_i) = 20\lg(1/\sqrt{5}) = -6.99\,dB$

Note that at $\omega = \omega_c$ the attenuation is $-3\,dB$.

13.3.2 *Multiple section filters*

The sharpness of the change in attenuation at the cut-off frequencies can be improved by cascading filter sections. To avoid the problem of each such section loading the preceding one with a different impedance, the sections have each an input impedance equal to its load impedance, i.e. the input impedance is equal to the characteristic impedance. A common form of such a cascaded filter has repeated series and shunt impedances (Fig. 13.8) and can be regarded as either a number of cascaded T-sections or alternatively cascaded π-sections.

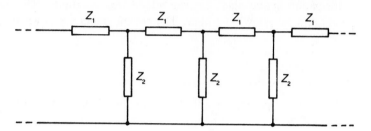

Fig. 13.8 A cascaded filter.

Figure 13.9 shows T-section low and high pass filters. At low frequencies the reactance of the inductor is insignificant compared with that of the capacitor and so most of the potential drop is across the capacitor. At high frequencies the converse occurs. Thus a low pass filter is produced with the inductors in series and the capacitor in parallel with the load while a high pass filter is with the capacitors in series and the inductor in parallel. The filter section is called a *constant k filter* since the product of total series and shunt impedances is a constant which is independent of frequency, i.e. LC.

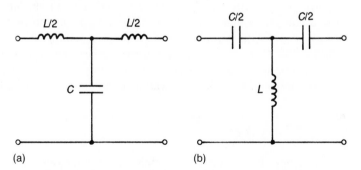

Fig. 13.9 T-section (a) low pass, (b) high pass filters.

The characteristic impedance Z_0 of the low pass filter is given by Equation (13.6), with $\frac{1}{2}Z_1 = j\omega\frac{1}{2}L$ and $Z_2 = 1/j\omega C$, as

$$Z_0 = \sqrt{(\tfrac{1}{4}Z_1^2 + Z_1 Z_2)} = \sqrt{[\tfrac{1}{4}(j\omega L)^2 + (j\omega L)(1/j\omega C)]}$$

$$= \sqrt{\left(\frac{L}{C} - \frac{\omega^2 L^2}{4}\right)} \tag{13.16}$$

Z_0 is a real quantity if $(L/C) > (\omega^2 L^2/4)$ and thus effectively purely resistive. Because all the elements in the T-section are reactive no power will be dissipated in the filter and thus all the input power will be transmitted through to the load, i.e. the attenuation is zero. Z_0 is an imaginary quantity if $(L/C) < (\omega^2 L^2/4)$ and thus purely reactive. When this occurs there is no power dissipation in the load, i.e. the attenuation is infinite. The T-section is thus operating as a low pass filter. The cut-off frequency is when $(L/C) = \tfrac{1}{4}\omega^2 L^2$, i.e.

$$f_c = \frac{1}{\pi\sqrt{(LC)}} \tag{13.17}$$

When the frequency ω is zero then Equation (13.16) gives for the characteristic impedance

$$Z_0 = R_0 = \sqrt{(L/C)} \tag{13.18}$$

This impedance is called the *design impedance R_0*.

For the high pass T-section filter in Fig. 13.9(b), the characteristic impedance Z_0 is given by Equation (13.6), with $\tfrac{1}{2}Z_1 = 1/j\omega 2C$ and $Z_2 = j\omega L$, as

$$Z_0 = \sqrt{(\tfrac{1}{4}Z_1^2 + Z_1 Z_2)} = \sqrt{[\tfrac{1}{4}(1/j\omega C)^2 + (1/j\omega C)j\omega L]}$$

$$= \sqrt{\left(\frac{L}{C} - \frac{1}{4\omega^2 C^2}\right)} \tag{13.19}$$

Z_0 is a real quantity if $(L/C) > (1/4\omega^2 C^2)$, i.e. $\omega^2 > 1/4LC$, and thus effectively purely resistive. Because all the elements in the T-section are reactive no power will be dissipated in the filter and thus all the input power will be transmitted through to the load. Thus the attenuation of the filter operating with $\omega^2 > 1/4LC$ is zero. Z_0 is an imaginary quantity if $(L/C) < (4\omega^2 C^2)$, i.e. $\omega^2 < 1/4LC$, and thus purely reactive. When this occurs there is no power dissipation in the load. Thus the attenuation is infinite. The T-section is thus operating as a high pass filter with a cut-off frequency given by

$$f_c = \frac{1}{4\pi\sqrt{(LC)}} \tag{13.20}$$

When the frequency ω is infinite then Equation (13.19) gives for the characteristic impedance

$$Z_0 = R_0 = \sqrt{(L/C)} \tag{13.21}$$

This impedance is called the *design impedance R_0*.

Similar calculations can be carried out for π-section filters.

Example   Design a high pass T-section filter to have a design impedance of $600\,\Omega$ and a cut-off frequency of 1 kHz.

The design impedance is given by Equation (13.20) as $R_0 = 600 = \sqrt{(L/C)}$. Thus $\sqrt{L} = 600\sqrt{C}$. The cut-off frequency is given by Equation (13.21) as

$$f_c = 1000 = \frac{1}{4\pi\sqrt{(LC)}} = \frac{1}{4\pi \times 600C}$$

Thus $C = 133\,\text{nF}$ and so a capacitor of $2C = 266\,\text{nF}$. This leads to $L = 600^2C = 47.9\,\text{mH}$.

13.3.3   Band pass and band stop filters

A *band pass filter* can be produced by the cascade connection of a low pass and a high pass filter. However, the more usual method is to use a T-section or π-section with the series and shunt branches being series or parallel resonant circuits. Thus a T-section will have for its series arms series resonant circuits and for its shunt arm a parallel resonant circuit. Both resonant circuits have the same resonant frequency. At frequencies below the resonant frequency the series arms of the T-section, which are series resonant circuits, have high impedances while the impedance of the shunt arm, a parallel resonant circuit, is low. Thus the output is low, i.e. the attenuation is high. At the resonant frequency the series arms have zero, or very low, impedance and the shunt arms have high impedance. Thus the output is high, i.e. the attenuation is low. At frequencies above the resonant frequency the series arms have high impedances while the shunt arm has a low impedance. The output is thus low, i.e. the attenuation is again high.

A *band stop filter* is produced by interchanging the series and shunt arms of the band pass filter.

13.4   Active filters

With low frequency filters the inductances required are fairly large and so the inductors are very bulky. Their use can be avoided by using operational amplifier circuits. Figure 13.10 shows circuits that can be used for low pass and high pass filters. The voltage gain of an operational amplifier circuit is (see Section 9.2 and Equation (9.2))

$$A_v = -\frac{Z_2}{Z_1} \tag{13.22}$$

where Z_1 is the input impedance to an inverting amplifier and Z_2 the feedback impedance. For the low pass filter (Fig. 13.10(a)) $Z_1 = R_1$ and Z_2 is the parallel arrangement of C_2 and R_2, i.e.

$$Z_2 = \frac{R_2(1/j\omega C_2)}{R_2 + (1/j\omega C_2)} = \frac{R_2}{1 + j\omega C_2 R_2}$$

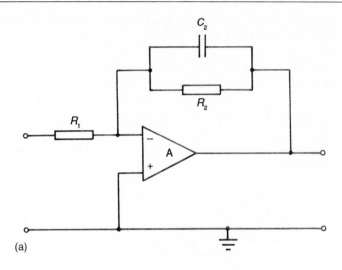

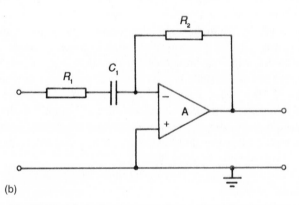

Fig. 13.10 (a) Active low pass filter, (b) active high pass filter.

Thus the voltage gain is

$$A_v = -\frac{R_2}{R_1(1 + j\omega C_2 R_2)} \tag{13.23}$$

The transition from the resistive term dominating to the reactive term dominating is when $\omega C_2 R_2 = 1$. Thus the cut-off frequency is

$$f_c = \frac{1}{2\pi C_2 R_2} \tag{13.24}$$

Similarly the high pass filter (Fig. 13.10(b)) has a voltage gain of

$$A_v = -\frac{j\omega C_1 R_2}{1 + j\omega C_1 R_1} \tag{13.25}$$

and a cut-off frequency of

$$f_c = \frac{1}{2\pi C_1 R_1} \qquad\qquad (13.26)$$

A band pass active filter can be obtained by combining the above low pass and high pass circuits so that the input impedance is that of R_1 in series with C_1 and the feedback impedance is that of C_2 in parallel with R_2. The band passed is between angular frequencies $1/R_1 C_1$ and $1/R_2 C_2$, the voltage gain between these frequencies being $-R_2/R_1$.

More complex operational amplifier circuits can be used to give sharper attenuation changes at the cut-off frequency.

Problems

1. A symmetrical attenuator has a current input of 100 mA, what will be the current output if the attenuation is -12 dB?

2. Three identical attenuator sections, each with an attenuation of -5 dB, are connected in cascade. What will be the overall attenuation?

3. Four identical attenuator sections are connected in cascade to give an overall attenuation of -20 dB. What is (a) the attenuation of each section and (b) the output of the final stage if the input to the first is 100 mV?

4. What is the characteristic impedance of a symmetrical T-section attenuator having series resistances of 100 Ω and a shunt resistance of 50 Ω?

5. What is the characteristic impedance of a symmetrical π-attenuator having a series resistance of 500 Ω and shunt resistances of 1 kΩ?

6. The voltage gain of a symmetrical π-section attenuator is given by

$$\text{voltage gain} = \frac{Z_0 + 2Z_2}{2Z_2 - Z_0}$$

where Z_0 is the characteristic impedance and $2Z_2$ the shunt impedance. Derive the equation.

7. What is the attenuation of a symmetrical π-section attenuator which has a series resistance of 100 Ω and a shunt resistance of 1000 Ω?

8. An RC filter has $R = 100$ Ω and $C = 10$ μF and is used as a low pass filter. What is the cut-off frequency and the attenuation in dB at that frequency?

9. What is the cut-off frequency and the design impedance for a constant k T-section low pass filter which has series inductances of 20 mH and a shunt capacitor of 0.1 μF?

10. Design a high pass T-section filter to have a design impedance of 600 Ω and a cut-off frequency of 3 kHz.

11. Derive an equation for the cut-off frequency and design impedance for a π-section low pass filter.

12. An operational amplifier circuit is as shown in Fig. 13.10(a) with $R_1 = 2\,\text{k}\Omega$, $R_2 = 10\,\text{k}\Omega$ and $C_2 = 0.1\,\mu\text{F}$. What is the cut-off frequency of the resulting low pass filter?

13. An active band pass filter has an operational amplifier with an input impedance of R_1 in series with C_1 and a feedback impedance is that of C_2 in parallel with R_2. Show that the band passed is between angular frequencies $1/R_1C_1$ and $1/R_2C_2$, and the voltage gain between these frequencies is $-R_2/R_1$.

Chapter 14

Digital Systems

14.1 Analogue and digital signals

In the work so far, we have been concerned with signals in an analogue form. Let us consider carefully what this means. We can take, as an example, the problem of producing a signal which will convey information regarding the depth of water in a tank. Problems such as this are quite common in practice and require the construction of a measurement system which, in some cases, must operate for long periods without attention, and with great reliability. One solution to the problem is illustrated in Fig. 14.1. The figure, which does not attempt to show the mechanical details of the measurement system, represents the water tank in which is mounted a potentiometer, the sliding contact being connected to a float, and which, therefore, alters its position as the water level changes. Obviously such a construction would require great mechanical refinement in order to produce a reliable system. Let us assume that at the maximum allowed water depth the float and slider just reach the top of the potentiometer travel. The output voltage V_o will then equal the supply voltage V. If also, at the lowest levels allowed, the float and slider have fallen to the bottom of the potentiometer, the output voltage V_o will equal zero. At any fraction x of the full water depth, the float and slider will rise to the same fraction x of the travel up the potentiometer and the output voltage will be

$$V_o = xV$$

The output voltage is thus proportional to the depth of water above the lowest allowed level and can be used as an indication of the depth. The output voltage is said to be an *analogue* of the water depth, and in an ideal system would vary in exactly the same way as the water depth. The relationship between the water depth and its ideal voltage analogue is shown in Fig. 14.2.

Many examples occur of the use of an analogue quantity to represent a physical variable. Thus in a motor car speedometer, the car speed is represented by the angular rotation of the speedometer pointer, while in an audio system, the intensity of the sound waves produced by the performers is converted by the microphone into a voltage analogue for amplification, recording and other processing. However, in life, systems are never ideal and there is a limit to the accuracy of the proportional relationship between the variable and its analogue.

Returning to the water depth system, we can point to various sources of inaccuracy. Thus, because of friction between the slider and the potentiometer wire, very small slow changes in water level may occur without the slider

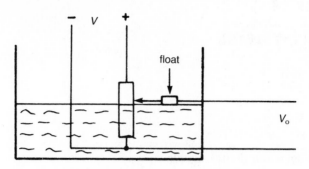

Fig. 14.1 Analogue float system.

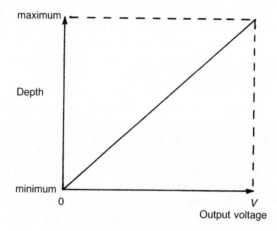

Fig. 14.2 Analogue relationship.

moving; the slider would, therefore, tend to respond to a slowly changing water depth in a series of small jerks. Further, because of corrosion, etc., on the surface of the potentiometer wire, the contact resistance between the contact and the wire will vary at different positions of the slider. Because of stray electromagnetic fields in the vicinity of the tank, perhaps due to adjacent mains electricity or other cables, various small induced alternating signals would also be present at the system output. These, and many other inherent factors, result in a *noise* voltage at the output which forms a limit to the smallest detectable change in water depth. There is, therefore, at any water depth, a range of uncertainty within which the true water depth is not known. This is illustrated in Fig. 14.3.

The amplitude of the noise voltage is, to some extent, independent of the value of the voltage V applied to the potentiometer. Some improvement in the depth uncertainty can, therefore, be obtained by using an increased supply voltage. A signal amplifier could also be used, especially if the depth indicator is to be mounted in a remote place, connected via lengthy lines to the data processing equipment. The use of such long lines will cause a reduction

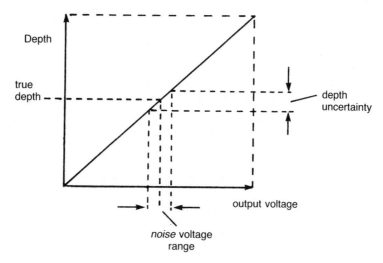

Fig. 14.3 Uncertainty range.

(attenuation) in the signal level reaching the end of the line, making the signal more easily affected by further noise voltages induced in the lines themselves. It must be remembered that the noise voltage production process is an inherent fact of life, which occurs everywhere, and which must be taken account of by good design techniques which minimize its effect. Further, once the minimum noise voltage has been achieved, by good design, it must be ensured that the amplitude of any signal used is always sufficiently greater than the amplitude of the noise, to given the required system accuracy.

Summarizing, an analogue signal is a signal which varies in exactly the same manner as the variable which is to be measured, and which can take any value within the allowable signal range. There is, therefore, theoretically, an infinite number of values which an analogue signal can take within an allowed signal range. In practice, as we have seen, there is, in fact, the smallest discernible change in signal value which is fixed by the unavoidable noise signals induced into the system.

A different solution to the water depth problem can be arranged by replacing the tank potentiometer with a multisegment switch, as illustrated in Fig. 14.4. The slider, moving with the float, now connects the supply voltage to one only of the eight output lines shown. Thus, in the figure, a voltage V will exist between line 5 and earth, while all the other lines will have, theoretically, zero voltage. Of course, there will be the unavoidable noise voltage on all the lines, but in order to detect which segment of the switch the float is connecting, we need only to be able to distinguish the presence of the signal voltage V, and this can be achieved even with noise voltages almost as large as the supply voltage V itself. This solution to the problem is a digital solution, and although the system proposed is a relatively crude one, it does illustrate the fundamental difference between analogue and digital systems. That is, that whereas in the analogue system we must be able to detect small changes in the analogue

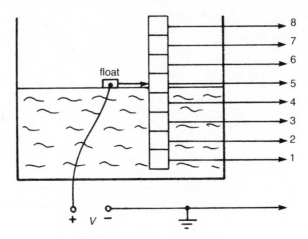

Fig. 14.4 Digital float system.

output voltage, in the digital system we are only interested in detecting the presence, or absence, of the voltage V.

In fact, since the magnitude of the voltage V contains no information (it must, of course, be larger than the interfering noise voltages), it is common practice to denote the presence of the voltage by the code 1 and the absence of the voltage by the code 0. The operation of the system can then be discussed without reference to the size of the voltage in any way. In fact, digital systems are quite common in which the signal is not a voltage but an air pressure or water pressure. The connecting wires are then, of course, pipes. However, whatever the nature of the actual signal, the use of the two codes 0 and 1 represents the absence or presence of the signal quantity.

The use of an eight segment switch in the water tank only allows the division of the full depth range into eight levels. The amplitude range is said to be *quantized*. In place of the depth uncertainty in the analogue system, due to noise, we have introduced a depth uncertainty, due to the quantizing error. Of course, we can reduce the quantizing error by using a switch with more contacts, but a limit to this improvement will be provided again by sticking friction between the slider and the switch contacts.

Having decided how many contact segments are to be used, we can draw up a table which lists the signals available on the output lines at each quantizing level. This is shown in Table 14.1. We can see that the depth range, from one half to five eighths of the full range depth above the zero level, is represented by the binary word

    0 0 0 0 1 0 0 0

This is said to be a *binary word* because the signal on each line can only take one of the two binary values 0 or 1. Each line signal is represented by one *binary digit* or bit. The word is, therefore, an eight bit word. The process of assigning a binary word to represent each possible output value is termed *encoding*. One obvious drawback to the system described is the fact that eight

Table 14.1 Digital output signals.

Depth range above minimum level	Signal on lines							
	Line 1	Line 2	Line 3	Line 4	Line 5	Line 6	Line 7	Line 8
$\frac{7}{8}-1$	0	0	0	0	0	0	0	1
$\frac{3}{4}-\frac{7}{8}$	0	0	0	0	0	0	1	0
$\frac{5}{8}-\frac{3}{4}$	0	0	0	0	0	1	0	0
$\frac{1}{2}-\frac{5}{8}$	0	0	0	0	1	0	0	0
$\frac{3}{8}-\frac{1}{2}$	0	0	0	1	0	0	0	0
$\frac{1}{4}-\frac{3}{8}$	0	0	1	0	0	0	0	0
$\frac{1}{8}-\frac{1}{4}$	0	1	0	0	0	0	0	0
$0-\frac{1}{8}$	1	0	0	0	0	0	0	0

signal lines are required to transmit the depth information. A system in which each bit of the code word is transmitted at the same instant is said to be a *parallel system*, and requires as many signal channels as there are bits in the complete word. A parallel data system is obviously the fastest way to convey the data, since the time required to transmit the complete word is the same as the time required to transmit one bit. An alternative cheaper method is to convert the parallel signal into a serial signal, i.e. the individual bits are transmitted one at a time, in time sequence, down a single channel. Although cheaper, the *serial system* is slower, because the time required to transmit the complete word is equal to n times the time required to transmit each bit, where n is the number of bits in the word. There is, therefore, a choice between speed and cost. If it is decided to use a serial transmission method with the water depth transducer, than a parallel to serial data converter would be required, the output from the transducer described being inherently a parallel signal.

Figure 14.5(a) shows one method by which the serial data signal can be produced. The eight output lines from the transducer are connected to the eight contacts on a rotary switch which is motor driven. As the rotating contact performs one complete revolution, it contacts in turn each of the eight lines, starting at line 1 and ending at line 8. The signal voltages on the eight lines (i.e. either 0 V or V V) appear in sequence upon the single output data line. Figure 14.5(b) shows, for each of the eight possible depth ranges, the resulting serial code word produced by the depth transducer. Notice that the duration of each of the time intervals in the serial code word is decided by the speed of rotation of the rotating switch. Each signal bit would have a duration of one eighth of the time of one revolution of the switch. The speed of transmission in such digital systems is often spoken of as the *bit rate*, i.e. the number of bits which are transmitted per second. Notice also that, in order that the signal receiver, which may be quite remote from the transducer/transmitter, can

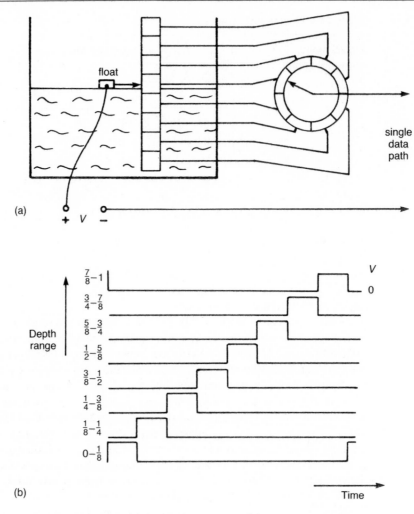

Fig. 14.5 (a) Serial data system, (b) serial code word.

make sense of the received signal, it must be synchronized with the transmitter timing, i.e. it must have a knowledge of the bit rate at which the serial signal is being transmitted. Let us consider an output from the transducer corresponding to a water depth in the range $\frac{7}{8}$ to 1 times full scale. The data word transmitted would be

$$0 \ \ 0 \ \ 0 \ \ 0 \ \ 0 \ \ 0 \ \ 0 \ \ 1$$

The word would exist on the data line as a 0 signal for seven consecutive bit periods, followed by a 1 signal for one bit period. Unless the receiver is aware of the length of a bit period, it will be unable to determine the number of 0 signals intended. For this reason digital systems are often timed by a system clock, which produces a regular sequence of pulses which are used to synchronize events. An additional data line is often used to distribute the clock signal throughout the system.

One other point which is important in serial systems is a knowledge, by the receiver, of the order in which the word is being transmitted. Consider again the transducer transmitting an output corresponding to a water depth in the $\frac{7}{8}$ to 1 times full scale range. If the direction of rotation of the rotary switch is now reversed, the transmitted serial word will change to

1 0 0 0 0 0 0 0

giving an ambiguity between the upper and lower depth ranges. We can thus see that, for the transmitted word to be meaningful to the receiver, the system as a whole must be aware of

(1) the transmitted bit duration, and
(2) the order of transmission of bits.

The inverse of the bit duration is known as the *bit frequency*, or bit rate, and is equal to the clock frequency of the system. For the eight bit system described, a total of eight bit periods would be required, in order to transmit one word. This finite word time would form a limit to the number of readings of water depth which could be made in any given time interval. A water depth transducer is, of course, inherently a relatively slowly operating system, and the speed of operation would be limited more by the mass, stiffness, etc., of the float potentiometer than by the speed of the data transmission system. However, if the serial digital data system is used with a very fast physical variable providing the data input, the ultimate limit to the speed will probably be provided by the time required to convert the value of the data variable into a digital word. A digital data system, because of the time required to transmit each coded data value, is inherently only capable of transmitting sample values of the data input. The basic principle is illustrated in Fig. 14.6, which shows an

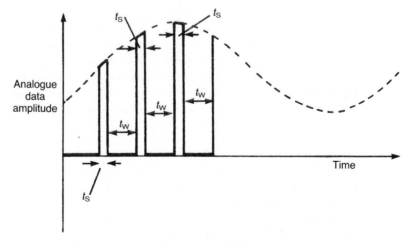

Fig. 14.6 Sampling.

analogue data signal being sampled, the time required to obtain a sample being t_s. Once the sample value has been obtained, the value can be converted into a digital word form, this taking the time t_w. The total conversion process thus takes a time

$$t_c = t_s + t_w$$

During the conversion time the analogue signal will, in general, change its value due to its normal variation. The conversion process is thus subject to a sampling error. It is obviously impossible to represent exactly a continuously varying quantity by fixed amplitude samples of finite duration.

Returning to the analogue system, is there a similar limit to the speed of obtaining readings? Remember that the output signal is a voltage whose magnitude is proportional to the water depth. Obviously again, the water level system is inherently a slowly operating system. However, if an analogue transducer is used with a very fast physical variable, the ultimate limit to the speed may again be provided by the speed of response of the data transmission system. In an analogue system this means the speed at which the signal voltage on the data line may be changed. This in turn will be decided by the properties of the line, i.e. the resistance, inductance and capacitance of the lines themselves. Because, however, in a digital system each transmitted digital word may consist of several changes of signal voltage, the digital system is inherently a slower method. In choosing to use a digital system, we have traded speed of response for relatively better immunity from electrical interference.

14.2   Number systems and codes

In the simple digital encoder described, eight bits were used to represent the eight amplitude segments into which the full range of the depth recorder was divided. The use of eight bits to represent eight sample values is very inefficient.

Bearing in mind that each binary digit can take one of two values, i.e. 0 or 1, then a digital word of n bits can represent 2^n different values. The power of 2 table shown in Table 8.2 indicates the range capability of digital words of various bit lengths. It can be seen that a 10 bit word has a capability of $2^{10} = 1024$ different amplitude samples, which will allow an accuracy of slightly better than 1 in 1000, i.e. of approximately 0.1%. The eight amplitude samples of the water depth transducer could have been represented by a three bit word, since

$$2^3 = 8$$

Table 14.3 shows the eight different values which may be represented by a three bit word. Listed in the order shown, the words form a code known as simple binary code. For use with the water depth transducer, the lowest segment would, perhaps, be represented by the binary word 000, while the highest segment, corresponding to the depth range $\frac{7}{8}$ to 1 times full scale could be represented by the binary word 111.

Table 14.2 Powers of 2.

No of bits in word n	Range capability $= 2^n$
1	2
2	4
3	8
4	16
5	32
6	64
7	128
8	256
9	512
10	1024
11	2048
12	4096

Table 14.3 Three bit word.

Decimal equivalent	3 bit binary word		
0	0	0	0
1	0	0	1
2	0	1	0
3	0	1	1
4	1	0	0
5	1	0	1
6	1	1	0
7	1	1	1

The use of a three bit word code would allow a considerable increase in the speed of transmission of the digitally coded data. However, the encoder used, i.e. the device which actually produces the digitally coded output word, would of necessity be slightly more complex.

The simple binary code can be expressed in a general form as

$$N = a_0 2^0 + a_1 2^1 + a_2 2^2 + a_3 2^3 + \ldots a_n 2^n \tag{14.1}$$

in which the constants $a_0 \ldots a_n$ can take either of the values 0 or 1. It is seen that the word is formed as the sum of a series of powers of 2 and is commonly written with the highest power of 2 at the left hand side, following the method used normally for decimal numbers. Thus the binary equivalent of decimal 15 is written as

1   1   1   1

i.e.

$$N = 1 \times 2^0 + 1 \times 2^1 + 1 \times 2^2 + 1 \times 2^3$$

$$= 15$$

The left hand bit is the *most significant bit* (MSB), since it involves the highest power of 2 in the word, while the right hand bit is the *least significant bit*.

As with decimal numbers, it is possible to deal with numbers of magnitude less than unity by using a binary point. Thus the binary word

    1 0 1 1 . 1 0 1

is understood as (starting with the MSB)

$$N = 1 \times 2^3 + 0 \times 2^2 + 1 \times 2^1 + 1 \times 2^0 + 1 \times 2^{-1} + 0 \times 2^{-2} + 0 \times 2^{-3}$$

$$= 8 + 0 + 2 + 1 + 0.5 + 0 + 0.125$$

$$= 11.625 \text{ decimal}$$

Although decimal and binary number systems are in quite common use, it is of course possible to develop number systems using any base. In general, the decimal equivalent of a number in a system of base r is

$$N = a_0 r^0 + a_1 r^1 + a_2 r^2 + \ldots a_n r^n \tag{14.2}$$

in which the constants $a_0 \ldots a_n$ may take any of the values 0 up to $(r - 1)$.

It is good practice, when dealing with numbers in different bases, to indicate the base applicable to the number as a subscript. Thus

$$1010_2 = 10_{10} = 12_8$$

Table 14.4 Hexadecimal system.

Hexadecimal value	Decimal equivalent
0	0
1	1
2	2
3	3
4	4
5	5
6	6
7	7
8	8
9	9
A	10
B	11
C	12
D	13
E	14
F	15

The great interest in the binary system, as far as digital system engineers are concerned, lies in the fact that only two levels, or characters, are used. In the decimal system, the ten characters 0 to 9 occur. In the binary system, however, only the characters 0 and 1 occur, and these are conveniently characterized by the presence or absence of a voltage, or the opening or closing of a switch.

Digital system engineers use, in addition to the binary system, the *octal system* (base eight) and the *hexadecimal system* (base 16). In writing down numbers in the hexadecimal system, the characters 0 to 9 are used together with six other characters. Any symbols, of course, may be used, but commonly the first six letters of the alphabet are employed. The hexadecimal code, and its decimal equivalent, are listed in Table 14.4.

14.2.1 *Conversion between bases*

(1) *Binary to decimal conversion.*
This process has already been considered. However, to recapitulate, the decimal equivalent is obtained by summing the decimal equivalent of each power of 2.

(2) *Decimal to binary conversion.*
This is best performed by a process of repeated division by 2.

(3) *Binary to octal conversion.*
This conversion is often used if a relatively long binary number is to be remembered, even for a very short time, for example, in order to enter the number by keyboard into a digital system. The conversion is very easy, because of the special relationship which exists between the bases 2 and 8 (i.e. $8 = 2^3$). The binary word is written with the bits separated slightly into groups of three bits. Note that if the word does not divide exactly into groups of three, zeroes may be added at the most significant bit end to complete the groups. Each group of three is now replaced by its octal equivalent, as given in Table 14.3. (Note that the octal and decimal equivalents of three bit binary numbers are identical.)

(4) *Binary to hexadecimal conversion.*
A similar relationship exists between the bases 2 and 16 ($2^4 = 16$). In this case the binary world is split into groups of four.

Example Convert 110110_2 to its decimal equivalent.

$$110110_2 = 0 \times 2^0 + 1 \times 2^1 + 1 \times 2^2 + 0 \times 2^3 + 1 \times 2^4 + 1 \times 2^5$$
$$= \quad 0 \quad + \quad 2 \quad + \quad 4 \quad + \quad 0 \quad + \quad 16 \quad + \quad 32$$
$$= 54_{10}$$

Example   Convert the binary word 110000011111_2 into its octal equivalent.

$$110 \quad 000 \quad 011 \quad 111$$

$$6 \qquad 0 \qquad 3 \qquad 7$$

therefore octal equivalent $= 6037_8$

Example   Convert 185_{10} to its binary equivalent

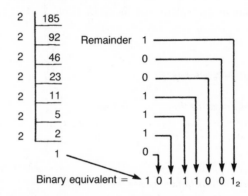

Example   Convert the binary word 0100111101_2 into its hexadecimal equivalent.

The binary word has ten bits. To divide it into groups of four bits, two zeroes must be added at the left hand (MSB) end.

$$0001 \quad 0011 \quad 1101$$

$$1 \qquad 3 \qquad D$$

hexadecimal equivalent $= 13D_{16}$

Remember that the 16 hexadecimal digits are the numbers 0 to 9 and the six letters A to F.

14.2.2   Arithmetic using the binary number system

The rules applied in performing simple arithmetic operations using the binary system are the same as those familiar from decimal arithmetic. However, any difficulties which arise from an inability to remember the 2 to 10 times tables in decimal arithmetic do not exist in binary arithmetic, as, of course, only the characters 0 and 1 exist.

(1) *Addition.* Remember, in performing additions, that $1 + 1 = 10$, which gives a 0 in the sum and a 1 to carry into the next higher column.

(2) *Subtraction*. In subtraction, where a borrow must be made from the next higher column, the borrow has the value of two.

(3) *Multiplication*. In multiplication, each partial product formed consists of 0 or 1 times the multiplicand. The partial products are, therefore, equal to zero, or to the multiplicand shifted left by the required number of places.

(4) *Division*. In division, the method is the same as for decimal long division. However, the divisor will divide into each section of the dividend either 0 times or 1 times.

Example   Perform the addition 101011 + 1001

$$
\begin{array}{ll}
101011 & \text{the augend} \\
+\ \ \ 1001 & \text{the addend} \\
\hline
110100 & \text{the sum}
\end{array}
$$

Example   Perform the subtraction 101101 − 11010

$$
\begin{array}{ll}
101101 & \text{the minuend} \\
-\ \ 11010 & \text{the subtrahend} \\
\hline
10011 & \text{the difference}
\end{array}
$$

Example   Perform the multiplication 110101 × 101

$$
\begin{array}{ll}
\ 110101 & \text{the multiplicand} \\
\times\qquad 101 & \text{the multiplier} \\
\hline
110101 & \\
000000 & \text{the partial products} \\
110101 & \\
\hline
100001001 & \text{the product}
\end{array}
$$

Example   Divide 100011 by 101

$$
\begin{array}{ll}
\qquad\ \ \ 111 & \text{the quotient} \\
101\overline{)100011} & \text{the dividend} \\
\ \ \ \ \underline{101} & \\
\ \ \ \ 00111 & \\
\ \ \ \ \ \ \underline{101} & \\
\ \ \ \ \ \ 0101 & \\
\ \ \ \ \ \ \ \underline{101} & \\
\ \ \ \ \ \ \ 000 & \text{the remainder}
\end{array}
$$

14.2.3   *Negative numbers*

Two methods are in common use whereby both positive and negative numbers may be represented in a binary code. In both methods, the same conventions are assumed:

(1) The MSB of the binary number is used as a sign indicator.
(2) The MSB value is to be 0 for a positive number and 1 for a negative number.

In the *sign + magnitude* method, the bits b_0 to b_6 give the magnitude of the number coded in simple binary. Thus we have:

b_7	b_6	b_5	b_4	b_3	b_2	b_1	b_0	
0	0	0	0	0	0	0	1	$+1_{10}$
0	0	0	0	0	0	0	0	$+0_{10}$
1	0	0	0	0	0	0	0	-0_{10}
1	0	0	0	0	0	0	1	-1_{10}
0	1	1	1	1	1	1	1	$+127_{10}$
1	1	1	1	1	1	1	1	-127_{10}

Although this is an obvious and simple method it has disadvantages. If the code for +1 is added to the code for −1, then we obtain:

```
+1        0 0 0 0 0 0 0 1
−1        1 0 0 0 0 0 0 1
          ‾‾‾‾‾‾‾‾‾‾‾‾‾‾‾
−2        1 0 0 0 0 0 1 0
```

i.e. it appears that the result is −2. In fact because the sign + magnitude code is not a weighted code (see Section 14.2) it is difficult to develop rules which allow arithmetic to be performed. This difficulty is overcome by the use of the *two's complement* method. For positive numbers, both methods are identical. However, the code for −1 is obtained by subtracting 1 from the code for 0. This method is continued, with each succeeding negative number being obtained by subtracting 1 from the preceding number. This removes the difficulty in the sign + magnitude method of having two codes to represent zero. Thus we have:

b_7	b_6	b_5	b_4	b_3	b_2	b_1	b_0	
0	0	0	0	0	0	0	1	$+1_{10}$
0	0	0	0	0	0	0	0	0_{10}
1	1	1	1	1	1	1	1	-1_{10}
1	1	1	1	1	1	1	0	-2_{10}
0	1	1	1	1	1	1	1	$+127_{10}$
1	0	0	0	0	0	0	0	-128_{10}

Using the same example again, i.e. adding +1 to −1,

```
+1              0 0 0 0 0 0 0 1
−1              1 1 1 1 1 1 1 1
                ‾‾‾‾‾‾‾‾‾‾‾‾‾‾‾
        1 ←     0 0 0 0 0 0 0 0
        carry
```

then, as long as a *carry* is allowed from the eight bit number into the non-existent ninth bit, the result obtained within the eight bits is correct. The two's complement method allows simple arithmetic because it is a weighted code; the weightings of the bits b_0 to b_6 are the simple binary weightings of 2^0 to 2^6. The weighting of the sign bit is 2^{-7}.

A simple rule for writing down the two's complement of a binary number is as follows. Write down the same binary number starting at the LSB (right hand) end, until and including the first logical 1 value. After the first 1, invert all values.

Example   Write down the two's complement of the 16 bit binary number

    0 1 1 0 1 0 1 1 0 1 1 0 1 0 0 0

The two's complement is

    1 0 0 1 0 1 0 0 1 0 0 1 1 0 0 0

14.2.4   Other possible binary codes

The simple binary code is not the only binary code which could be used for encoding the amplitude levels. In fact, any arrangement of the bit characters would be usable provided that each amplitude level has a unique coding and, of course, also provided that a knowledge of the code used is available at both the transmitter and the receiver.

The simple binary code is an example of a *weighted* code, by which it is meant that each bit in the code has a definite weighting assigned to it. Thus, each bit in the simple binary code has a weighting which is two times the weighting of the bit on its right hand side. The choice of code used will, however, affect the design of the encoder, and perhaps of the data transmission circuitry, and certain codes may have definite advantages, in cost or in reliability. A code which is often used in such applications is shown in Table 14.5. This code, known as *Gray binary code*, is listed in the figure in four bit form. If required in three bit form, for example, to code the eight levels of the water depth transducer, then the most significant bit will not be required. Code listings such as this are best remembered by looking for the pattern which exists in the code. Thus the least significant bit pattern consists of a sequence of four zeroes and four ones, but starting with two zeroes. The number of zeroes and ones in the pattern increases for each bit as a power of two.

The Gray binary code is an example of a special class of codes which are cyclic and which also change between adjacent code characters in only one bit position. The use of this type of code gives advantages in certain types of transducer.

Figure 14.7 shows a shaft position encoder which uses an array of light sensitive detectors to produce the electrical output, each detector having its own output line. The digitally coded output is thus available in parallel form. The light sensitive detectors are illuminated by an equivalent set of lamps which are directed through a transparent cylinder. An encoding pattern is

Table 14.5 Gray code.

Equivalent decimal value	4 bit Gray code			
0	0	0	0	0
1	0	0	0	1
2	0	0	1	1
3	0	0	1	0
4	0	1	1	0
5	0	1	1	1
6	0	1	0	1
7	0	1	0	0
8	1	1	0	0
9	1	1	0	1
10	1	1	1	1
11	1	1	1	0
12	1	0	1	0
13	1	0	1	1
14	1	0	0	1
15	1	0	0	0

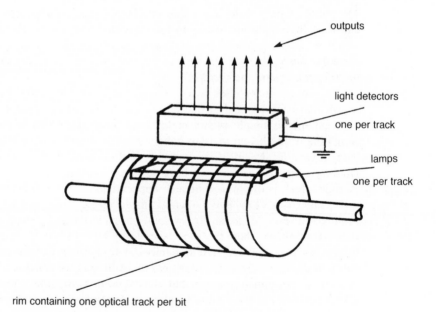

outputs

light detectors
one per track

lamps
one per track

rim containing one optical track per bit

Fig. 14.7 Shaft position encoder.

formed photographically upon the cylinder, obscuring certain detectors in each position of the cylinder. The use of a four bit code, giving only $2^4 = 16$ different code patterns would, of course, only allow an angular discrimination of 360/16 = 22.5°.

However, the use of a ten bit code, requiring ten lamps and detectors, will allow an angular discrimination of 360°/1024, i.e. of approximately one third of one degree. What are the advantages of the Gray code over simple binary code for such an application? Consider first the method of operation of the encoder. In any position of the shaft certain of the light detectors will be illuminated. As the shaft is turned the output word will change to a new value. If all the bits do not change to their new values at exactly the same instant, a set of false output words will occur before the output settles to give the new correct value. Examination of the Gray code shows that between each adjacent code, in only one bit is a change valid. Multiple output changes, which would thus occur in a practical design, may, therefore, be ignored by the system. The code also is cyclic, in that there is only one bit difference between the first and last code character. Comparison with the equivalent change in the simple binary system will show that in returning from the last to the first code character all bits in the code are required to change.

A further common form of binary code is the *binary coded decimal* (BCD) code. This code is often used where the decimal form of the coded number is to be retained. The decimal characters 0 to 9 are coded using simple binary code. These characters are then used to replace the decimal characters. The decimal number 346_{10} would then be coded as

0011	0100	0110
3	4	6

Four binary bits are required for each decade in the decimal number. The code is, therefore, less efficient in the use of bits than is the simple binary, for twelve bits are required to represent numbers up to 999_{10}. Taking a specific example, to represent the decimal number 346 in simple binary,

i.e. $346_{10} = 11011010_2$

eight bits are required, while 12 bits are required if binary coded decimal representation is used.

14.3 Boolean algebra

The original work of the mathematician George Boole, describing propositions whose result is either true or false, has been applied to signals in which the variable can take one of two values. The Boolean algebra is now extensively applied to binary coded signals and forms a major tool for engineers employed in digital system design. The basic rules and identities are easily described, with reference to the opening or closing of sets of switch contacts. Thus the two binary values 0 and 1 can be used to represent a switch open (non-conducting) or closed (conducting), or the absence or presence of a voltage, or the absence or presence of a current. The interpretation can be extended to include the absence or presence of all the necessary conditions for the success of a particular process.

14.3.1   *Boolean variables, operations and identities*

Let us consider the simple circuit of Fig. 14.8. This consists of a battery, a switch A and a lamp. If we denote the action of lighting the lamp by the Boolean variable F, which will take the value 1 if the lamp is lit and 0 if the lamp is unlit, then we can write

$$F = A \qquad\qquad (14.3)$$

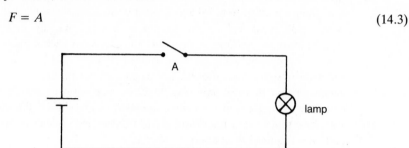

Fig. 14.8 $F = A$.

where A is another Boolean variable, which takes the value 0 if the switch is open, and the value 1 if the switch is closed. An extension of this method defines the complement or negation of a variable. Thus the negation of the variable A is written as $\bar{A}$ (sometimes A'). If the variable A has the binary value 0 then its complement $\bar{A}$ has the value 1 and vice versa.

Two basic operations must also be clearly defined. Consider the circuit of Fig. 14.9, in which the operation of the lamp is controlled by two switches, A and B, in series. Again, denoting the lighting of the lamp by the variable F, then we can write

$$F = A \text{ AND } B$$

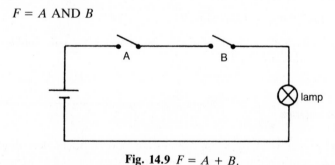

Fig. 14.9 $F = A + B$.

It is clear that both switches must be closed for the lamp to light. The logical operation AND is denoted in Boolean equations by a dot, although in common practice the dot is often omitted. Thus the equation would be written

$$F = A \cdot B$$

or more commonly

$$F = AB \qquad\qquad (14.4)$$

When a Boolean equation written in this manner is read it must be read as F equals A AND B.

An alternative operation is denoted by the electrical circuit of Fig. 14.10, which shows a lamp controlled by two switches A and B connected in parallel. In this case, closure of either of the switches will result in the lamp lighting. The circuit is a realization of the logical OR function and the logical equation is

$$F = A \text{ OR } B$$

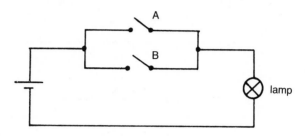

Fig. 14.10 $F = AB$.

The OR function is usually written

$$F = A + B \tag{14.5}$$

where the + sign denotes the OR function. Again it is necessary when reading the equation to read the + sign as OR and to make sure that no confusion with the arithmetical version of the + sign occurs. The OR and the AND are the basic logical functions and quite complex switching circuits may be represented in Boolean form using these functions, together with the system variables in their normal and complemented forms.

It is also common practice to use brackets to link together sections of equations which are to be ANDed together. Thus the equation

$$F = A + BC$$

represents the circuit of Fig. 14.11(a), while the equation

$$F = A(B + C)$$

represents the circuit of Fig. 14.11(b). In Fig. 14.11(a) either A OR both B AND C are required to be closed for the lamp to operate, while in Fig. 14.11(b) switches A AND either B OR C are to be closed.

Using these basic functions, we can now specify some basic identities

$$
\begin{array}{ll}
A + 0 = A & (1) \\
A + 1 = 1 & (2) \\
A \cdot 0 = 0 & (3) \\
A \cdot 1 = A & (4) \\
A + A = A & (5) \\
A \cdot A = A & (6)
\end{array}
$$

$$A + \bar{A} = 1 \qquad (7)$$
$$A \cdot \bar{A} = 0 \qquad (8)$$
$$\bar{\bar{A}} = A \qquad (9)$$

(14.6)

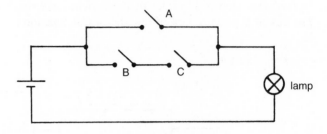

(a)

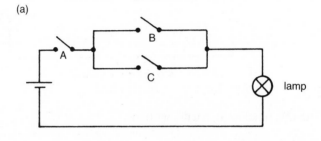

(b)

Fig. 14.11 (a) $F = A + BC$, (b) $F = A(B + C)$.

The validity of these identities may be checked by checking their truth for all possible values of the variables. It also helps to visualize the meaning by drawing the equivalent electrical circuit, remembering that the value 0 represents an open circuit, while 1 represents a short circuit. The equivalent electrical circuit for each identity is shown in Fig. 14.12.

For identities involving a larger number of variables, it becomes useful to list the possible values of the variables in an ordered manner to facilitate checking each possible combination. Such an ordered listing of all possible values of the system variables is called a *truth table*. For any logical equation, therefore, the equality may be checked by drawing its truth table. It is very convenient, when listing the combinations of variables, to do so according to the simple binary code.

Consider the Boolean equation

$$A(B + C) = AB + AC \qquad (14.7)$$

The truth table for the variables A, B and C is shown in Table 14.6. For the two sides of the equation to be identical the two columns representing the two sides, i.e. the columns $A(B + C)$ and $AB + AC$, should be identical for all combinations of variables. This is seen to be true.

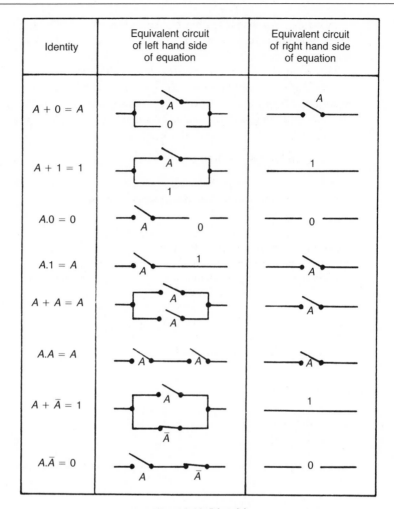

Fig. 14.12 Identities.

Table 14.6 Truth table $A(B + C) = AB + AC$.

Variables			Left hand side of equation		Right hand side of equation		
A	B	C	$(B + C)$	$A(B + C)$	AB	AC	$AB + AC$
0	0	0	0	0	0	0	0
0	0	1	1	0	0	0	0
0	1	0	1	0	0	0	0
0	1	1	1	0	0	0	0
1	0	0	0	0	0	0	0
1	0	1	1	1	0	1	1
1	1	0	1	1	1	0	1
1	1	1	1	1	1	1	1

Two functions whose truth tables are identical, i.e. functions which have the same Boolean value for any combination of the input variables, are identical functions.

This example also illustrates that the technique of removal of brackets by *multiplying out*, familiar from simple arithmetic, is also applicable to Boolean equations. There are, however, steps which are valid in the manipulation of Boolean equations which have no counterpart in simple arithmetic.

Consider another example, the Boolean equation

$$A + BC = (A + B)(A + C) \tag{14.8}$$

The two columns headed $A + BC$ and $(A + B)(A + C)$ in Table 14.7 are identical, and the validity of the equation is proved.

Table 14.7 $A + BC = (A + B)(A + C)$.

Variables			Left hand side of equation		Right hand side of equation		
A	B	C	BC	$A + BC$	$A + B$	$A + C$	$(A + B)(A + C)$
0	0	0	0	0	0	0	0
0	0	1	0	0	0	1	0
0	1	0	0	0	1	0	0
0	1	1	1	1	1	1	1
1	0	0	0	1	1	1	1
1	0	1	0	1	1	1	1
1	1	0	0	1	1	1	1
1	1	1	1	1	1	1	1

Comparison of the equations proved in these two examples is interesting

$$A(B + C) = AB + AC$$

$$A + BC = (A + B)(A + C)$$

These two equations illustrate the dual nature of the two Boolean operations, the logical AND and the logical OR. The second equation is obtained from the first, by replacing each AND operation with an OR operation and also by replacing each OR operation with an AND operation. The removal of the brackets by multiplying out in the first equation also has its dual in the second equation. Thus the right hand side may be obtained by *ANDing out* the left hand side of the equation, i.e. by ANDing together the logical OR of the variable A with each of the variables in the second term in turn. This operation has no counterpart in simple arithmetic, but the use of the dual nature of the logical operations AND and OR in Boolean arithmetic is of great value. Although the use of the Boolean operation may seem strange at first, familiarity is soon obtained with a little practice in the manipulation of equations.

One further very useful manipulation tool will be introduced at this point. This is the theorem due to *De Morgan*. This may be stated in equation form as

$$\overline{A + B} = \bar{A}\bar{B} \tag{14.9}$$

$$\overline{AB} = \bar{A} + \bar{B} \tag{14.10}$$

The truth of these two forms of De Morgan's theorem should be checked using the truth table method. In words, the theorem states that to negate a logical expression of two or more variables each variable should be negated and the logical operation changed (i.e. AND becomes OR and vice versa). The method will be illustrated by examples.

Example   Negate the function $F = A\bar{B}$

Applying De Morgan's theorem directly

$$\bar{F} = \overline{A\bar{B}} = \bar{A} + B$$

Example   Negate the function $F = AB + CD$

Here De Morgan's theorem must be applied twice. First consider AB and CD to be single variables. Then

$$\bar{F} = \overline{AB + CD} = \overline{AB}.\overline{CD}$$

Applying De Morgan a second time

$$\bar{F} = \overline{AB}.\overline{CD} = (\bar{A} + \bar{B}).(\bar{C} + \bar{D})$$

Example   Negate the function $F = A + \bar{B} + \bar{C}D$

Applying De Morgan's theorem

$$F = \overline{A + \bar{B} + \bar{C}D}$$
$$= \bar{A}.B.\overline{\bar{C}D}$$

The theorem may be applied yet again to the two variable term $\bar{C}D$ to give

$$\bar{F} = \bar{A}.B(C + \bar{D})$$

14.4   Electronic gates

The Boolean arithmetic has been explained in terms of the interconnection of various switch contacts. In modern digital systems, however, the use of moving mechanical contacts would result in systems with a very slow maximum speed of operation. As a result of the requirement for faster operation, electronic switches, or gates as they are usually termed, have been developed which are capable of operating in times as short as several nanoseconds. These have enabled the design and production of the very fast and complex digital systems of today, such as the digital computer. Over recent years, various technologies have been used in the manufacture of electronic gates, resulting in the availability of several series or families of gates, the members of each family being compatible with each other in terms of the required power supplies and the voltage or current levels at the inputs and outputs of the gate. Simple interconnections between arrays of gates will, therefore, quite easily allow the realization of very large switching systems.

Although it will be instructive to consider briefly the principles of one or two of the various gate families, the internal construction of each gate may be very complex. When using the gates the mechanism of their internal operation is of little interest to the systems designer, whose aims are affected only by their external characteristics. Thus the user will wish to know the specified power supply voltages, the time required for each gate to operate correctly, the FAN OUT, i.e. the number of similar gates which may be safely driven from the output of the gate, and the FAN IN, i.e. the number of similar gate outputs which may be safely connected to the gate input. The gates are available in integrated circuit form in units which may contain up to eight separate gate circuits in one package. Each individual gate circuit may be an extremely complex design whose price is very low only because the units are made and sold in enormous quantities. The design and construction of the gates is, therefore, a matter for the large semiconductor manufacturing companies only. With the advent of the large scale integration (LSI) techniques, it is commonplace for complex arrays of gates to be produced and sold on one chip, so providing the designer with large segments of systems which may be interconnected to realize the complete system.

14.4.1   Positive and negative logic

Before proceeding to examine various gate circuits, we must decide the meaning of the terms logic 0 and logic 1 when applied to actual circuits. In the previous sections we have discussed the logic operations in a general mathematical way, without reference to the actual voltage or current levels represented by the logic 0 or 1. When looking at actual gate circuits, however, the 0 and 1 levels are specified in terms of the actual voltage or current which is required at the gate input in order to cause the gate to switch correctly.

Switching systems may be described in terms of either positive or negative logic, depending whether the logic 1 level is more positive or more negative than the logic 0 level. Figure 14.13(a) shows several typical logic voltage levels, each of which is a positive logic system, while Fig. 14.13(b) shows possible negative logic voltage levels. Notice that in a positive logic system both the 0 and 1 logic voltages may be negative with respect to the system earth or zero line; however, the 1 level will be less negative (more positive) than the 0 level.

Instead of the terms logic 1 or logic 0, the terms HIGH and LOW are often

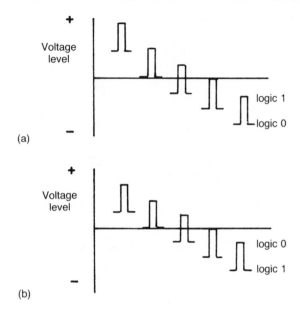

Fig. 14.13 (a) Positive logic levels, (b) negative logic levels.

used. Thus, in changing from the logic 0 to the logic 1 level, a gate output is said to go from LOW to HIGH. Further, if the normal resting state of the output is in the LOW state, and if transition to the HIGH state initiates some further step in the overall system, the gate is said to have an active HIGH output. With the LSI circuits using the MOS technology, which are now widely available, and which will be described in a later section, the active signal level is often the low state.

14.4.2   Diode logic

The earliest form of electronic gate used the unidirectional conducting property of a diode to produce two distinct circuit conditions. Figure 14.14 shows a positive logic diode AND gate with three inputs. Its action may be explained as follows. With all inputs at or near the earth potential, all the input diodes will be held in the conducting state, with the total current flowing to earth, via the diodes being limited by the resistor R. The output voltage is thus held at a potential more positive than the logic 0 level of the input by a voltage equal to the potential across the diodes in the conducting condition. This potential is normally of the order of 0.5 to 0.7 V. If one input is taken to the HIGH or logic 1 state, the diode connected to that input will become non-conducting, while the output potential will still be held in the low state by the remaining diodes, which are still conducting. Thus, only if all the inputs are taken to the logic 1 level will the output level rise positively to the logic 1 level. The circuit thus performs the logical AND function for positive logic inputs.

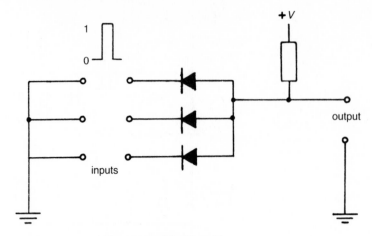

Fig. 14.14 Diode AND gate.

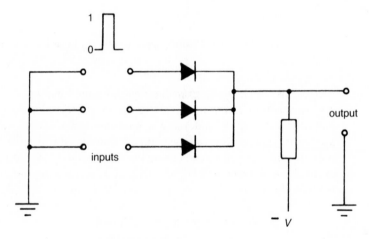

Fig. 14.15 Diode OR gate.

Figure 14.15 shows a positive logic diode OR gate in which, with all the inputs in the 0 level, the output potential is held slightly less positive than the input 0 level, due to the voltage drop across the diodes which are all conducting. If any input is raised in potential to the logic 1 level, the output potential will also rise to a value slightly less positive than the input logic 1 level, all the other input diodes becoming reverse biased. The gate thus performs the logical OR function.

One interesting point can be noted from Fig. 14.16, which shows the diode AND gate of Fig. 14.14 with a negative logic input. In other words the logic 0 level of Fig. 14.14 is called the logic 1 level in Fig. 14.16. Note that the actual signal voltages are not altered, only the logic level name assigned to it. In Figure 14.16, if all the inputs are at logic 0 (which in a negative logic system means they are all at their most positive value), then the output potential will

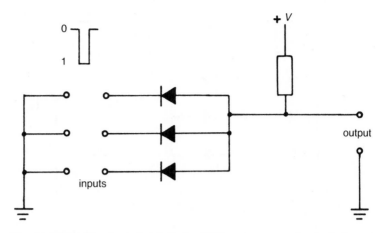

Fig. 14.16 Positive logic level diode AND gate − negative logic inputs.

also be at its most positive value, i.e. at logic 0. If, however, any input is taken negative to its logic 1 value, the output voltage will also be taken negative to its logic 1 value. The gate now is functioning as an OR gate. Thus, the function performed by any circuit depends also upon the definition of the logic levels assigned at its inputs and outputs.

What then are the disadvantages of the simple diode gates described? The main disadvantage derives from the use only of passive devices, i.e. devices which have no power gain. Without gain, the signal power at the output of the gate must be less than that at its input, due to the unavoidable power loss in the gate. As can be observed from a consideration of the method of operation of the gate, the low and high voltage levels are not well defined, and vary between input and output by almost one volt, due to the voltage drop in the conducting diode. This problem is made rapidly worse if similar gates are cascaded in order to perform more complex Boolean functions, a process which is obviously necessary in any but the most simple digital system. The result is that in practice the low and high voltage levels rapidly converge until there is little discrimination between them, resulting in very unreliable operation. A satisfactory family of gates would, in practice, have well defined logic 0 and 1 levels, whereby the manufacturer is able to specify reasonably fine limits about the nominal levels within which correct switching of the gate is guaranteed. Another obvious disadvantage of the gates described is the need for the provision of both positive and negative power supplies.

The inclusion of an active device in the circuit, to give satisfactory power gain, leads also to a major modification in the switching theory already described. The majority of amplifying devices, in their simpler forms, also give signal inversion as well as amplification. Figure 14.17 shows a natural extension to the diode logic gate, in which a common emitter mode transistor amplifier is added to give power gain. The amplifier also gives signal inversion. In the figure, the AND gate is formed by the diodes and the resistor R_1. The transistor inverting amplifier is coupled to the AND gate with the resistors R_2

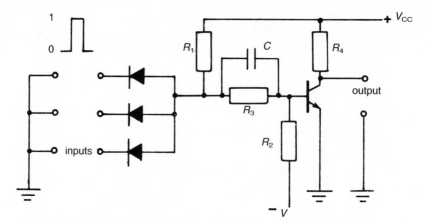

Fig. 14.17 Positive logic AND gate with inverting amplifier.

and R_3, the negative supply voltage being included to ensure that the transistor is completely non-conducting unless the AND gate output is in the HIGH state. The capacitor C is included to improve the speed of operation of the gate.

Let us now examine the logical function performed by an AND gate, followed by an inverting amplifier. Considering a two input system for simplicity, the truth table is as follows. The function performed is the NOT AND or NAND function, which is defined, for a three input gate as

$$F = \overline{A.B.C.} \tag{14.11}$$

The output from a positive logic NAND gate is thus HIGH unless all the inputs are HIGH, when the output is LOW (Table 14.8). The addition of a similar inverting amplifier to the OR gate results in the NOT OR or NOR function. The NOR function is defined, for a three input gate, as

$$F = \overline{A + B + C} \tag{14.12}$$

The truth table for a two input NOR function is as shown in Table 14.9. The output from a positive logic NOR gate is thus LOW unless all its inputs are low, when the output is HIGH.

Table 14.8 NAND gate.

Inputs		Output from AND gate	Inverted output from AND gate
A	B		
0	0	0	1
0	1	0	1
1	0	0	1
1	1	1	0

Table 14.9 NOR gate.

Inputs		$A + B$	Output
A	B		$\bar{A} + \bar{B}$
0	0	0	1
0	1	1	0
1	0	1	0
1	1	1	0

14.4.3   More complex logic families

The four logic functions already described, the AND, the OR, the NAND and the NOR, form the basis of the logic families commonly in use at the present time. The circuits of the gates are, however, more complex than those shown in Figs 14.14 to 14.17.

Figure 14.18 shows the circuit of a TTL (transistor–transistor logic) NAND gate. This family of gates employs a specially produced multiple emitter transistor for the inputs. The operation of the circuit is as follows. With all the inputs (emitters of transistor T_1) in the logic 1 state, the current in resistor R_1 flows through the base–collector diodes which are forward biased. The resulting current flows into the base of transistor T_2, saturating the transistor. Transistors T_3 and T_4 form an output buffer stage, providing the gate with a very low output impedance in both the LOW and HIGH states. The voltage developed by the emitter current of T_2 across the resistor R_3 causes transistor T_4 to be saturated, while the low voltage at the T_2 collector ensures that transistor T_3 is switched off, i.e. into its high impedance state.

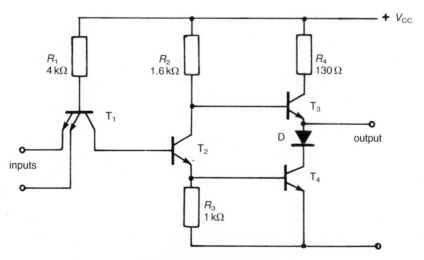

Fig. 14.18 TTL NAND gate.

When the input at any emitter of transistor T_1 is taken into the LOW (logic 0) state, transistor T_1 is switched into the conducting state. Provided the input voltage is sufficiently low the collector voltage of T_1 will also be sufficiently low to prevent transistors T_2 and T_4 from conducting. The collector of T_2 will hence be at a high potential, ensuring that T_3 is in the saturated low impedance state. The output potential thus switches between the LOW and the HIGH states. When any input emitter of T_1 is taken into the LOW state, the driving circuit must be capable of accepting (sinking) the emitter current. This means that the emitter current must be able to pass to earth, via the driving stage output impedance. The gate described is typical of the 7400 family of TTL gates which are a standard in use at the present time. These gates operate from a single 5 V power supply and have a logical 0 output impedance of $12\,\Omega$, and a logical 1 output impedance of $70\,\Omega$ for the standard gates. The gates change state as the changing input voltage passes through a threshold value of approximately 1.4 V. The output voltage is typically 3.3 V in the logical 1 state, and 0.2 V in the logical 0 state. The output can, therefore, tolerate typically 1.9 V of negative going noise in the 1 state, and 1.4 V of positive going noise in the 0 state, before falsely triggering the gate it is driving. Switching time for these gates is of the order of 10 nanoseconds.

The gates are available in dual in line (DIL) packages, with either 14 or 16 pins. Depending upon the number of pins required per gate, one or more gates may be provided in one package. Thus, allowing 2 pins for the power supply connections, four 2 input NAND gates (3 pins each) can be provided on a 14 pin package. Alternatively, one 8 input NAND can be provided on a 14 pin package. This family of TTL gates is produced by several different manufacturers. The standard gates in this family have a FAN OUT of the order of 10, i.e. the output will drive correctly the input of 10 similar gates.

The TTL gates are an example of saturated transistor logic in that the transistors used therein are used in either the saturated or the cut-off states. Transition between those two states is arranged to be as rapid as possible. This is the most efficient method when looked at from the power dissipation point of view, because the cut-off and the saturated states are the states in which the power dissipation in the device is a minimum. However, from the point of view of speed of operation, an improvement may be made by preventing the transistors from switching between these two extreme levels. The operating voltage swing is restricted to a very small level, so minimizing the amount of charge which must be supplied to, or extracted from the device. Emitter coupled logic gates use this technique to provide a very fast system (propagation delay times typically of the order of 2 nanoseconds or less), at the expense of increased power dissipation. Use of this family of gates is, therefore, normally restricted to the application requiring the high switching speed.

The emitter coupled logic gate circuit is based upon the circuit of the differential amplifier shown in Fig. 14.19. The circuit is completely symmetrical so that, with equal input and reference voltages, the potentials at the two transistor collectors are equal. When used as a differential amplifier, this is the normal mode of operation of the circuit. If, however, the input voltage is

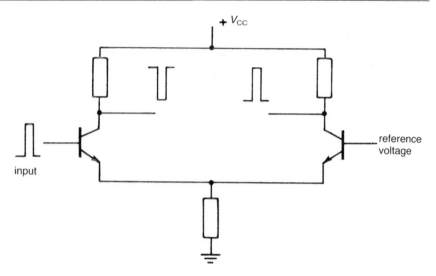

Fig. 14.19 Emitter coupled logic gate.

raised above that of the reference, the collector voltage of the input transistor will increase. If the input voltage is reduced below the reference voltage, the opposite effects occur. The current in the emitter resistor remains virtually constant. The application of a step input voltage change will, therefore, result in voltage steps at the transistor collectors, the changes at the collectors being in antiphase.

To allow multiple inputs to the circuit, several input transistors are connected in parallel. The gate input, therefore, performs the OR function, and the collector voltages provide both the OR and the NOR (= $\overline{\text{OR}}$) outputs.

In Figure 14.20 a typical three input ECL gate is shown. This circuit is basically the same as that of Fig. 14.19, with added refinements to improve its operation. Three input transistors are operated in parallel, while the output from each collector is taken via an emitter follower circuit, whose functions are to give a reduced output impedance, so giving an improved FAN OUT, and also to assist in making the output voltage levels equal to the input levels, this of course being necessary if the gates are to be cascaded. The reference voltage input is also provided with an emitter follower circuit, together with a diode whose function is to improve the temperature stability of the gate.

In a typical commercially available gate, the supply voltage is 5.2 V, while the logic 0 and logic 1 voltages are 2.95 V and 3.70 V, a logic voltage swing of 0.75 V. Other names sometimes given to this type of gate are *current steering logic* (CSL) or *current mode logic* (CML).

A more recent development, which is now readily available and in widespread use, is the MOS (metal-oxide-semiconductor) family of gates. These devices are extremely simple and their great advantage lies in their applicability to integrated circuit techniques. Large scale integration techniques have allowed the production of a wide range of complex circuits, making available a wide variety of digital functions.

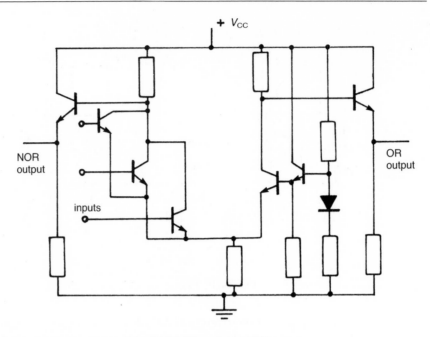

Fig. 14.20 Three input ECL gate.

A very successful family of gates is based upon the complementary properties of p channel and n channel devices. This family, known as the complementary MOS (CMOS) family, is competing most successfully with the TTL family at frequencies up to 25 MHz. The basis of the CMOS gate is the complementary pair inverter, which consists of a p channel device and an n channel device, connected as shown in Fig. 14.21. The power supply voltage range is not critical, the devices operating satisfactorily from 3 V to 18 V (compared with the allowable range for TTL gates of 4.75 V to 5.25 V).

The principle of operation is extremely simple. If a high input voltage is applied to the inverter, the n channel device turns on, giving a low impedance from the output to earth. The p channel device is off, giving a high impedance to V_{DD}, the supply voltage. For a low input voltage the opposite conditions apply; the output voltage is virtually at the V_{DD} value. The main advantage of the CMOS family is the low power dissipation. The d.c. current drain for a CMOS inverter is in the low nanoampere range. The noise immunity of the family is of the order of 30 per cent of the value of the supply voltage V_{DD}. The devices are available in dual in line (DIL) packages of up to 40 pins, although the smaller gate circuits are supplied in the common 14 or 16 pin forms.

The relative characteristics of the three families of gates are summarized in Table 14.10. The product

speed (nanoseconds) × power dissipation (mW)

with the unit of picojoules, is taken as a figure of merit for a gate system.

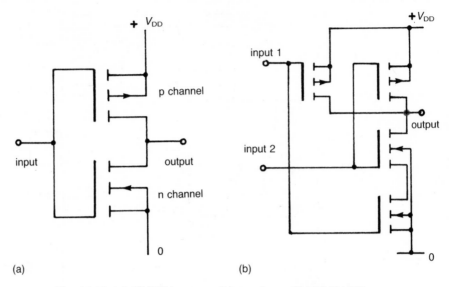

Fig. 14.21 (a) CMOS invertor, (b) two input CMOS NAND gate.

Table 14.10 Gate types.

Logic type	Relative cost	Typical propagation delay ns	Power dissipation mW	Speed–power product pJ	Noise margin V	Fan out
TTL	Medium	6	20	120	1.0	10
ECL	High	2	150	300	0.4	25
CMOS	Very low	15	0.0001	0.005	$0.3\,V_{DD}$	50

14.5 The realization of switching functions

Once the desired output function for a system has been obtained, the circuit can be constructed by interconnecting the required gates selected from the gate families described in the previous section. Depending upon the family chosen, the design will have to be implemented mainly in the more readily available NAND or NOR logic.

14.5.1 Gate symbols

In drawing gate interconnection diagrams, symbols are used to represent the individual gates. Unfortunately, a single universal series of symbols has not emerged in common use. Figure 14.22 shows the two main systems used.

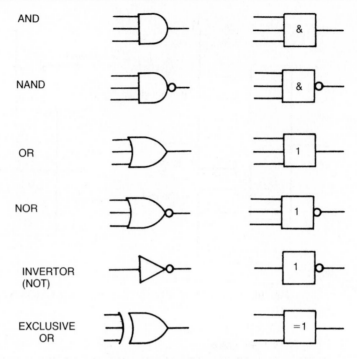

Fig. 14.22 Gate symbols.

14.5.2 AND/OR logic

Let us consider the realization of the Boolean expression

$$F = AB + CD$$

using simple AND and OR gates. The term AB may be realized directly, using a two input AND gate, as can the second term CD. The outputs from the two AND gates will then form the inputs to a two input OR gate whose output is the required function. This realization is shown in Fig. 14.23. The function is thus directly realizable using AND gates as the input gates followed by OR gates as the output gate. The Boolean expression is said to be in AND/OR form.

The same expression may be obtained in the different OR/AND form as follows. The complement of the expression F is obtained by applying the De Morgan theorem.

$$\bar{F} = \overline{AB + CD} = \overline{AB} + \overline{CD}$$

$$= (\bar{A} + \bar{B})(\bar{C} + \bar{D})$$

$$= \bar{A}\bar{C} + \bar{A}\bar{D} + \bar{B}\bar{C} + \bar{B}\bar{D}$$

The original function may be obtained again by applying the De Morgan theorem a second time

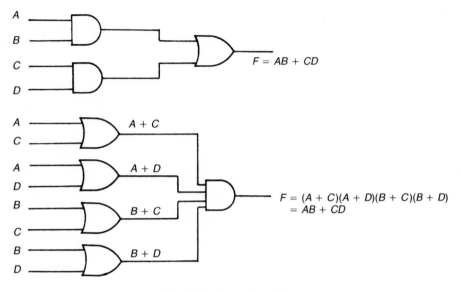

Fig. 14.23 $F = AB + CD$.

$$\bar{\bar{F}} = F = \overline{\overline{AC} + \overline{AD} + \overline{BC} + \overline{BD}}$$

$$= \overline{\overline{AC}} \cdot \overline{\overline{AD}} \cdot \overline{\overline{BC}} \cdot \overline{\overline{BD}}$$

$$= (A + C)(A + D)(B + C)(B + D)$$

The function is now directly realizable using OR gates as the input gates followed by an AND gate, as the output gate, as shown in Fig. 14.23.

It can be seen that there are several ways in which a function may be realized, and, in general, some realizations will require fewer gates than others. In fact, since the cost of gates is to a great extent dependent upon the number of pin connections, the number of gate inputs used in a realization is a good guide to the relative cost of different circuits. To realize the function $F = AB + CD$ will require six gate inputs for the AND/OR realization, but 12 gate inputs for the OR/AND realization.

The process of simplification or reduction of Boolean expressions to different forms is often termed minimization; the property of the function which is being minimized will, however, depend upon the actual design problem.

Example   Realize the function

$$F = A\bar{B} + \bar{A}B$$

This is directly realizable in AND/OR form as shown in Fig. 14.24(a). The function can be converted to the OR/AND form by the application of De Morgan

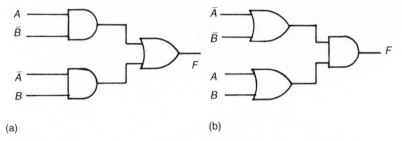

Fig. 14.24 (a) $F = A\bar{B} + \bar{A}B$, (b) $F = (\bar{A} + \bar{B})(A + B) = A\bar{B} + \bar{A}B$.

$$\bar{F} = \overline{A\bar{B} + \bar{A}B}$$

$$= \overline{A\bar{B}}.\overline{\bar{A}B} = (\bar{A} + B)(A + \bar{B})$$

This may be simplified by multiplying out the brackets.

$$\bar{F} = A\bar{A} + \bar{A}\bar{B} + AB + B\bar{B}$$

This may be simplified, remembering that

$$A\bar{A} = 0 \quad \text{and} \quad B\bar{B} = 0$$

$$\bar{F} = \bar{A}\bar{B} + AB$$

Since F represents the lamp being lit, then the complement $\bar{F}$ must represent the lamp being unlit. The lamp is unlit if neither of the switches A and B are operated, or, alternatively, if both switches are operated.

The original function may now be obtained by a further application of De Morgan

$$F = \bar{\bar{F}} = \overline{\bar{A}\bar{B} + AB} = \overline{\bar{A}\bar{B}}.\overline{AB}$$

$$= (A + B)(\bar{A} + \bar{B})$$

The function can now be realized directly in OR/AND form, as shown in Fig. 14.24(b).

In these realizations, the inputs A and B are required in both the normal (A) and the complemented ($\bar{A}$) forms. Invertors would, therefore, have to be included if the complemented variables are not directly available from the system inputs. The two realizations, in this case, are seen to require the same number of gate inputs. As previously mentioned, the function

$$F = \bar{A}B + A\bar{B}$$

is frequently used in digital systems, and is known as the EXCLUSIVE OR function. It is so commonly used that it is made available as a gate in its own right in most logic gate families. The symbol representing the *exclusive or* gate is also shown in Fig. 14.22.

Example Simplify the function

$$F = A + \bar{A}B$$

The method of solution of this simplification is not obvious, but the example is included as this expression is often met in more complex problems.

$$F = A + \bar{A}B = A(B + \bar{B}) + \bar{A}B$$

The addition of the factor $B + \bar{B}$ does not change the expression because

$$B + \bar{B} = 1$$

and

$$A.1 = A$$

The new form of the expression may be expanded to give

$$A + \bar{A}B = AB + A\bar{B} + \bar{A}B$$

$$= AB + A\bar{B} + \bar{A}B + AB$$

The addition of the extra term AB does not change the expression because

$$AB + AB = AB$$

The expression may now be factorized to give

$$F = A + \bar{A}B = AB + A\bar{B} + \bar{A}B + AB$$

$$= A(B + \bar{B}) + B(A + \bar{A})$$

$$= A + B$$

Thus

$$F = A + \bar{A}B = A + B \tag{14.13}$$

This identity is a useful simplifying tool for use in more complex problems.

14.6   The Karnaugh map

The Karnaugh map provides an alternative method for listing the output value of a Boolean expression for all possible combinations of the Boolean input variables. In this, the Karnaugh map acts in a very similar manner to the truth table, in that expressions that are identical have identical Karnaugh maps also. The map method, however, also forms the basis of a method for simplifying Boolean expressions in a systematic manner, so removing many of the difficulties which are met in simplifying Boolean expressions by the algebraic methods previously described.

14.6.1   *Plotting Boolean expressions on a Karnaugh map*

A Karnaugh map consists of a set of boxes, each box representing one possible combination of the Boolean input variables. The box is marked with either a 1 or a 0 to indicate the value of the Boolean expression for the particular combination of the inputs that the box represents.

For one variable, (A), therefore, only two boxes are required, one to represent the input value A, and one to represent the value $\bar{A}$. For two variables $(A$ and $B)$, there are four possible combinations of variables ($\bar{A}\bar{B}$, $\bar{A}B$, $A\bar{B}$, AB). In general, for n input variables, the map will require 2^n boxes. There are, therefore, exactly as many boxes as there are entries in the equivalent truth table. Figure 14.25 shows the Karnaugh map arrangement of the boxes for 1, 2, 3 and 4 input variables. The maps can be extended for

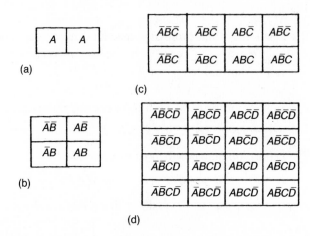

Fig. 14.25 Karnaugh map arrangements with (a) 1, (b) 2, (c) 3 and (d) 4 input variables.

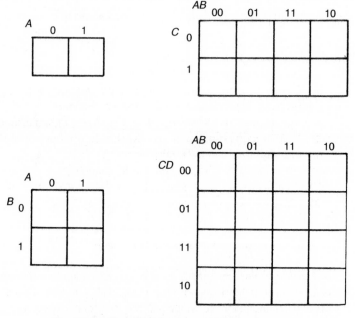

Fig. 14.26 Normal labelling of Karnaugh maps.

greater numbers of variables, but owing to the difficulty in actually drawing the maps in a useful way for greater than seven variables, the method is rarely used for expressions of this size.

In Fig. 14.25, the maps have been labelled in each box with the particular combination of variables represented by the box. The usual method for labelling the maps is shown in Fig. 14.26, with the individual rows and columns marked at the top or left hand side of the map. Comparison of Figs 14.25 and 14.26 will show that the two methods of labelling give the same result. The maps are used by placing a 1 in each box for which the combination of variables makes the expression take the logical value 1. It is not the usual practice to enter a 0 into the remaining boxes, although, of course, these represent the combinations of the input variables which make the expression take the value 0.

Example   Plot on a Karnaugh map, the Boolean expression

$$F = A\bar{B}C\bar{D} + ABCD + A\bar{B}\bar{C}\bar{D}$$

The solution is shown in Fig. 14.27.

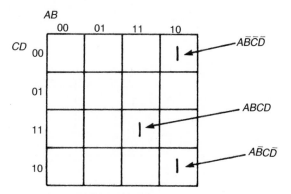

Fig. 14.27 Example.

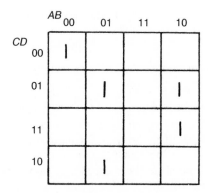

Fig. 14.28 Example.

Example   Plot on a Karnaugh map the Boolean expression

$$F = \bar{A}\bar{B}\bar{C}\bar{D} + \bar{A}\bar{B}C\bar{D} + A\bar{B}\bar{C}D + A\bar{B}CD + \bar{A}BC\bar{D}$$

The solution is shown in Fig. 14.28.

Example   Plot on a Karnaugh map, the Boolean expression

$$F = \bar{A}\bar{B}\bar{C}D + BC + \bar{C}\bar{D}$$

This expression is again a four variable expression. However, not all the variables are included in each term. For example, the term BC indicates that all squares in which both B and C are included must contain a 1. The solution to this example is given in Fig. 14.29, from which it can be seen that there are, in all, four squares whose labels contain the variables B and C together. Similarly, the term $\bar{C}\bar{D}$ also represents four squares on the map, while the term $\bar{A}\bar{B}\bar{C}D$ represents only one square. It can be concluded that, in a four variable problem, a term containing four variables represents only one square, a term containing three variables represents two squares, a term containing two variables represents four squares, while a term containing only one variable represents eight squares.

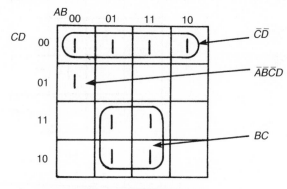

Fig. 14.29 Example.

Example   Plot on a Karnaugh map, the Boolean expression

$$F = \bar{A} + \bar{C}\bar{D} + A\bar{B}\bar{C}$$

The solution to this example is shown in Fig. 14.30. The three terms of the function consist of a group of eight, a group of four and a group of two squares. However, note that the terms overlap each other on the map. This is quite reasonable since there are, for example, two squares on the map which contain both $\bar{C}\bar{D}$ and $\bar{A}$. These are the two squares in which $\bar{C}\bar{D}$ and $\bar{A}$ overlap.

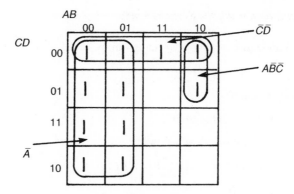

Fig. 14.30 Example.

Example Determine the Boolean function, represented by the Karnaugh map of Fig. 14.31.

The map consists of four squares containing 1s. The function may be read off as

$$F = \bar{A}\bar{B}CD + AB\bar{C}\bar{D} + ABCD + A\bar{B}CD$$

or, alternatively, by grouping together two 1s to give a three variable term

$$F = \bar{A}\bar{B}CD + AB\bar{C}\bar{D} + ACD$$

We have produced from the map two solutions, both correct, one being in a more simple form than the other. This is the basis of the use of the map for the minimization of functions.

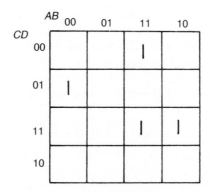

Fig. 14.31 Example.

14.6.2 Minimization using the Karnaugh map

Minimization, using a Karnaugh map, is based upon the Boolean identity

$$A + A = 1 \qquad (14.14)$$

The function

$$F = ABCD + ABC\bar{D}$$

may be minimized as follows

$$F = ABCD + ABC\bar{D} = ABC(D + \bar{D})$$

$$= ABC$$

This function of two terms is plotted on a Karnaugh map in Fig. 14.32.

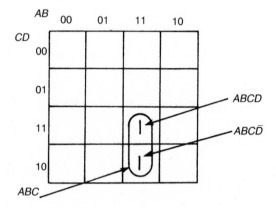

Fig. 14.32 $F = ABCD + ABC\bar{D}$.

The minimization is possible because it is possible to read the function from the map as either two single square terms or one double square term. This property of the Karnaugh map is possible because of the method used to label the squares. The method used is chosen because squares next to each other in the same row or column differ in their labelling by only one variable. Thus, in Fig. 14.32, the two adjacent 1s are $ABCD$, and $ABC\bar{D}$, and differ only in the value of the variable D. The rows and columns of the map have been labelled according to the Gray binary notation, which is, of course, a binary code in which adjacent characters differ only in one variable.

Consider an extension of this principle. Figure 14.33 shows a Karnaugh map of the function

$$F = AB\bar{C}\bar{D} + ABC\bar{D} + \bar{A}\bar{B}CD + A\bar{B}CD$$

which minimizes as

$$F = AB\bar{D}(\bar{C} + C) + \bar{B}CD(A + \bar{A})$$

$$= AB\bar{D} + \bar{B}CD$$

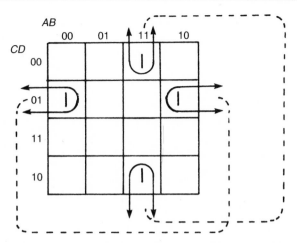

Fig. 14.33 $F = AB\bar{C}\bar{D} + ABC\bar{D} + \bar{A}B\bar{C}D + A\bar{B}\bar{C}D$.

The four terms thus reduce to two 3 variable terms.

From this example we see that the idea of adjacent squares is extended so that the top edge of the map is the same as the bottom edge of the map, as if the map had been rolled round a cylinder. Similarly, between the left and right hand edges there is also the same relationship as if the map had been rolled round a vertical cylinder. It is obviously difficult to visualize the map rolled round two cylinders at the same time. The map is, however, in effect a closed surface. For ease of drawing, it is, therefore, drawn in square form.

Minimization on a Karnaugh map is thus only a task of looking for groups of adjacent ones which can be grouped together. Notice that ones can only be grouped together in powers of 2. On a 4 variable map of 16 squares, we can join together groups of 2, 4 or 8 ones. A group of 16 would, of course, mean that all squares are ones which would mean that the function is itself always equal to 1. Figure 14.34 summarizes the possible types of groupings for a four variable function.

Example　Minimize, using the Karnaugh map, the function

$$F = \bar{A}\bar{B}\bar{C}D + \bar{A}\bar{B}CD + \bar{A}B\bar{C}D + \bar{A}BCD$$

The function is plotted in Fig. 14.35. We see that the function consists of a group of four adjacent ones. This suggests that we should be able to reduce the function to one term only. Using algebraic methods

$$F = \bar{A}\bar{B}\bar{C}D + \bar{A}\bar{B}CD + \bar{A}B\bar{C}D + \bar{A}BCD$$

$$= \bar{A}\bar{B}D(C + \bar{C}) + \bar{A}BD(C + \bar{C})$$

$$= \bar{A}\bar{B}D + \bar{A}BD$$

$$= \bar{A}D(B + \bar{B})$$

$$= \bar{A}D$$

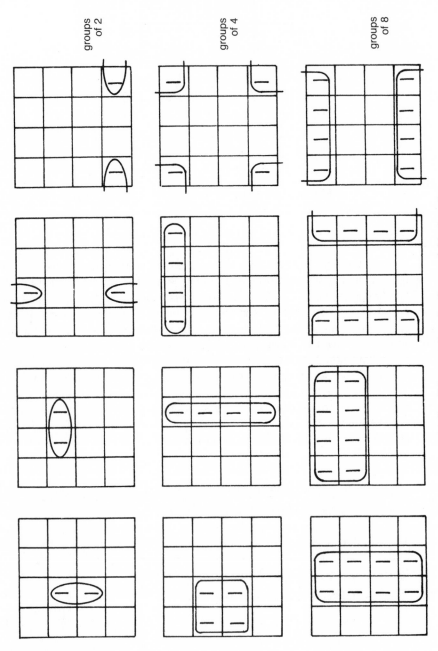

Fig. 14.34 Possible groupings for four variable function.

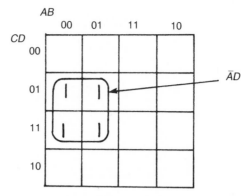

Fig. 14.35 Example.

Example Minimize the function

$$F = ABCDE + ABCD\bar{E}$$

Obviously this will minimize algebraically to

$$F = ABCD(E + \bar{E}) = ABCD$$

The function is plotted in Fig. 14.36.

For a five variable problem we need extra squares; in fact we need $2^5 = 32$ squares. The map consists of two sets of 16 squares, as shown in Fig. 14.36. The two ones in the map are also to be considered adjacent. This can be better visualized in the three dimensional drawing of Fig. 14.37, in which the two 4 variable maps have been positioned, one on top of the other. Similar squares on each map are now adjacent to each other.

Many actual problems can, however, be solved, using up to six variables. Minimization, using algebraic methods, can be quite awkward to apply in some cases. The Karnaugh map method will, with a little practice, reduce the task to one of trivial proportions.

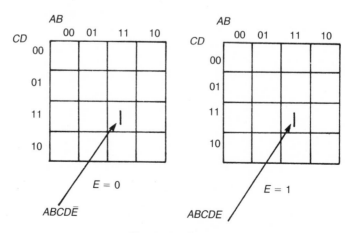

Fig. 14.36 Example.

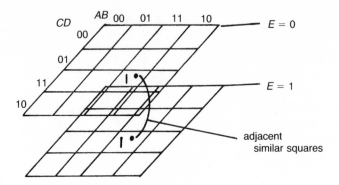

Fig. 14.37 Example.

Example Minimize the Boolean function

$$F = \bar{A}\bar{C}\bar{D} + \bar{A}B\bar{C}D + BCD + BC\bar{D}$$

The function is plotted in Fig. 14.38 and minimizes to

$$F = \bar{A}\bar{C}\bar{D} + \bar{A}B + BC$$

Note, in this example, that it is perfectly correct, in grouping together adjacent ones, to use a one in more than one grouping. A circuit to realize this function would thus require

1	3 input	AND	gate
2	2 input	AND	gate
1	3 input	OR	gate

Inverters may again be required if the negated forms of the input variables are not already available.

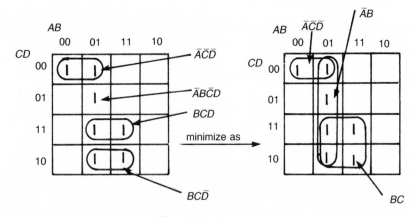

Fig. 14.38 Example.

14.6.3 *Don't care conditions*

In certain problems, involving a number of input variables, certain combinations of the variables may never occur. If a certain input combination is known never to occur, the network output may be allowed to take any output value for this combination of the inputs. In certain cases, this technique can allow simpler network realizations. Such an input variable combination is described as a *don't care* condition. The technique is, perhaps, best explained by means of an example.

Example   Minimize the Boolean function

$$F = \bar{A}B\bar{C}D + AB\bar{C}D + \bar{A}BCD \tag{14.15}$$

The input combination ABCD will never occur. A Boolean function, including don't care conditions, is often written

$$F = \bar{A}B\bar{C}D + AB\bar{C}D + \bar{A}BCD + \{ABCD\}_{x \text{ or } \phi} \tag{14.16}$$

where the terms included in the bracket are the don't care terms. The bracket usually has the symbol x or ϕ against it. When the function is plotted on a Karnaugh map, the don't care terms are again marked with either an x or with the ϕ symbol. A plot of the function is shown in Fig. 14.39.

If the don't care term is ignored, the minimization will require two two term loops, giving

$$F = B\bar{C}D + \bar{A}BD$$

However, if the network output is allowed to be 1 for the don't care term (which, of course, will never occur), a minimization can be used of one four term loop, giving

$$F = BD$$

This technique can produce significant savings in the realization of functions.

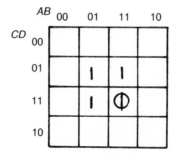

Fig. 14.39 Example.

14.7 Realizations using NAND or NOR gates

We have already considered the problem of the realization of a Boolean function, using only AND or OR gates. Any Boolean function may be realized in this way. The majority of gates available in the standard logic gate families are, however, of the NAND or NOR type and some method must be obtained which will enable the realization of any Boolean function using these more complex gates with their inherent inverting action. Let us consider first whether it is indeed possible to realise all Boolean functions using the NAND/NOR gates.

Figure 14.40 shows two circuits, Fig. 14.40(a) representing a two input NAND gate driving into a single input NAND gate, while Fig. 14.40(b) represents a two input NAND gate with a single input NAND gate in series with each input.

Consider first the circuit of Fig. 14.40(a). The two input NAND gate has an output given by

$$F_1 = \overline{AB}$$

This provides the input to the single input NAND gate whose output is

$$F = \overline{F_1} = \overline{\overline{AB}} = AB$$

A single input NAND gate operates simply as an inverter. The two NAND gates of this circuit thus perform together the logical AND function.

Consider now the circuit of Fig. 14.40(b). Each input is connected to a single input NAND gate; the outputs of these gates are thus $F_1 = \overline{A}$ and $F_2 = \overline{B}$. The outputs of the two inverters form the inputs to the two input NAND whose output function is, therefore,

$$F = \overline{F_1 F_2} = \overline{\overline{A}\,\overline{B}} = A + B$$

The circuit of Fig. 14.40(b) will thus perform the logical OR function.

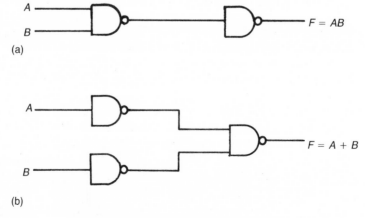

(a)

(b)

Fig. 14.40 (a) $F = AB$, (b) $F = A + B$.

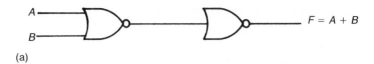

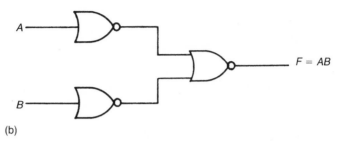

Fig. 14.41 (a) $F = A + B$, (b) $F = AB$.

Because we can use NAND gates to perform both the OR and the AND function it is perfectly possible to realize any Boolean function using NAND gates alone. Similarly, it can be shown that the same is true using NOR gates alone. Figure 14.41 shows the NOR gate realization of the simple AND and OR functions. However, realizations performed by replacing AND or OR gates directly with the circuits of Figs 14.40 and 14.41 would be very inefficient indeed.

14.7.1   Realizations with NAND gates

Let us examine the action of a NAND gate more closely. Figure 14.42(a) shows the truth table for a two input positive logic NAND gate, with the inputs and outputs marked as H (high) or L (low) rather than as 1 or 0. Used in a positive logic system we have the table in (b).

If this same gate was used in a negative logic system, in which the level H would represent a 1 and the level L would represent a 0 (for both inputs and outputs), then the output would be at logic 1 only when both the inputs were at logic 0, i.e. the table in (c). This is true of the NOR function. We can conclude that the function performed by the gate is determined not only by the action of the gate, but also by the way in which the logic levels at its terminals are defined.

Figure 14.42(b) to (e) summarizes the action of the positive input NAND gate with all possible definitions of the logic levels. This shows that, provided we can define, at will, the levels at the gate terminals, we can use the gate to perform any of the four basic logic functions.

Inputs		Output
A	B	F
L	L	H
L	H	H
H	L	H
H	H	L

(a)

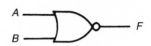

Inputs		Output
A	B	F
0	0	1
0	1	1
1	0	1
1	1	0

input   logic positive
output logic positive
(b)   function   NAND

Inputs		Output
A	B	F
1	1	0
1	0	0
0	1	0
0	0	1

input   logic negative
output logic negative
(c)   function   NOR

Inputs		Output
A	B	F
0	0	0
0	1	0
1	0	0
1	1	1

input   logic positive
output logic negative
(d)   function   AND

Inputs		Output
A	B	F
1	1	1
1	0	1
0	1	1
0	0	0

input   logic negative
output logic positive
(e)   function   OR

Fig. 14.42 The action of a positive logic NAND gate.

Example Realize the exclusive OR function using NAND gates only. The function $F = A\bar{B} + \bar{A}B$ was realized in simple AND/OR gate form in Section 14.5.2. The AND/OR gate network produced is repeated in Fig. 14.43(a).

The output OR gate in this figure may be replaced by a NAND gate directly if we can define the logic levels at its inputs as negative logic. We can also use NAND gates to replace the input AND gates because, having defined the logic at their outputs as negative logic, NAND gates will actually perform the AND function. Figure 14.43(b) shows the resulting NAND gate realization. Bear in mind, however, that as we have defined the input and output logic as positive logic, but the logic between the gates as negative, the Boolean function obtained at points between the gates will be the logical inversion of that obtained at the same point in the circuit of Fig. 14.43(a).

It is common practice to call the output gate in such a network the level 1 gate, the gates which drive the level 1 gates are called the level 2 gates and so on, working backwards towards the input. The circuits of Fig. 14.43 would thus be described as two level networks.

A simple rule may now be stated for obtaining a NAND realization of a Boolean function: Obtain the required Boolean function in AND/OR form, and draw the network required, using AND and OR gates. The output gate will be an OR gate. Replace all gates with NAND gates. The logic level at the inputs to all ODD level gates is negative logic and all new inputs to these gates must be inverted.

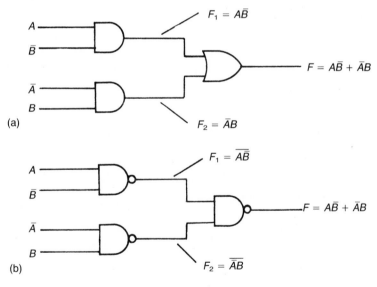

Fig. 14.43 Example.

Example Realize the function

$$F = AB = C(D + E)$$

using NAND gates.

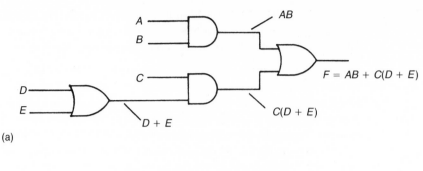

(a)

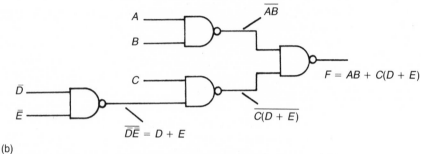

(b)

Fig. 14.44 Example.

The function is directly in AND/OR form and is realized in simple gate form as shown in Fig. 14.44(a). Notice that this is a three level realization and that the type of gate alternates between OR and AND as we proceed backwards through the levels from the output gate. Because the output gate is an OR gate we can replace all the gates with NAND gates giving the final realization shown in Fig. 14.44(b). At the level 3 gate, however, because at the inputs to all ODD levels the logic is inverted, the input variables required are $\bar{D}$ and $\bar{E}$.

14.7.2 Realizations with NOR gates

Figure 14.45 examines the action of a positive logic NOR gate for all the possible definitions of the logic levels at its terminals. Again, by a suitable choice of the logic level definitions, we can make the NOR gate perform any of the four simple logical operations. In particular, with a positive logic output and negative logic input, the NOR gate will perform the AND function. With negative logic output and positive logic input the function performed is the logical OR. NOR gates can be used, therefore, to replace directly the gates in a simple OR/AND network, provided that the output gate is an AND gate. Again, care must be taken with inputs to odd level gates, because the logic definition at such inputs is inverted.

(a)

Inputs		Output
A	B	F
L	L	H
L	H	L
H	L	L
H	H	L

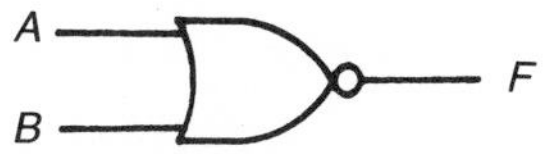

(b)

Inputs		Output
A	B	F
0	0	1
0	1	0
1	0	0
1	1	0

input   logic positive
output logic positive
function   NOR

(c)

Inputs		Output
A	B	F
1	1	0
1	0	1
0	1	1
0	0	1

input   logic negative
output logic negative
function   NAND

(d)

Inputs		Output
A	B	F
0	0	0
0	1	1
1	0	1
1	1	1

input   logic positive
output logic negative
function   OR

(e)

Inputs		Output
A	B	F
1	1	1
1	0	0
0	1	0
0	0	0

input   logic negative
output logic positive
function   AND

Fig. 14.45 The action of a positive logic NOR gate.

Example Realize the exclusive OR function, using NOR gates only.

The exclusive OR function is

$$F = A\bar{B} + \bar{A}B$$

As shown in Section 14.5.2, this AND/OR function can be converted into the OR/AND form of the same function to give

$$F = (A + B)(\bar{A} + \bar{B})$$

A simple gate realization of this function is given in Fig. 14.46(a). The gates can then be all replaced directly by NOR gates to give the solution of Fig. 14.46(b).

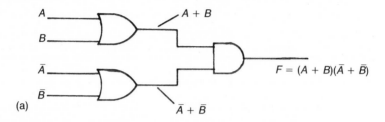

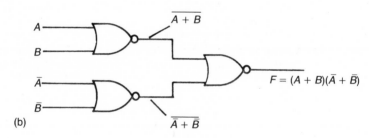

Fig. 14.46 Example.

The rule for obtaining a NOR gate realization of a Boolean function may be stated as: Obtain the required Boolean function in OR/AND form and realize the function using AND and OR gates. The output gate will be an AND gate. Replace all gates with NOR gates. The logic levels at the inputs to all ODD levels are inverted.

Example Realize the function

$$F = AC + \bar{A}BC$$

in both NAND and NOR forms.

The function is already in AND/OR form, and may be realized using NAND gates to give a two level solution as shown in Fig. 14.47(a). However, the function may be minimized (either algebraically, using the identity $X + \bar{X}Y = X + Y$, or, alternatively, using the Karnaugh map method) to give the result

$$F = AC + BC$$

This will allow a two level NAND realization, with fewer gate inputs, as shown in Fig. 14.47(b).

To obtain the NOR realizations, the function may be factorized

$$F = C(A + B)$$

This form of the function allows the NOR realization using only four gate inputs, although, of course, it may be necessary to include an inverter to produce the inverted input $\bar{C}$ for the input to the level 1 gate. The NOR realization is shown in Fig. 14.47(c).

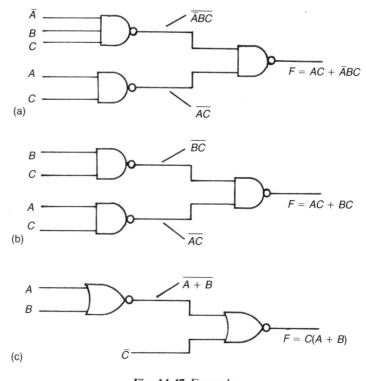

Fig. 14.47 Example.

14.8   Alternative gate symbols

A difficulty often arises when dealing with relatively large digital system designs in deciding the function for which a particular gate is included when only the standard logic gate symbols are used. For this reason an alternative set of symbols is often used (Fig. 14.48). This system obviates this problem by

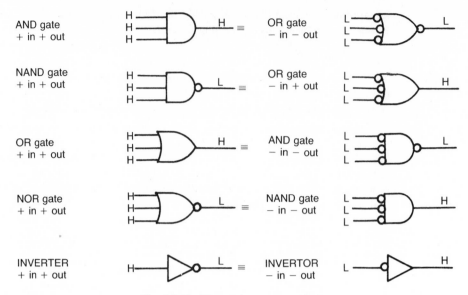

Fig. 14.48 Alternative gate symbols.

maintaining the characteristic AND gate shape for gates used basically in an AND function, and the characteristic curved OR shape for gates used in a basic OR function. The system thus allows identical gates, even gates on the same chip, to be represented using different symbols. In the system, an input or output marked with a small circle is taken as *active LOW*, while a connection not so marked is taken as *active HIGH*.

Figure 14.49 shows a NAND gate realization of the exclusive OR function, using the two different symbol schemes. Using the standard symbols, no indication is given of the fact that the output gate is performing basically an OR function upon the outputs of the other two NAND gates, which are included to perform basically an AND function upon their inputs. In trying to trace a signal through a large network drawn in this way, great difficulty would be met. With the alternative symbols, the symbol shapes tell that the two input gates are performing an AND function upon their inputs, and that if both A and $\bar{B}$ are HIGH, then the top gate output will go LOW, while if both $\bar{A}$ and B are HIGH, the lower gate will go LOW. The output gate shape tells us that it is

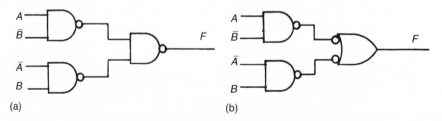

Fig. 14.49 (a) Standard symbols, (b) alternative symbols.

performing an OR function, and that if either of its inputs goes LOW, its output will go HIGH.

Example   Determine the method of operation of the network of Fig. 14.50.

The network consists of three similar segments, together with some output gating. Each segment consists of two input AND gates whose outputs drive an OR gate. Because one of the input AND gates is driven via inverters, one AND gate will go to active LOW if the inputs A_1 and A_2 are both HIGH, and

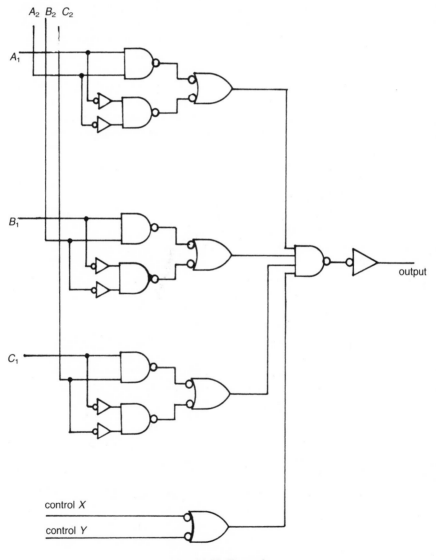

Fig. 14.50 Example.

the other AND gate will go to active LOW if the inputs A_1 and A_2 are both LOW. If either of these events occurs, the OR gate output will go HIGH. The OR gate outputs will thus indicate, by going HIGH, if the inputs A_1 and A_2 are in the same state. The three OR gate outputs, together with the output from the fourth control input OR gate, drive an AND gate. The output will thus go HIGH if the three bit words $A_1B_1C_1$ and $A_2B_2C_2$ are identical, and an active LOW, control X or control Y, is received.

14.9 Use of a multiplexer as logic function generator

A different solution to the realization of combinational logic functions is provided by the multiplexer device. Several such devices are available from various manufacturers within the 7400 series of logic chips. A multiplexer is illustrated in Fig. 14.51 and has a number of data inputs, typically four or eight, and an equivalent number of address inputs. A binary coded input address will select one only of the inputs, so that the data value present at that input appears at the device output. Figure 14.51 shows an eight input, three address multiplexer with TRUE and COMPLEMENTED data outputs. A further input, $\bar{E}$, is provided which is an enable input and which must be held in the logic LOW state to activate the chip. In effect the device operates as an eight input data selector so that with, for example, an input address of

$$ABC = 000$$

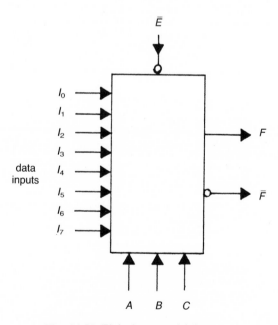

Fig. 14.51 Eight input multiplexer.

the data value present at input I_0 will be transferred to the output F, while with the input address of

$$ABC = 111$$

the data value present at input I_7 will be transferred to the device output.

Example   Realize the logic function $F = CD + \bar{A}C + ABC + A\bar{C}\bar{D}$ using a sixteen input, four address multiplexer device.

This logic function is a function of four variables, $ABCD$. The function should first be expanded to its canonical form; in other words, each term in the function should contain each of the four variables. This process is the reverse of minimization and is, perhaps, most easily accomplished using a Karnaugh map. The function should be plotted on a four variable map as shown in Fig. 14.52. The expansion can be completed by reading the combination of variables for each square in which the function output is 1. Thus

$$F = \bar{A}\bar{B}C\bar{D} + \bar{A}\bar{B}CD + \bar{A}BC\bar{D} + \bar{A}BCD + A\bar{B}C\bar{D} + A\bar{B}CD + AB\bar{C}\bar{D}$$

$$+ ABC\bar{D} + ABCD$$

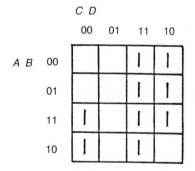

Fig. 14.52 $F = CD + \bar{A}C + ABC + A\bar{C}\bar{D}$.

It can be seen that for nine of the 16 possible squares, an output $F = 1$ is required. This can be arranged by connecting a 1 input to each input selected by these nine combinations of the input variables, while connecting a 0 input to the remaining seven. This is shown in Fig. 14.53.

Example   Realize the logic function of the previous example using an eight input, three address multiplexer.

From the four input variables specified in the function, three can be chosen as the address inputs to the multiplexer. Let us assume the inputs A, B and C are to be used as the address inputs. The function can be rewritten as follows:

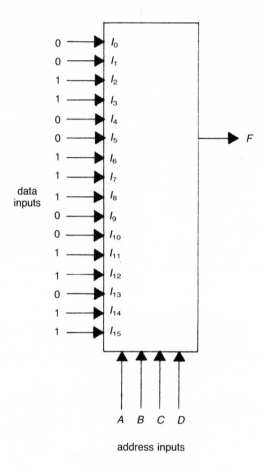

Fig. 14.53 Example.

$$F = \bar{A}\bar{B}\bar{C}(0) + \bar{A}\bar{B}C(D + \bar{D}) + \bar{A}B\bar{C}(0) + \bar{A}BC(D + \bar{D}) + A\bar{B}\bar{C}(\bar{D})$$

$$+ A\bar{B}C(D) + AB\bar{C}(\bar{D}) + ABC(D + \bar{D})$$

It can be seen that the terms including the eight possible combinations of the inputs A, B and C have been collected together. The rewritten function can be checked by multiplying out again the brackets to produce the original function. The values in the brackets are termed the *residues* and are the data inputs required for each of the eight input connections to the multiplexer.

Input	Term	Residue
I	$\bar{A}\bar{B}\bar{C}$	*0*
I	$\bar{A}\bar{B}C$	$D + \bar{D} = 1$
I	$\bar{A}B\bar{C}$	*0*
I	$\bar{A}BC$	$D + \bar{D} = 1$
I	$A\bar{B}\bar{C}$	$\bar{D}$

$$
\begin{array}{lll}
I & A\bar{B}C & D \\
I & AB\bar{C} & \bar{D} \\
I & ABC & D + \bar{D} = 1
\end{array}
$$

The final solution is shown in Fig. 14.54.

Note that if other than inputs A, B and C had been chosen as the address inputs to the multiplexer, a different but equally valid solution would have been obtained.

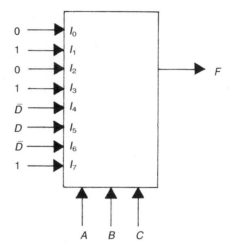

Fig. 14.54 Example.

14.10   Read only memories as logic function generators

Memory devices are discussed in some detail in Chapter 15. However, the read only memory chip, or ROM, provides a convenient way to realize combinational logic functions, especially where the required output is a function of many input variables and also where perhaps several different output functions of the same input variables are required.

A memory device is an array of individual memory cells, each cell with the ability to store a data value, 0 or 1. The cells are arranged so that the data values may be read as a series of output words of n values or bits, with each individual word being selected by a number of word address inputs. Figure 14.55 shows diagrammatically a memory device consisting of 16 words of four bits each. The device is represented in the figure as programmed with four bit Gray code. The stored binary words are accessed by the address inputs to the device, the four address inputs shown in the figure being decoded to select any one of the $2^4 = 16$ stored words to be output.

The read only memory or ROM is available in various forms and sizes, and for large scale production use would be obtained ready programmed with the desired data pattern from the manufacturer. Several forms exist, however, which allow individual user programming and reprogramming and these

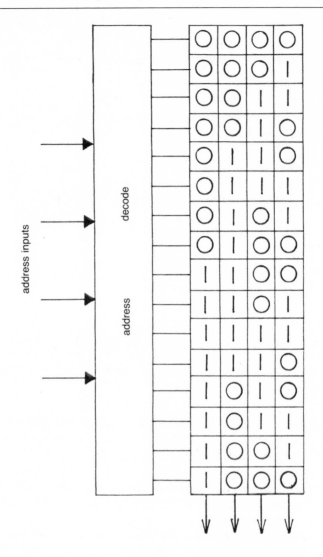

Fig. 14.55 Memory device.

devices form a useful means of realizing multi-input multi-output functions. Such programmable ROM, or PROM, require the use of a PROM pro-grammer for reprogramming purposes. This equipment selects the word to be programmed by driving the address inputs; the desired stored word is supplied to the data connections and a programming control input on the PROM is driven by a defined programming voltage.

In some devices such programming produces permanent, irreversible changes, but in other forms the data cells may be prepared for reprogramming an almost unlimited number of times, by shining an intense ultra-violet light through a quartz window in the device. ROM chips are available in many sizes and may contain many thousands of individual cells, allowing the production of very complex functions. For example, a device is available containing 32 768 individual cells, organized as an array of 4096 words, each of eight bits. Such an array would require an input address of 12 bits to access the $2^{12} = 4096$ individual words. To realize a logic design using a ROM device, each function should be expanded to its canonical form.

Example Using a ROM chip, realize the following logic function of four input variables.

$$F = AB + \bar{B}C\bar{D} + AD$$

$$F = A\bar{B}C\bar{D} + AD + BCD$$

$$F = ABCD + \bar{A}B\bar{C} + \bar{B}\bar{D}$$

$$F = A\bar{C}\bar{D} + BCD + \bar{B}$$

Figure 14.56 shows the expansion to canonical form using Karnaugh maps and the resultant memory pattern for the function. The solution requires the use of a ROM chip of 64 bits, organized as 16 words, each 4 bits in length.

14.11 Programmable logic arrays

The ROM solution to logic functions is inefficient where the logic equation contains a few terms of many variables. Consider the extreme example of

$$F = ABCDEFGH + \bar{A}\bar{B}\bar{C}\bar{D}\bar{E}\bar{F}\bar{G}\bar{H}$$

In a function of eight input variables, using a ROM with eight address inputs, there are a possible $2^8 = 256$ different input combinations, i.e. the ROM will have a 256 word storage capability. In the extreme case of this function, only two of these words are required to generate a logic 1 output. The space available on the chip is thus not used effectively.

Yet another, different, solution is provided by the programmable logic array (PLA) which, like ROM devices, is obtainable in a manufacturer programmed form, or in a user programmable form, the field programmable logic array (FPLA). In a PLA, not only is the output pattern resulting from each input address specified, but also the required addresses themselves are specified. For a given size of device, therefore, it is possible to have more input variables than with a normal ROM. A two input, two output, three term PLA is shown schematically in Fig. 14.57.

The two inputs A and B are made available in TRUE and COMPLEMENTED forms by the buffer circuits shown and are distributed on horizontal busbars in the figure. The three vertical busbars each represent one product term selected from the input variables, the input variables being ANDed together to give the required term. In an unprogrammed device, all

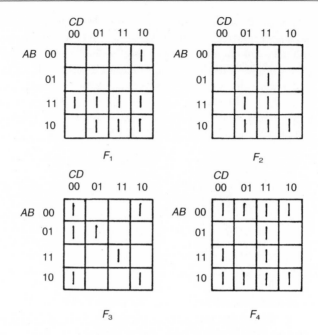

| Inputs | Outputs | | | |
A B C D				
0 0 0 0	0	0	1	1
0 0 0 1	0	0	0	1
0 0 1 0	1	0	1	1
0 0 1 1	0	0	0	1
0 1 0 0	0	0	1	0
0 1 0 1	0	0	1	0
0 1 1 0	0	0	0	0
0 1 1 1	0	1	0	1
1 0 0 0	0	0	1	1
1 0 0 1	1	1	0	1
1 0 1 0	1	1	1	1
1 0 1 1	1	1	0	1
1 1 0 0	1	0	0	1
1 1 0 1	1	1	0	0
1 1 1 0	1	0	0	0
1 1 1 1	1	1	1	1

F_1 F_2 F_3 F_4

Fig. 14.56 Example.

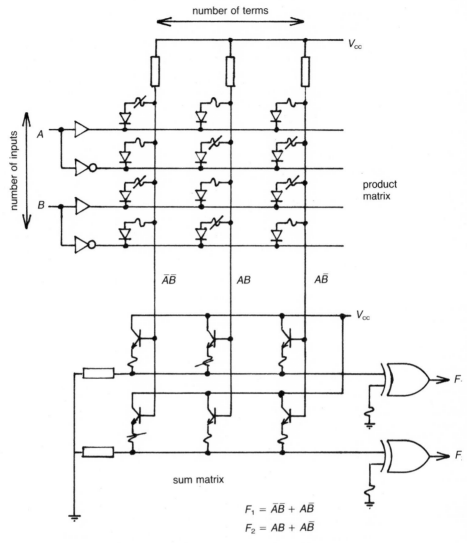

number of terms

number of inputs

V_{cc}

product
matrix

$\overline{A}\overline{B}$ AB $A\overline{B}$

V_{cc}

sum matrix

$F_.$

$F_.$

$$F_1 = \overline{A}\overline{B} + A\overline{B}$$
$$F_2 = AB + A\overline{B}$$

Fig. 14.57 Two input, two output, three term programmable logic array.

the input variables are ANDed to each vertical busbar, being connected by
small fusible links as shown in the figure. Selection of the input variables
required is accomplished during programming by opening, i.e. by fusing, the
links for those variables not required. In the figure, those links not required
are shown *slashed*, leaving the three vertical busbars to represent the terms
$\overline{A}\overline{B}$, AB and $A\overline{B}$. Since these are product terms of the input variables, this
section of the PLA is termed the *product matrix*.

   In order to produce the desired outputs, all that is required is to OR
together those of the product terms needed in the function. Each output can
have included each of the product terms, the selection again requiring the

breaking of fusible links for those terms not desired. In Fig. 14.57 the functions are shown being output via EXCLUSIVE-OR gates, each with one gate input grounded via a fusible link. This arrangement allows the output function to be obtained in ACTIVE HIGH or ACTIVE LOW form, since if the fusible link is opened, the EXCLUSIVE-OR gate will function as an inverter.

PLA devices are available in a wide variety of sizes. For example, a typical device may allow 16 input variables, with their complements, and generate up to eight output functions, each with up to 48 product terms.

14.12   Integrated circuits

The logic circuits described so far in this chapter can all be constructed using standard electronic components, i.e. resistors, capacitors and discrete transistors. However, circuits constructed in this way, especially in relatively large systems, suffer from major disadvantages.

(1) The labour required to produce such circuits is prohibitively expensive.
(2) The reliability of circuit connections made manually is relatively poor.
(3) The physical size and weight of the resulting circuit is excessive.

The first major reduction in circuit size and increase in its reliability occurred with the introduction of the transistor in the late 1940s, replacing the thermionic valve as the active component in electronic circuits. A typical thermionic valve operates with 200 to 300 V power supplies and consumes power of the order of 2 W to heat its cathode. A major drawback of large value systems therefore was the size of the power supplies required and the large amount of waste heat which had to be removed from the equipment.

The next major circuit improvement came with the introduction in the early 1960s of the first commercial integrated circuits. An integrated circuit is manufactured by a process which allows not only the active circuit components, such as transistors, but also the passive components, the resistors, capacitors and their interconnections, to be formed upon a tiny silicon slice. As a result, an integrated circuit is a complete circuit in itself, requiring normally only the addition of the correct power supplies to produce a functioning system. An integrated circuit can be defined therefore as 'the physical realization of a number of electrical elements, inseparably associated on or within a continuous body of semiconductor material to perform the functions of a circuit'.

The first integrated circuits, with from 1 to 10 transistors per chip, are now designated as of SSI (small scale integration) technology. Improved manufacturing methods eventually produced MSI (medium scale integration) and then LSI (large scale integration). VLSI (very large scale integration) technology devices are now commercially available and newer devices will soon enter the SLSI (super large scale integration) era. The different technologies may be roughly classified as follows:

SSI	1 to 10 transistors per chip
MSI	10 to 100 transistors per chip

LSI	100 to 10 000 transistors per chip
VLSI	10 000 to 50 000 transistors per chip
SLSI	Over 50 000 transistors per chip

Several different technologies have been used in the production of integrated circuits.

A large proportion of the small and medium scale integration devices, including the ever popular 7400 TTL logic family is based upon the bipolar technology described in Section 6.3. This is one of the fastest technologies available today but suffers from the disadvantages of relatively large power consumption and relatively low packing density. It is not, therefore, in its basic form, particularly suitable for LSI devices.

Current LSI systems are based mainly upon the MOS (metal-oxide-semiconductor) technology described in Section 6.9.2. The first LSI devices were based upon the PMOS technology in which a p channel transistor is formed in an n type substrate. This is a well understood and reliable process but has the disadvantage that, being based upon the mobility of positive charge carriers to provide conduction, PMOS devices are relatively slow in operation.

Later devices are based upon the NMOS technology described in Section 6.9.2. and, using the more mobile negatively charged electrons rather than the positive holes as the charge carriers, are capable of about a twofold increase in speed. However, even the NMOS devices are approximately ten times slower in operation than the bipolar devices.

The CMOS (complementary MOS) technology, described in Section 14.4.3, uses a combination of a p channel transistor and an n channel transistor, and thus achieves speeds between those of the PMOS and NMOS devices. The CMOS device has excellent noise immunity and very low power consumption, making it suitable for use in electrically-noisy environments or where low power consumption is very important. The CMOS process was indeed developed directly for the aerospace industry, where extreme operating conditions are common.

Problems

1. Convert the following decimal numbers into their simple binary equivalents:

   4596_{10} 322_{10} 49.875_{10}

2. Convert the following into their decimal equivalents:

   2710_8 11010110_2 1101_3

3. Complete the following additions:
   (a) $11001_2 + 01111_2$
   (b) $0111_2 + 1100_2 + 1001_2$
   (c) $457_8 + 112_8$

4. Complete the following subtractions:
   (a) $11000_2 - 00111_2$
   (b) $651_8 - 377_8$

5. Express 349_{10} in its equivalent form in base 16 using A B C D E F as the base 16 characters corresponding respectively to the decimal characters 10 to 15.

6. Write the Boolean expressions for (a) a four input AND gate, (b) a four input OR gate.

7. Write the Boolean expression and truth table for the logic diagram given in Fig. 14.58.

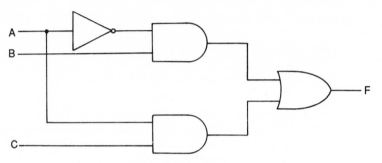

Fig. 14.58 Problem 7.

8. By constructing the truth tables for the functions, show that

$$B + D = \bar{A}B + B\bar{D} + AD + \bar{A}\bar{B}D$$

9. By the method of De Morgan, negate and simplify the following functions:

$$A + \bar{B}, \bar{C}D, AB + \bar{A}\bar{B}, A(B + \bar{C})$$

10. Three members of a panel game are asked to record their opinion of a record using ON/OFF switches. Instead of recording individual preferences, however, data processing is required, such that a HIT lamp is lit if a majority of the panel like the record, while a MISS lamp is lit if a majority dislike the record. Construct a truth table for the problem and develop and minimize the logical equations to the lamp signals.

11. Design a logic system that has three inputs A, B and C and which gives a high output only when the majority of the inputs are high.

12. Design a logic system which will activate a buzzer when either of the following conditions occur with a car: the headlights are on while the ignition is off, the driver's door is open while the ignition is on.

13. Design a logic system which will allow a door to be opened when the correct combination of four push buttons A, B, C and D is pressed. An incorrect combination will result in an alarm sounding. Take the correct combination to be A and B pressed.

14. What gates would be needed to (a) realize the following function, (b) realize the function after simplification?

$$\overline{AB} + \overline{AC} = F$$

15. Use Karnaugh maps to simplify
    (a) $\bar{A}\bar{B}\bar{C} + \bar{A}B\bar{C} + A\bar{B}C + ABC = F$
    (b) $\bar{A}\bar{B}\bar{C}\bar{D} + \bar{A}B\bar{C}\bar{D} + A\bar{B}C\bar{D} + A\bar{B}CD + ABC\bar{D} + ABCD = F$
    (c) $\bar{A}\bar{B}\bar{C} + \bar{A}BC + A\bar{B}\bar{C} + A\bar{B}C + ABC = F$
    (d) $(AC + A\bar{C}D)(AD + AC + BC) = F$

16. How can NOR gates be used to produce OR, AND, NAND and invert gates?

17. Draw the circuit using only NOR gates to obtain the function

    $(\bar{A} + B)(\bar{C} + D) = F$

18. Draw the circuit using only NAND gates to obtain the function

    $B(\bar{A} + \bar{C}) + \bar{A}\bar{B} = F$

Chapter 15
Sequential Logic Systems

15.1 Sequential networks

The logic networks so far considered are examples of combinational logic networks. In a *combinational network*, the output is always determined by the combination of input variables present at that time. This is not exactly true, in that real gates take a finite time to change state, and invalid transient outputs may briefly occur. However, the statement is true once all transients have died away. It is possible, however, to create a gate network whose output is not decided only by the particular input logic values at any time.

Consider the network of Fig. 15.1. This is again only a two NAND gate network. In order to examine this network further, let us assume that the output F is at logical 0, and that both the inputs A and B are at logical 1. The output function from each NAND gate is given by

$$F_1 = \overline{BF} = \bar{B} + \bar{F}$$

and

$$F = \overline{AF_1} = \bar{A} + \bar{F}_1$$

Thus, with the values we have assumed, $\bar{F} = 1$, and hence $F_1 = 1$, giving $\bar{F}_1 = 0$, and since $\bar{A} = 0$, then $F = 0$. The assumed values are, therefore, perfectly consistent.

Alternatively, let us assume, initially, that the output F is at logical 1, and that both the inputs A and B are at logical 1. Thus both $\bar{B}$ and $\bar{F}$ are at 0, giving $F_1 = 0$ and hence $F = 1$. These assumed initial values are also perfectly consistent.

We have, therefore, shown that with $A = B = 1$ the output F may be either 0 or 1, i.e. the output value is no longer specified solely by the values applied at any time to the inputs. This property of the network arises from the feedback connection from the output F back into an input of one of the gates. The connection allows the state of the output to modify the input variables of the network. It can be shown that the output value is now a function of the values of the input variables, but is also a function of the sequence in which the input variables are applied. In effect, the extra dimension of time has been added to the operation of the network, because, as the output value is now a function of the sequence of application of the inputs, it can be used to indicate the sequence which occurred, long after the sequence has ended. The network can be used, in fact, as a *binary memory*.

Let us consider an application for such a memory. Assume that a certain

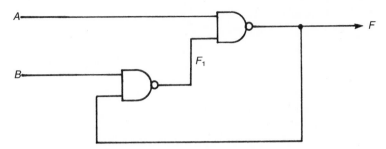

Fig. 15.1 A binary memory network.

process is to be started when each of two remote stations have signalled their readiness. It is unlikely that both signals will occur at the same instant. Each station could signal by placing a logical 1 on its signal line. If both lines are then connected to the inputs of a two input AND gate, the gate output will go to logical 1, when both stations are ready, and this can start the process. This circuit is a combinational logic solution to the problem.

If, however, the operator at each station signifies his readiness by pushing momentarily a push button, only a short duration transient logical 1 will occur on the line. Some memory device must be included to remember that a transient 1 has occured, and to provide a steady logical 1 output to drive the AND gate. This solution to the problem would be a *sequential logic* solution. The problem is, of course, trivial. However, the great majority of problems involving digital solutions are solved by sequential methods. The basic memory element in sequential systems is provided by one of several *bistable circuits*, which are so called because the circuit output will take one or two different, but quite stable, values and can be arranged to switch between these values in order to memorize the occurrence of some event.

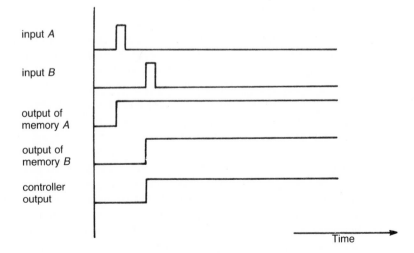

Fig. 15.2 Timing diagram.

A more commonly used name for this type of circuit is the *flip flop*. The operation of a sequential circuit is often more clearly understood with the aid of a timing diagram. Figure 15.2 shows the timing diagram for the process control problem. In drawing this timing diagram, we have again assumed perfect devices. In practice, of course, propagation delays would occur before each element assumes its new state. In a simple circuit, the propagation delays would have little effect and would result only in a delay of a few nanoseconds before the output $F = 1$ is obtained. In certain circumstances, however, in more complex systems, such delays may cause malfunctions, or *hazards* as they are termed, and the possible effects of delays upon any circuit should be examined, using a timing diagram, before any design is finalized.

15.1.1   *The Set Reset bistable (flip flop)*

A *Set Reset (SR) flip flop* is a bistable circuit which has two outputs, one being the logical inversion of the other, usually labelled Q and $\bar{Q}$. It also has two inputs, the set input S being used to set the Q output to 1 ($\bar{Q}$ to 0) and the reset input R being used to reset the flip flop to the Q = 0 state.

The circuit diagram of an SR bistable using transistors is shown in Fig. 15.3(a). The principle of operation of the circuit may, perhaps, be more clearly understood if it is redrawn as in Fig. 15.3(b).

With the exception of the two input connections, S and R, and the input components R_7, R_8, D_1 and D_2, the circuit is seen to consist of two simple transistor switching circuits; i.e. transistor T_1, with its collector load resistor R_1, and input resistor R_4, whose output is fed into a second similar circuit made up from T_2, R_2 and R_3. The two capacitors C_1 and C_2 are included to improve the switching speed of the circuit, and are known as *speed up* capacitors. The output of the second circuit is, in turn, connected directly back into the input of the first. The transistors normally operate in either the cut-off or saturated *bottomed* states, the resistor values being chosen so that if T_1 is in the cut-off state, the high voltage resulting at the collector of T_1 holds the transistor T_1 in the ON or saturated state; in turn the very low voltage existing at T_2 collector or in the saturated state holds transistor T_2 in the cut-off state. The conditions in the circuit are, therefore, quite stable. They are also quite stable in the reverse condition, i.e. with T_1 the saturated transistor holding T_2 in the cut-off state. These two stable possibilities form the bistable states of the circuit, either of the two states being obtainable by means of suitable triggering input signals.

Transition between the two stable states occurs very rapidly since, as each transistor starts to change from the cut-off state to the conducting state, or vice versa, it enters the high gain active region of operation. In the transitional state, therefore, the circuit consists of two high gain amplifiers in cascade, with the output connected back into the input. This is a highly unstable condition, and the opposite stable state is very rapidly reached. The input circuits shown will allow the application of a positive control signal. Assuming positive logic, the application of a logical 1 voltage to the SET input will ensure that T_1 is

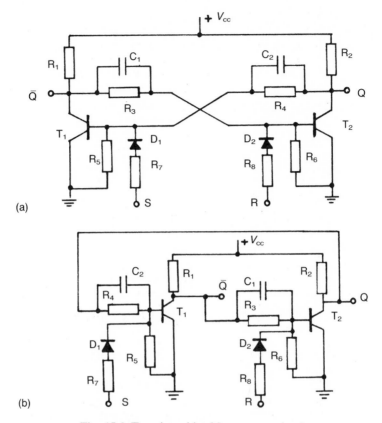

Fig. 15.3 Transistor bistable memory circuit.

switched ON and hence also that T_2 is switched OFF. Similarly, the application of a logical 1 voltage to the RESET input will switch T_2 ON and also T_1 OFF. A SET input will, therefore, result in a low voltage at T_1 collector and a high voltage at T_2 collector. The Q output connection is, therefore, taken from the collector of transistor T_2 and the $\bar{Q}$ output from the collector of T_1.

The logical operation of the circuit is described by the truth table of Fig. 15.4, which also shows the logic circuit symbol used to represent the network. Again, the circuit symbol makes no attempt to represent the actual circuit components used to create the flip flop. The box symbol, however, shows the two input connections S and R, and the two output connections, Q and $\bar{Q}$. The truth table for a sequential network is more commonly termed the *state table*. The state table, as in the case of the truth table, for combinational circuits, lists the output changes which occur for all possible combinations of the input variables. In fact, because, as we have just considered, the specification of the inputs to a sequential circuit does not specify uniquely the value of the output, each set of input variable values must be considered for both possible states of the output.

Taking an example from Fig. 15.4, the state table shows that if the input

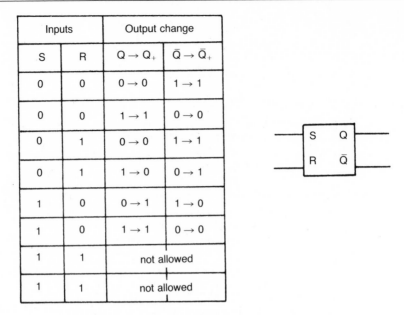

Inputs		Output change	
S	R	$Q \rightarrow Q_+$	$\bar{Q} \rightarrow \bar{Q}_+$
0	0	$0 \rightarrow 0$	$1 \rightarrow 1$
0	0	$1 \rightarrow 1$	$0 \rightarrow 0$
0	1	$0 \rightarrow 0$	$1 \rightarrow 1$
0	1	$1 \rightarrow 0$	$0 \rightarrow 1$
1	0	$0 \rightarrow 1$	$1 \rightarrow 0$
1	0	$1 \rightarrow 1$	$0 \rightarrow 0$
1	1	not allowed	
1	1	not allowed	

Fig. 15.4 State table and logic symbol for SR flip flop.

values $S = 0$ and $R = 0$ occur while the output $Q = 0$, then it will stay at 0. Also, if the same input values occur while the output $Q = 1$, then it will stay at 1. The state table for the two input flip flop thus has eight entries.

Notice, in the state table, the use of the + subscript to denote the passage of time. Thus, Q_+ is the value of the variable Q after the change, caused by the inputs, has finished. Similarly, $\bar{Q}_+$ is the new value of $\bar{Q}$. It is normally not necessary to list, in the output column, the value of the output $\bar{Q}$ as well as the value of the output Q because, of course, $\bar{Q}$ is always the logical negation of Q (except, perhaps, during the actual switching transient).

The operation for the SR flip flop may be summarized as follows:

(1) With $S = 0$ and $R = 0$, the output is not affected, and remains at its previous value.
(2) With $S = 1$ and $R = 0$, the output Q will change to $Q = 1$, if it was previously at 0, and will remain at 1 if it was previously at 1. The output value $Q = 1$ is termed the SET condition. Notice also that the term HIGH is again often used to describe the 1 condition at any terminal.
(3) With $S = 0$ and $R = 1$, the output Q will go LOW (to $Q = 0$) if it was previously HIGH (at $Q = 1$) and will remain at LOW if it was previously LOW.
(4) The input condition $S = 1$, $R = 1$ is not allowed with this flip flop. With both S and R inputs at 1, both transistors are made fully conducting, resulting in 0 values at both collectors which form the outputs Q and $\bar{Q}$. Obviously, both Q and $\bar{Q}$ cannot take the logic value 0 at the same time; this combination of the input variables must, therefore, be prevented from occurring.

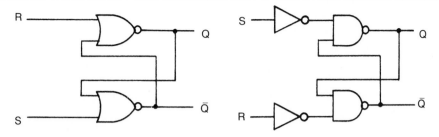

Fig. 15.5 SR flip flops using cross coupled gates.

An SR flip flop may be conveniently constructed from two two-input NOR gates, or, alternatively, two two-input NAND gates, with input inverters as shown in Fig. 15.5.

15.1.2 The trigger flip flop

The trigger (T) flip flop is another form of bistable circuit. Like other flip flops, it has two outputs, Q and $\bar{Q}$, but only one input T. The logic symbol and the state table for the T flip flop are shown in Fig. 15.6. The T flip flop changes state on each T input signal and remains in that state as long as the T input remains LOW.

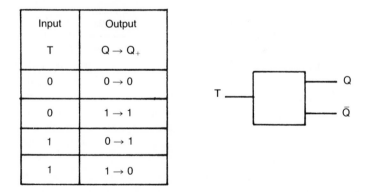

Input	Output
T	$Q \rightarrow Q_+$
0	$0 \rightarrow 0$
0	$1 \rightarrow 1$
1	$0 \rightarrow 1$
1	$1 \rightarrow 0$

Fig. 15.6 State table and logic symbol for T flip flop.

15.1.3 The JK flip flop

The SR flip flop and the T flip flop were developments of circuits commonly constructed using discrete transistors. The advent of integrated circuit techniques led to the development of the JK flip flop, which, since it can realize the truth tables of both the SR flip flop and the T flip flop, is the circuit in most common use in current practice. Figure 15.7 shows the state table and the logic symbol for the JK flip flop. The state table is identical to that of the

Inputs		Output
J	K	Q → Q$_+$
0	0	0 → 0
0	0	1 → 1
0	1	0 → 0
0	1	1 → 0
1	0	0 → 1
1	0	1 → 1
1	1	0 → 1
1	1	1 → 0

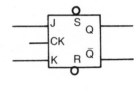

Fig. 15.7 State table and logic symbol for JK flip flop.

SR flip flop, with the exception that the input condition J = 1, K = 1 is allowed, the flip flop acting as a T flip flop for these inputs. The J input can thus be considered as the SET input, while the K input acts as a RESET input.

The JK flip flop is intended to be used in a clocked (synchronous) mode of operation. The J and K inputs do not themselves initiate a change of logic state in the device, but are used as control inputs to determine the change of state which is to occur at a time decided by a pulse input to the CLOCK terminal. This mode of operation allows the precise timing of the state changes in a sequential circuit. The signal applied to the clock terminal is of pulse form, and the change of state of the flip flop will occur normally coincident in time with the rear edge of the pulse. Different designs of JK flip flop are available, however, which are clocked by a clock pulse edge of either polarity.

The JK flip flops of the TTL logic family are available in several forms, in 14 and 16 pin packages. The forms are dictated by the number of pins required to make the connections to the circuit. One or two flip flops are enclosed in a package; these may have single J and K inputs or multiple J and K inputs to allow the logical ANDing together of input signals. They may also be provided with one or both of SET and CLEAR inputs which can be used to set output Q to 1 or clear output Q to 0 at any time. In the logic symbol of Fig. 15.7, the SET and CLEAR inputs, marked S and C, are shown with a small circle against the input. This is to indicate that these inputs are activated by a LOW input, i.e. in a positive logic system by a 0 level input and in fact, during the normal operation of the flip flop, these inputs must be held at the logic 1 level.

15.2 Registers and counters

We have seen that the bistable circuits we have considered may be used to remember that an event has happened, i.e. to remember that a certain signal has occurred at the input terminals. It can be arranged relatively easily that the occurrence of a logic 1 level on a certain line will set the flip flop so that its Q output is also at logic 1. The flip flop, in this way, will function as a single bit memory, memorizing the fact that the logic 1 level has occured. For a binary signal of length n bits, i.e. an n bit binary word, n bistable circuits would be required to construct a word memory. When used in this manner, the bistables together are commonly called a *register*.

Different circuit arrangements are used to enter the value of each bit into the flip flops forming the register, depending upon the application.

If the n bit binary word is available in parallel form, i.e. available on n lines with one line for each bit, each flip flop in the register may be entered separately, usually at the same time. This is the fastest method, but since there are as many data input connections to the register as there are bits in the binary word, this is also a relatively expensive method. This form is known as a *parallel entry register*.

Alternatively, the binary word may be in serial form, i.e. available on one data line in time sequence. The values are entered serially, via the one data input connection, and would be timed into the register by a system clock. In this form the register is known as a *serial entry* or *shift register*. The term shift register arises from the action of entering one data bit into the first flip flop, which we can call A, on the first clock timing pulse; on the second clock pulse, the data contents of the first flip flop are shifted into the second flip flop (B), and the second data bit is entered into the first flip flop (A). On the third clock pulse the value in B is shifted into flip flop C, the value in A is shifted into B, and the value of the next bit, bit 3, is entered into flip flop A. The serial entry register thus requires as many operations (shift and store) as there are bits in the binary word to be stored.

Registers of the types described are available with either serial or parallel entry, and with either serial or parallel read out. Yet another type of register is the counting register, which is a number of flip flops arranged to store a binary word which represents the number of pulses applied to the input. Such counting registers can be arranged to count in any code, although, of course, counting according to the decimal code or the simple binary code is very common.

Using n flip flops, a total count of 2^n may be made. A four bit counter could, therefore, count up to 16 counts before any stored word would occur twice. The number of input pulses or counts needed to cycle a counter through its full range, and which would return the counter state to its starting value, it called the MODULUS of the counter, often abbreviated to MOD. Thus a MOD 8 counter would have a count cycle of eight counts.

Sequential networks are of two basic types:

(1) *Synchronous networks*, in which the state changes in the various parts of the network occur at the same instant, timed by the system clock.
(2) *Asynchronous networks*, in which the state changes in the various parts of the network do not occur at the same instant, and in fact may occur at a time decided by a state change in another part of the same system.

Figure 15.8 shows the state table and the timing diagram for a three bit counting register (counter), which counts according to the simple binary code. The three flip flops used to realize the counter are labelled, for convenience, A, B and C, with flip flop A representing the least significant bit of the binary coded signal. Examination of the state table shows that each flip flop is required to change state when the next less significant flip flop changes from the 1 state to the 0 state. Using JK flip flops, 1 to 0 voltage steps at each flip flop Q output may, therefore, be used to clock the next more significant flip flop, if the control inputs J and K are both held continually in the logic 1 state. The JK flip flops are thus being used as T flip flops in this application. The resulting circuit is also shown in Fig. 15.8.

Notice carefully that in this network the flip flop A has to respond to the input pulse before it provides an output level change to trigger the next flip

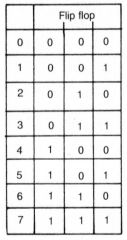

State table

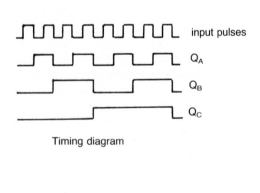

Timing diagram

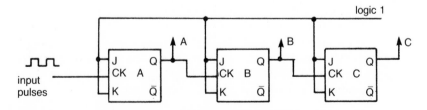

Fig. 15.8 Asynchronous three bit binary counter.

flop. In particular, when in state 7 with $Q_A = Q_B = Q_C = 1$, each flip flop will change state in turn on the receipt of the next input pulse. Thus flip flop A will change from 1 to 0, triggering flip flop B to change from 1 to 0, triggering flip flop C to change from 1 to 0. The network is, therefore, asynchronous in nature and is often termed a *ripple counter*, because of the way in which the change of state *ripples* along the flip flop chain. Care must be taken in reading out the counter state because, of course, between the state 7 (111) and the next state (000) a series of very short duration transient states will occur. The sequence will, in fact, be

$$111 - 011 - 001 - 000$$

An alternative synchronous realization of this state table is shown in Fig. 15.9.

Examination of the state table shows that flip flop A is required to change state at every input pulse. If the input pulses are fed to the CLOCK input, a JK flip flop A will change state at every clock pulse, if the control inputs J_A and K_A are connected to logic 1. Similarly, flip flop B is required to change state on every input pulse which occurs while flip flop A is at logic 1. The control inputs J_B and K_B are, therefore, connected to the output of flip flop A, Q_A. Flip flop C is required to change state on every input pulse when both A AND B are at logic 1. The control inputs J_C and K_C are, therefore, driven by a logic function obtained by ANDing together the outputs Q_A and Q_B. This may be done by using AND gates, or, alternatively, using JK flip flops with multiple J and K inputs, as shown in the figure. Each flip flop in the figure has three J and three K inputs. Unused inputs may be left open as they automatically assume a logic 1 state in the open-circuit condition.

The basic principle of this type of synchronous circuit is that the logic levels available at the outputs Q and $\bar{Q}$ of each flip flop are used as inputs to logic gate networks to produce the required control signals for the J and K inputs of each flip flop, so as to ensure that all flip flops change state correctly, according

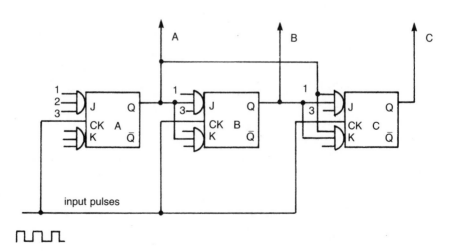

Fig. 15.9 Synchronous three bit binary counter.

to the desired state table when the input clock pulses occur. Because all the flip flops change state at the same time, the difficulties due to the transient states in the asynchronous ripple counter do not occur.

Example   Realize a synchronous MOD 10 counter using JK flip flops.

Because of the common use of the decimal system, the MODULUS 10 counter is often required. Figure 15.10 shows the state table for such a counter, counting in the simple binary code. The flip flops may be directly programmed as follows. Flip flop A is required to change state on every input pulse. Its J and K inputs may, therefore, be left open, i.e. in the 1 state. Flip flop B is required to change state on every second input pulse, i.e. every time A is at logic 1, except that the change from 0 to 1, which would occur after state 9, must be inhibited. The inputs J_B and K_B may be connected directly to the

State	D	C	B	A
0	0	0	0	0
1	0	0	0	1
2	0	0	1	0
3	0	0	1	1
4	0	1	0	0
5	0	1	0	1
6	0	1	1	0
7	0	1	1	1
8	1	0	0	0
9	1	0	0	1
10 = 0	0	0	0	0

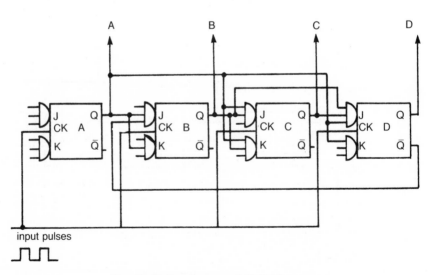

Fig. 15.10 Synchronous MOD 10 counter.

output of flip flop A, with the extra J_B input from $\bar{Q}_D$. The input functions are thus

$$J_B = A\bar{D} \qquad K_B = A$$

Flip flop B will thus set to 1 only when A is at 1 and D is at 0. Flip flop C is required to change state every time both A and B are at logic 1. Its input functions are

$$J_C = K_C = AB$$

Flip flop D is required to set to 1 when all other flip flops are at logic 1. It is required to reset once in the count cycle, when changing from state 7 to state 8. However, if $K_D = A$ it will attempt to reset on every clock pulse which occurs while A is at logic 1. However, since it is only set to 1 itself once in the cycle, at state 8, it will, therefore, only be able to reset in state 9. The input functions for flip flop D are thus

$$I_D = ABC \qquad K_D = A$$

The resulting network is also shown in Fig. 15.10.

Example Realize a shift register using JK flip flops.

The shift register was described earlier as a serial entry n bit word memory in which the data stored in one flip flop is shifted to the next flip flop on the command of the clock pulse. A shift register may be constructed using JK flip flops as shown in Fig. 15.11.

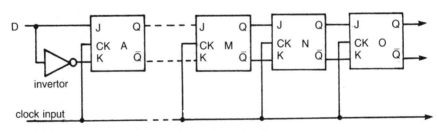

Fig. 15.11 Shift resister using JK flip flop.

Consider flip flop N as representative of any flip flop in the register. If the previous flip flop M is in state 1, then the inputs to flip flop N will be

$$J_N = 0 \qquad K_N = 1$$

These will result in flip flop N resetting to state 0 on the next clock pulse (or staying at state 1 if it is already reset). Similarly, if flip flop M is in state 1, then the inputs to flip flop N will be

$$J_N = 1 \qquad K_N = 0$$

On the next clock pulse, flip flop N will set to state 1, or will stay set to 1 if it is already in that state.

For the first flip flop in the register, these same input conditions can be arranged by providing the control signal K_A from the data input D via an inverter. Thus

$$J_A = \text{DATA VALUE}$$

$$K_A = \overline{\text{DATA VALUE}}$$

15.3 Formal method for synchronous counter design

The design of synchronous counters, using the methods outlined in the previous section, relies heavily upon the experience of the designer and a more formal method is obviously required. Not all counters have successive states which follow the sequence of a simple code, as does the counter illustrated in Fig. 15.10. The successive states are often shown in a state diagram, as in Fig. 15.12, which represents the same counter. Each state is shown as a circle, with the logic values of the individual flip flop stages shown inside the circle. An arrow shows the state change which occurs when the counter is clocked by its input pulse. A state diagram, although perhaps rather obvious in the case of a simple counter, often helps the operation of a synchronous circuit to be understood and, in more complex designs, is an essential part of the design process.

The state table of Fig. 15.10 is only a tabular version of the state diagram. The first step in a formal design procedure will therefore be to produce a state diagram of the state changes which are to be realized; once completed, this can be redrawn as the state table.

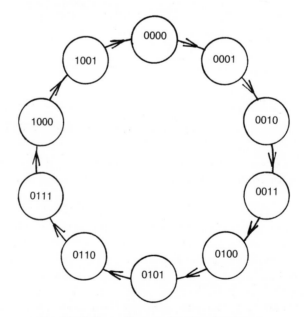

Fig. 15.12 State diagram for decade counter.

From the state table it can easily be determined which changes are required in logic levels in each flip flop to achieve the changes of state listed. For example, from Fig. 15.10 it can be seen that, when in state 0, an input pulse is required to change the counter into state 1, and that this will require flip flop D to remain at 0, C to remain at 0, B to remain at 0 and A to be set to 1. The state table for the JK flip flop, shown in Fig. 15.7, is more usefully drawn in a different form which shows the control J and K inputs required to produce a specified state change. From Fig. 15.7 it can be seen that two sets of inputs will result in any particular change of output. For example, for both inputs $J = 1$, $K = 0$ and $J = 1$, $K = 1$ the flip flop will set from 0 to 1 when clocked. These changes are best displayed in a JK excitation table as shown in Fig. 15.13.

State change $Q \rightarrow Q_+$	Inputs J K
$0 \rightarrow 0$	0 Ⓞ
$0 \rightarrow 1$	1 Ⓞ
$1 \rightarrow 0$	Ⓞ 1
$1 \rightarrow 1$	Ⓞ 0

Fig. 15.13 Excitation table for JK flip flop.

This shows that to achieve, for example, a state change of 0 to 1, the required inputs are $J = 1$ and $K = 0$ or $K = 1$. The K input is a *don't care* condition, indicated in the figure by the Ⓞ symbol. Using this table, it is now possible to determine the overall control functions for the inputs of each flip flop in turn. The states of the decade counter of Fig. 15.10 can be represented on a Karnaugh map as shown in Fig. 15.14.

	BA 00	01	11	10
DC 00	0	1	3	2
01	4	5	7	6
11	Ⓞ	Ⓞ	Ⓞ	Ⓞ
10	8	9	Ⓞ	Ⓞ

Fig. 15.14 State map for decade counter.

The state 0, i.e. when $D = 0$, $C = 0$, $B = 0$, $A = 0$, is represented by the top left hand square of the K map. Similarly, all other states of the counter are represented by one square on the map. With four state variables, A, B, C and D, there are obviously 16 state squares available, but as only ten are required for this counter, the remaining six squares can be treated as don't care states *as*

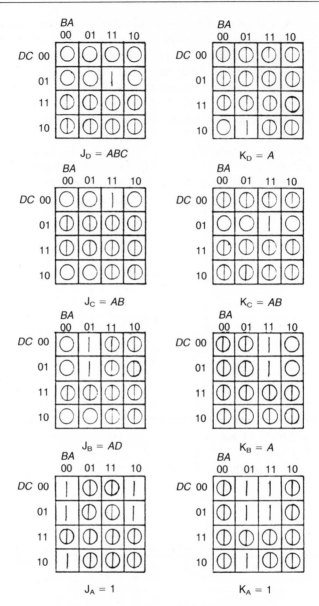

Fig. 15.15 Control maps for JK flip flops.

long as the states are not entered. Assuming we can ensure that this provision is met, then the design can be simplified.

In Fig. 15.15 Karnaugh maps are shown for the control signals for each of the inputs of the four flip flops. Note that don't care inputs have been assumed for each of the don't care states. The maps have been derived as follows.

Considering maps J_D and K_D: when in state 0 (top left hand corner) the state table of Fig. 15.10 shows that D must be prepared to change from 0 to 0. The excitation table shows that this change is achieved by inputs of $J = 0$, $K = 0$; these values are therefore entered into the equivalent squares of the maps. Each square of each map is treated in the same way. Thus when in state 8 (bottom left-hand square of maps), flip flop D must change from 1 to 1, $J_D = 0$, $K_D = 0$, flip flop C must change from 0 to 0, $J_C = 0$, $K_C = 0$, flip flop B must change from 0 to 0, $J_B = 0$, $K_B = 0$ and flip flop A must change from 0 to 1, $J_A = 1$, $K_A = 0$. When all squares have been completed, the functions are minimized to produce the results shown in Fig. 15.15, which can be seen to agree with those obtained intuitively in the previous section. The method described is rather tedious to perform but has the advantage that, provided the steps are followed by rote, the design will be achieved. The assumption however that states 10 to 15, the don't care states, will never be entered must be justified. When power is first applied to the counter circuit, the flip flops will set randomly into states, and it is certainly possible for a don't care state to be entered. One solution would be to reset each flip flop to zero before commencing counting, by connecting briefly each flip flop RESET input to ground. Counting would then commence from state 0.

Example   Determine the control functions for a counter using JK flip flops, to count according to the state diagram of Fig. 15.16(a).

The counter sequence has a cycle of six states and will therefore require three flip flops, giving a total of eight available states. The two available states not in the cycle could be treated as don't care states as in the previous example. However, the state diagram specifies a different treatment. If either of the two states is entered on power-up, the first input pulse is used to transfer into one of the valid counter states, in this case state 000. Obviously one count would be lost; whether this is important or not depends upon the use for which the counter is intended, and this we are not told in the specification. The state table for the design is shown in Fig. 15.16(b). Note that the state numbers shown do not follow the simple binary sequence, again for a reason not specified. However, this unusual form of coding means that successive states in the count sequence do not appear in adjacent squares in the state map (Fig. 15.16(c)). Consequently it often assists to include a further column in the state table, as has been done in this case – the column showing the *next state* for any *present state*.

Using the JK excitation table, repeated in Fig. 15.16(d), the Karnaugh maps for the control functions for each flip flop are completed as shown in Fig. 15.16(e).

15.4   The multivibrator

The astable multivibrator is a basic pulse generator which is often used as a clock pulse generator for sequential networks. It is similar in form to the bistable memory circuit of Fig. 15.3, and consists of two transistor switching

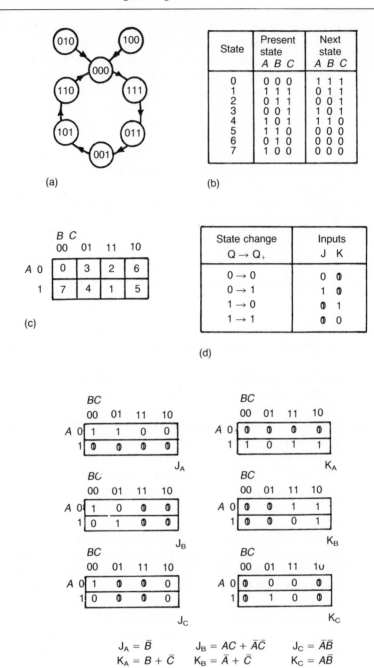

(a)

(b)

(c)

(d)

(e)

Fig. 15.16 Example, (a) state diagram, (b) state table, (c) state map, (d) JK excitation table, (e) control maps.

circuits coupled in cascade. The coupling is, however, made via capacitors, rather than via resistors, with the result that each stable state of the circuit is only stable for a finite time, at the end of which the circuit returns to the active region of operation, and is rapidly switched to its other temporarily stable state. The circuit of a two transistor multivibrator is shown in Fig. 15.17.

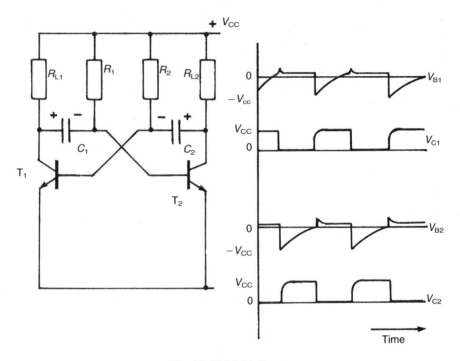

Fig. 15.17 Multivibrator.

In order to understand the mode of operation of the circuit, consider initially that transistor T_2 is in the cut-off state, with its collector potential almost equal to V_{CC}, being held thus by the charge on capacitor C_1. Transistor T_1 is heavily conducting, and is in the saturated state, with its collector potential very low. Transistor T_1 base will be at a potential of about $0.6\,V$; the capacitor C_2 will be charged therefore, in the polarity shown, to a voltage equal to the collector supply voltage V_{CC} less $0.6\,V$.

The potential at transistor T_2 base is rising positively from a negative value, as the capacitor C_1 charges via resistor R_1 from the supply voltage V_{CC}. Eventually the base potential goes positive with respect to the 0 line. Transistor T_2 commences to conduct, its collector potential starting to fall. There can be no instantaneous change of voltage across capacitor C_2, because to change a capacitor voltage requires the movement of a finite quantity of charge. As T_2 collector potential falls, therefore, the fall in voltage is transferred also to the base of T_1. It is driven rapidly negative, so cutting off T_1. The potential of T_1 collector rises rapidly to the $+V_{CC}$ value, this rise being transferred also to the base of T_2, so reinforcing the rapid switching ON of T_2.

It is this possitive feedback, when the circuit is in the active region of its characteristic, which gives the rapid switching action between the two extreme states. The timing cycle then repeats. Figure 15.17 also shows, on a timing diagram, the waveforms of the collector and base voltages. It can be seen that the collector waveforms are approximately square in nature, and have a $1:1$ mark to space ratio if the two halves of the circuit are identical.

The duration of each half period of the waveform may be estimated as follows. When one of the transistors is rapidly switched ON, for example T_1, its collector potential falls from $+V_{CC}$ to almost zero. This voltage change is transferred via the coupling capacitor to the other transistor (T_2) base, which is, therefore, driven negative to a voltage of very nearly $-V_{CC}$ volts. This holds the transistor cut off until the capacitor C_1 charges from the collector supply voltage $+V_{CC}$ via the resistor R_1 to a voltage at T_2 base of zero volts. Effectively, therefore, the capacitor is charging from a voltage of $-V_{CC}$ towards a voltage of $+V_{CC}$ (i.e. through $2V_{CC}$ V) and the half period ends when it has charged through V_{CC} V. Thus

$$v = V_{CC} = 2V_{CC}(1 - e^{-\frac{t}{C_1 R_1}})$$

or

$$t = 0.69 C_1 R_1 \tag{15.1}$$

The total period is, therefore,

$$T = 0.69(C_1 R_1 + C_2 R_2) \tag{15.2}$$

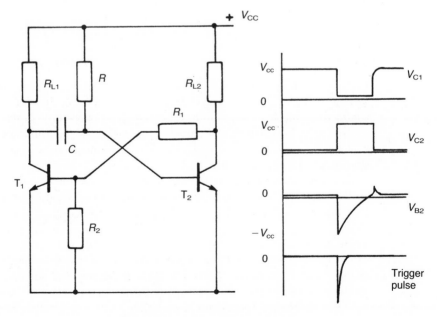

Fig. 15.18 Monostable multivibrator.

Figure 15.18 shows a further modification of the same basic circuit, in which one coupling between the switching transistors is via a resistor, while the other coupling is via a capacitor. This results in a network with one completely stable state and one temporarily stable state. The circuit thus waits in its stable state until triggered into its temporarily stable state; after a fixed timing period it then automatically reverts into its original stable state.

It is of interest to note that the operation of this circuit is the origin of the name *flip flop*, i.e. the circuit is *flipped* into its temporary state, and eventually *flops* back into its stable state. The circuit is called the *monostable multivibrator*, one use being to produce output pulses of predetermined duration from input trigger pulses of variable duration.

The duration of the temporary state is estimated as in the case of the astable multivibrator, and is given by

$$T = 0.69CR \tag{15.3}$$

The circuit may be triggered by a negative going trigger pulse applied to the base of transistor T_2 or a positive going trigger pulse applied to the base of transistor T_1.

Problems

1. Draw the logic circuit diagram of an SR flip flop using (a) NOR gates and (b) NAND gates, explaining their action and giving the state table.

2. Draw the logic circuit diagram of a four bit asychonous 8–4–2–1 binary counter and explain its action.

3. How many flip flops would be required to build a binary counter that can be used to count from 0 to 1023?

4. Draw the logic circuit diagram of a four bit synchronous decade counter and explain its action.

5. Explain the use of the term modulus when used in connection with a counter.

6. A counter has a clock signal of 256 kHz and gives an output from the last flip flop of 2 kHz. What is the modulo number?

7. Produce a state table for a count of 7 (modulo 7) counter using the simple binary code. Develop the logical equations to the input signals for the JK flip flops required to realize the sychonous counter to perform the state table.

8. Draw the logic circuit diagram of a four stage serial in/parallel out shift register constructed from JK flip flops.

9. What is the frequency of a two stage transistor multivibrator, circuit diagram as in Fig. 15.17, if $C_1 = C_2 = 0.001\,\mu F$, $R_1 = 2.2\,k\Omega$ and $R_2 = 100\,k\Omega$?

Chapter 16

Microprocessors

16.1 Digital computers

Without doubt, the digital computer has had a profound effect upon modern life and continues to hold out exciting prospects for the future. The basic idea of a calculating machine interested scientists as early as the 17th century and several attempts were made to produce working machines using mechanical devices. The modern digital computer, although a very complex digital system, consists only of combinations of the same basic logical circuits which we have considered in Chapters 14 and 15; i.e. AND, OR, NAND, NOR gates and shifting and counting registers.

The production of commercially viable machines awaited the development of the manufacturing techniques which allowed the production of large numbers of reliable gate and register circuits. Modern digital computers fall into two categories: general purpose and special purpose machines. General purpose computers are designed to be programmed to perform many different and quite unrelated tasks, while a special purpose computer is designed to operate on one specific problem and its circuits are minimized to provide only the facilities required for the particular task. As a consequence, a special purpose machine can be smaller and cheaper than a general purpose machine.

A digital computer operates by performing upon its input data a sequence of relatively simple arithmetical or logical operations. The available operations are, in general, quite limited in scope and consist of operations such as add, subtract, complement, shift left, shift right, etc. In order to perform a task a method must be devised whereby the task is broken down into a sequence of the operations which the particular computer is able to perform; i.e. an *algorithm* is devised which will allow the task to be performed by the computer. The required sequence of operations for the task is the computer *program* and this is stored in the computer *memory*.

The structure of a basic computer is illustrated in Fig. 16.1. It consists of five fundamental units. The central unit is the central processing unit, usually referred to as the CPU. This unit itself has two parts; the control unit (the CU) and the arithmetic and logic unit (the ALU). The function of the ALU is to perform the basic computer arithmetic and logic operations, i.e. the add, complement, shift operations etc., upon the data presented to it. The function of the control unit is to FETCH from memory, in turn, the individual sequence instructions, to DECODE the instructions and to generate the resulting sequence of control signals which will arrange for the correct data flow to and from the ALU and which will synchronize the EXECUTION of the

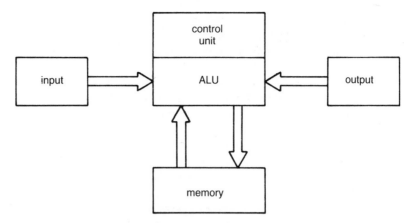

Fig. 16.1 Basic computer structure.

instruction by the ALU. The computer operation thus consists of a sequence of FETCH, DECODE, EXECUTE cycles under the control of the control unit. The memory is shown as a separate unit of the computer. Its function is to store until required any data which may be used by the computer. It is used to store the program which is to be executed and will, in general, also be used to store data which are required for the program. The size of the memory provided for a computer will depend upon the task or tasks for which the machine is to be programmed; a special purpose machine may require only a short program with little input/output data, while a general purpose machine may store several very long programs at the same time, together with extensive data tables. The remaining two units shown in Fig. 16.1 are the input and output units which provide the input and output interfaces to the external world, by means of which new data and new programs can be entered into the machine and the results of the computation obtained. Data signal flow between the basic units of the computer takes place upon sets of signal lines which are called buses. The three standard buses in a computer are the data bus, the address bus and the control bus.

The data bus provides the highway upon which data are passed between the various units of the computer, the data signal flow being controlled by signals produced by the control unit and distributed upon the control bus. The address bus provides the highway for signals, produced by the control unit, which select the correct store location within the memory unit. Each memory store location within the memory unit is given a unique address, by means of which the control unit can select any individual location, within the memory, as the source or destination of data or as the source of the next program instruction.

The application of modern LSI manufacturing techniques to computer circuits has now brought about a revolution in the use of computing techniques. The LSI microprocessor, in which the functions of the CPU of a digital computer have been realized upon a single silicon chip, has enabled the production of complete digital processing facilities at a low cost. This has enabled the use of digital computing methods to be applied to a wide range of applications, from industrial process controls to musical door chimes and has changed the philosophy of the design of digital systems.

The development of the microprocessor device, as it exists today, grew out of research into LSI devices intended for the calculator industry. The first general purpose microprocessor chip, the 8008, was introduced in 1972 by the Intel Corporation. Other manufacturers followed and the development of microprocessor chips continues with more powerful microprocessors being produced almost yearly.

A microprocessor is an LSI device in which the functions of an ALU and its associated control unit are produced on a single chip. It is possible to include, on the same chip, a clock generator circuit and also sufficient memory to allow the device to be described correctly as a single chip computer, and which would be adequate in power to perform small programs. However most micro-processor systems require two or perhaps three LSI devices; in addition to the LSI central processing unit itself, clock generator circuits, memory devices and serial and parallel input/output devices are all available in LSI form.

16.2 The power of a CPU

The power of a CPU, i.e. its ability to process binary data, is dependent upon three main factors. These are:

(1) The order code (or instruction code) of the CPU. This is the list of arithmetic and logic operations which it can perform. To perform any given task, a program must be devised which will achieve the required result, the program being a sequence of the operations provided by the CPU. Thus a CPU which has an extensive and well-designed order code may be able to perform the work with a shorter program.

(2) The speed of operation of the CPU. This is determined by the technology by which it is manufactured. Thus the Motorola 6800 CPU, manufactured in NMOS technology achieves instruction times of the order of 2 to 5 microseconds.

(3) The word length of the CPU. This is the number of binary digits (bits) which it can process at any one time. Microprocessors typically have word lengths of 4, 8, 16 or 32 bits. Common examples of 8 bit microprocessors are the Intel 8080 and 8085, the Motorola 6800 and 6502 and the Zilog Z80. Examples of 16 bit microprocessors are Intel 8086 and 8088, Motorola 68 000 and 65 816 and Zilog Z8000. The Intel 80 386, Motorola 68 020 and Zilog Z80 000 are examples of 32 bit devices.

Table 16.1 shows the data accuracy which can be obtained with various word lengths. In Section 14.2 the accuracy of encoded data was discussed. It can be seen that an 8 bit CPU would be able to process data with an accuracy of 1 part in 256, i.e. with a data accuracy of better than $\frac{1}{2}\%$.

If an accuracy better than this value is required, using an 8 bit CPU, this can be arranged, but only by using more than one CPU word to represent the data.

Table 16.1 Data accuracy of different word lengths.

Number of bits	2^n	Data accuracy $\dfrac{1}{2^n} \times 100\%$	Known as
4	16	6.250	Nibble
8	256	0.391	Byte
10	1 024	0.098	
16	65 536	0.002	

The program could be written to express the data in, for example, two 8 bit words which can be considered to be placed side by side, to give effectively 16 bit accuracy. However the use of *double precision* methods, in this manner, requires longer programs and consequently the rate of data processing will be reduced.

It is common practice to refer to a binary word of 4 bits, as a *nibble* and a word of 8 bits as a *byte*. No other specific word is in common use for other lengths.

16.3 Memory organization

The memory of a computer system is used to store both the programs which are to be executed by the processor and the data which are to be operated on by the processor. The memory can be thought of as a number of memory locations, each of the same word length as the CPU itself. Each location will store one binary word and, in order that the location can be accessed to store or retrieve data at will, each store location is given a unique address by which it may be referenced. The memory addresses are, indeed, binary words themselves. An 8 bit CPU therefore, if using one 8 bit word as the memory address, can reference only 2^8 (256) locations. A memory of only 256 locations would severely restrict the processor to small programs. It is therefore common practice for 8 bit microprocessors to use two data words, i.e. 16 bits for the memory location addresses. The size of memory which can be uniquely addressed using a 16 bit address is thus 2^{16} (65 536) locations in length. But.

$$2^{16} = 2^6 \times 2^{10}$$
$$= 64 \times 1024$$

The memory is said to be 64 K in size, where

1 K (kilobyte) = 1024 locations.

A memory address is therefore a 16 bit binary word. Such a 16 bit string of 0s and 1s is inconvenient and clumsy to deal with, being difficult to remember and lengthy to write. It is common practice however to express the 16 bit binary string in its equivalent hexadecimal form, using the conversion techniques described in Section 14.2.1. The 16 bit word is thus contracted to four hexadecimal characters which, with a little practice, can easily be converted back into binary whenever required.

For example, the 16 bit binary word

$$0110 \quad 1010 \quad 0110 \quad 0011$$
$$6 \qquad A \qquad 6 \qquad 3$$

converts to the hexadecimal form 6A63. Bear in mind that the use of the hexadecimal form in this way is only as a shorthand method of writing down long binary words i.e. the microprocessor itself operates purely in binary.

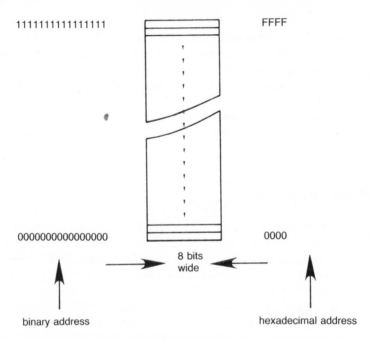

Fig. 16.2 Representation of 16 bit addresses on a memory map.

Figure 16.2 shows a representation of the computer memory in which all of the memory locations are together shown as a vertical column, containing, as we have seen, $2^{16} = 65\,536$ separate locations. The binary addresses are shown at the left hand side, with the equivalent hexadecimal versions of the addresses shown at the right hand side. Notice that each hexadecimal address requires two bytes, i.e. each hexadecimal character requires one nibble. The memory addresses extend therefore from address 0000_{16} to address $FFFF_{16}$. A representation of the system memory addresses in this way is termed a *memory map*.

In practice, the main memory of a microprocessor system is normally implemented by one or more LSI MOS memory chips with, in a special purpose design, only sufficient memory connected to perform the required task. Thus the main memory may range in size from, say, only 128 bytes to the maximum of 64 kilobytes.

16.3.1 Microprocessor memory devices

Two main types of memory devices are used in microprocessor systems: *RAM – random access memory* and *ROM – read only memory*.

The RAM is essentially a read/write memory, i.e. the word stored at any location may be read or alternatively a new word may be stored or written into the location. The main disadvantage of a RAM is that it is *volatile*, which means that whenever the power supply is removed from the RAM device, the stored contents of the chip are lost. When power is applied to a RAM, all memory locations will assume completely random stored values. Because of their volatility, complete programs are seldom stored in RAM, for if the power were to be removed from the chip, by accident or due to supply failure, the programs would be destroyed and would have to be re-entered into the memory before execution could begin again.

RAM is therefore used essentially to store data, which are of a temporary nature, or alternatively it is used for the temporary storage of programs which are under development and which therefore are being edited or changed. It is possible to purchase RAM devices which are provided with battery back-up to provide a retention of stored data under power fail conditions. In order to reduce the battery drain, memory devices used with battery back-up are usually of the CMOS technology. For example, 4 K memory cards are available which specify a retention time of 30 days starting with a fully charged battery. RAM devices are available in various configurations. The Motorola 6810 RAM is a 1 K bit chip, i.e. contains 1024 single bit cells, which is organized for byte oriented systems and has 128 memory locations each 8 bits wide. One such chip will therefore be all that is required for very small systems. Alternatively 1 K bit devices may be obtained, organized to have say 256 locations, each 4 bits in width. Thus two such chips will be required to provide a 256 location 8 bit memory. The latest memory devices available commercially are 64 K bit devices, holding out the prospect of large memories with very few chips.

A ROM is a memory which, once programmed, can only have its contents read: i.e. the stored contents are permanent. Thus a ROM is non-volatile and is used mainly for the storage of programs.

There are several types of ROM in common use. The mask-programmed ROM is programmed permanently during manufacture. The required pattern is supplied by the purchaser to the manufacturer who prepares a mask which will establish the required bit pattern. Because of the cost of making the programming mask, the process is only economically feasible where quantities of the order of 1000 or more similar devices are to be purchased. These devices are thus intended for large scale manufacturing industry.

The programmable read only memory, the PROM, is supplied by the manufacturer with all its memory calls in the same state. It can be user programmed by use of a special equipment, a PROM programmer. Each memory cell in the chip is provided with a fusible link which can be *fused* or *blown* using the PROM programmer. However the programming of such a device is a once-only operation; a fusible link once blown cannot be replaced. A further, very commonly used, user programmable PROM is known as an EPROM, or erasable PROM. The most common form of EPROM is a device which is manufactured with a quartz window on the top of the chip. The memory cells may be set all to the same logic value by exposing the chip to ultra-violet (UV) light for a period of 5 to 20 minutes. Once cleared in this way the EPROM is reprogrammable using the PROM programmer. This type of memory chip, known as a UV erasable PROM, is well suited for use in the development of programs, since, once the program has been developed and edited in RAM, it may be programmed or *burnt* into EPROM and checked for operation in its final form. The EPROM thus provides a quick, easily programmed form of ROM for development purposes or for use in manufactured systems where the number of ROMs required does not warrant the cost of a mask-programmed ROM. The clearing and reprogramming process may be repeated a large number of times.

The memory in a typical microprocessor system will in general be made up of both ROM devices and RAM devices. The time of access of data from the memory into the CPU is in the range 200 to 600 nanoseconds.

Example   In an 8 bit microprocessor system memory, three 1 K bit RAM memory devices are addressed so that they occupy the lowest addresses on the memory map. What are the addresses located in each device?

Figure 16.3 shows the three devices configured in the lowest addresses of the memory map. Each memory chip contains 1024 memory bit cells. When used

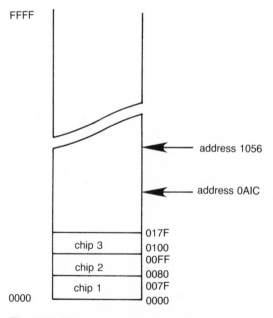

Fig. 16.3 Memory map for examples.

on an 8 bit system, each chip will provide $1024/8 = 128$ memory locations. In hexadecimal notation, each device provides

$$128_{10} = 80_{16} \text{ locations}$$

The addresses, therefore, in chip 1 will be the first 80_{16} locations, i.e. addresses 0000 to 007 F. In chip 2, the addresses will be 0080 to 00 FF. In chip 3, the addresses will be 0100 to 017 F.

Example On the memory map of Fig. 16.3, two memory addresses are shown. These are locations 0A1C and 1056. How many memory locations are there between these two addresses?

This result may be obtained by subtracting the lower address from the higher, but the subtraction must be performed in hexadecimal mathematics. To assist in this, Table 16.2 shows a hexadecimal subtraction table.

$$\begin{array}{r} 1056 \\ -0A1C \\ \hline 063A \end{array}$$

The difference is $063A_{16}$ locations.

Table 16.2 Hexadecimal subtraction table, $X - Y = ?$

$X =$

	0	1	2	3	4	5	6	7	8	9	A	B	C	D	E	F
1	F	0	1	2	3	4	5	6	7	8	9	A	B	C	D	E
2	E	F	0	1	2	3	4	5	6	7	8	9	A	B	C	D
3	D	E	F	0	1	2	3	4	5	6	7	8	9	A	B	C
4	C	D	E	F	0	1	2	3	4	5	6	7	8	9	A	B
5	B	C	D	E	F	0	1	2	3	4	5	6	7	8	9	A
6	A	B	C	D	E	F	0	1	2	3	4	5	6	7	8	9
7	9	A	B	C	D	E	F	0	1	2	3	4	5	6	7	8
$Y =$ 8	8	9	A	B	C	D	E	F	0	1	2	3	4	5	6	7
9	7	8	9	A	B	C	D	E	F	0	1	2	3	4	5	6
A	6	7	8	9	A	B	C	D	E	F	0	1	2	3	4	5
B	5	6	7	8	9	A	B	C	D	E	F	0	1	2	3	4
C	4	5	6	7	8	9	A	B	C	D	E	F	0	1	2	3
D	3	4	5	6	7	8	9	A	B	C	D	E	F	0	1	2
E	2	3	4	5	6	7	8	9	A	B	C	D	E	F	0	1
F	1	2	3	4	5	6	7	8	9	A	B	C	D	E	F	0

For results below the stepped line borrow a one from the next higher order digit.

16.4 Registers

The term *register* is used for small memories that are part of the control section. They are the working stores used for the manipulation and short-term storage of data, and to keep track of the internal operation of the microprocessor. Common types of registers are:

Program counter: This contains the address of the next progam instruction. Remember that the CPU performs a sequence of FETCH–DECODE–EXECUTE operations. Each operation consists of a FETCH of the binary contents of the address in the memory which is currently held in the program counter. The program counter contents are then incremented, i.e. increased by 1, to point to the next memory address. The binary word obtained in the FETCH is then decoded by the CPU. The function of the program counter is thus to keep track of the address in memory of the next byte of the program.

Accumulators: Accumulators are temporary storage registers used during most arithmetic and logic operations to hold one of the operands. Thus if a number X, stored in memory location L_1 is to be added to a number Y, stored in memory location L_2, with the result of the addition to be stored in location L_3, the operation sequence would be performed, using one of the accumulators, as follows:

Step 1: Load the number Z from location L_1 into accumulator A.
Step 2: Add to the contents of accumulator A the contents of location L_2. The result of this addition will be the new contents of accumulator A.
Step 3: Store the contents of accumulator A in the memory location L_3.

General purpose registers: These registers may serve as temporary storage for data or addresses. They are not assigned a specific task and the programmer may, for example, be able to assign them as accumulators.

Index register: This is used to hold the address of an operand in the main memory when the indexed mode is used. In some microprocessors, e.g. Motorola 68 000, general purpose registers are used as index registers.

Stack pointer register: It is often convenient to make use of a section of read/write memory in the main memory for use as further temporary storage of data. Such a section of memory is termed a stack. In order to keep track of the position of the top of the stack within the memory, the stack pointer is used to point to the top unused memory location in the stack.

Status or condition code register: This holds 1 bit indicators, called flags, that represent the state of conditions inside the CPU. They are used in the same manner as the monarch's flag, the Royal Standard, on Buckingham Palace. Whenever the monarch is in residence at the Palace the flag is raised to indicate the fact. Each of the flags in the condition register is similarly set to logic 1 to indicate the occurrence of a certain event within the microprocessor system. The flags are then used as the basis for decisions taken by the CPU.

Different microprocessors have different numbers and types of flags. Among the common flags are:

CARRY: This is used in arithmetical and logical operations as the means of indicating that there is a 1 to carry over from operations performed in the accumulators. Thus, for example, if with an 8 bit accumulator two 8 bit numbers are added and the result would be a 1 in the ninth bit, then since there is no such bit the flag is used to indicate a carry over of 1. For example, adding the two following numbers

$$1 1 0 1 1 0 0 1$$
$$\underline{1 0 0 0 0 0 0 0}$$
Carry 1 $0 1 0 1 1 0 0 1$

OVERFLOW: The number range covered by an 8 bit word using the two's complement method extends from -128 to $+127$ decimal. Obviously it is quite likely that the result of an arithmetic operation upon numbers which are valid will result in an answer which lies outside the allowable number range. For example, the decimal addition $126 + 126 = 252$ cannot be accommodated in an 8 bit two's complement range. In binary the addition becomes

$$\begin{array}{ll} 0\ 1\ 1\ 1\ 1\ 1\ 1\ 0 & = 126_{10} \\ \underline{0\ 1\ 1\ 1\ 1\ 1\ 1\ 0} & = 126_{10} \\ 1\ 1\ 1\ 1\ 1\ 1\ 0\ 0 & = -4_{10} \end{array}$$

The result is obviously incorrect since the sum of the two positive numbers appears to give a negative answer, i.e. the result has a 1 in the sign bit. This effect is due to a *carry* propagating within the word into the most significant bit, so changing the sign of the result. Whenever this overflow occurs, the overflow flag is set to 1.

ZERO: This flag is set to 1 whenever the results of an operation, arithmetic or logical, in either an accumulator or a memory location is zero.

SIGN or NEGATIVE: Whenever an operation results in a 1 in the most significant bit of an accumulator of a memory location, this flag will also be set to 1. In arithmetic operations, the most significant bit is taken to be the sign bit and a 1 in this bit would indicate a negative result.

INTERRUPT MASK: This flag is used by the processor in dealing with interrupt programs. It is set to 1 if an interrupt is allowed, 0 if not. See Section 16.9.

HALF-CARRY: This flag is used when dealing with binary coded decimal characters. The BCD code was described in Section 14.2.3. Each BCD character requires 4 bits. An 8 bit word will therefore represent two BCD characters. When performing BCD arithmetic in the binary processor, an addition, for example, might generate a carry from bit 3 of the result into bit 4, i.e. from the first BCD digit into the second BCD digit. This would result in an incorrect answer and must be corrected. The HALF-CARRY

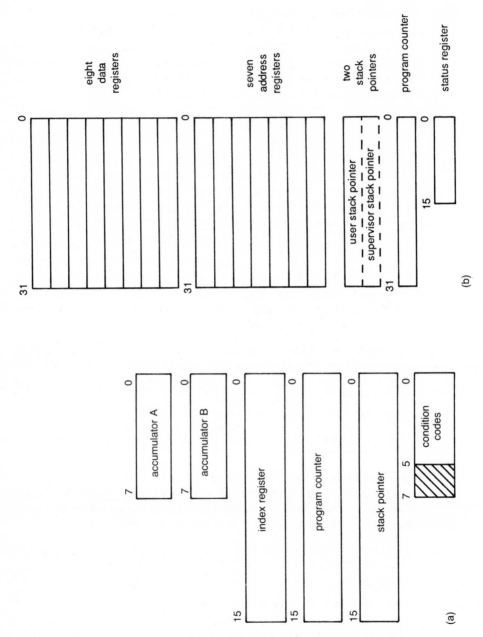

Fig. 16.4 Registers, (a) Motorola 6800, (b) Motorola 68000.

flag therefore sets to 1 to indicate a carry from bit 3 into bit 4 in the result of a binary addition.

The register sets and sizes of a microprocessor are conveniently displayed on a diagram. Figure 16.4(a) shows the set for the Motorola 6800 with figure 16.4(b) showing the set for the Motorola 68 000. The 6800 can handle addresses which are 8 bits wide, the 68 000 32 bits wide. The 6800 has two accumulators and no general purpose registers. The 68 000 has no specific register corresponding to the accumulator since any of the eight data registers can function as accumulators. The 68 000 has also no specific index registers since any of the address registers can be used in this way.

16.5 Instruction sets

The instruction set is a list of the operations a microprocessor can perform and includes such operations as data transfer, arithmetic and logical operations, branching instructions and sub-routine instructions. Thus, for example, the data transfer instructions might include: load, store, move, input and output. The arithmetic instructions could include: add, subtract, increment, decrement, compare and negate. Logical operations might be: AND, OR, EXCLUSIVE OR and NOT. Branching instructions could include: unconditional branch, branch if zero, branch if not zero, branch if equal, branch if not equal, branch if positive and branch if negative. The sub-routine instruction is to make the progam being followed jump to a special group of instructions to carry out some specific task. Such types of instruction are generally referred to as a CALL or JUMP instruction.

Different ways of describing the same instruction tend to occur with the different microprocessors and recourse must be made to the code list issued with a specific microprocessor.

16.5.1 Addressing methods

There are many techniques used to fetch the desired operand during the execution of an instruction, these techniques being called *addressing modes*. These include:

Inherent or implied addressing mode: In this mode the address containing the operand is already included in the operation code, these operation codes thus requiring no additional data other than the operation code. Inherent mode codes are thus 1 byte in length.

Immediate addressing mode: In this mode the data on which the operation is to be performed immediately follow the instruction. Immediate mode instructions require 1 byte for the instruction together with 2 or 3 bytes for the data.

Direct or absolute addressing mode: Whereas the immediate mode instructions provide the processor with the actual numbers which are to be operated on by the instruction, the direct mode provides the processor with a memory address which contains the data to be operated on or where the data are to be stored.

Relative addressing mode: In normal operation, each step in the program is performed in sequence. At certain points in the program decisions may have to be taken, usually as a result of certain tests, to alter the course of the program, i.e. to branch into a different programming sequence. The first byte of the instruction is the operation code. The second byte, called the offset, gives the number of memory locations to be moved by the program counter relative to its present position.

Indexed addressing mode: This mode is similar to the direct mode in that an address is specified from which the data required by the operation can be obtained. In the direct mode, however, the specified address is given as a constant written into the program and which cannot be varied during the running of the program. The indexed mode of addressing specifies the address of the operand as being given by the contents of the index register. As the contents of the index register can be altered, under program control, then the instruction operand can be varied while the program is running.

16.5.2 *Machine code programming*

The instruction set, or order code, supplied with a microprocessor lists the instructions. Many of the instructions are provided with more than one addressing mode, which alters the manner in which the CPU obtains the data on which to perform the operation. The CPU recognizes each instruction as a binary code and a program can be written and presented to the processor as a sequence of such binary words. A program written in this manner is said to be a *machine code program* because it is written in the binary language which the CPU itself understands. This is the lowest level of programming language and although it is necessary to understand the principles of machine code programming, the use of this level of language for other than small programs would be very tedious. A higher level of language uses mnemonics to represent the individual instructions and this, together with other important facilities, makes *assembler* language programming, as it is called, a much more convenient process.

16.5.3 *M6800/6802 examples*

To illustrate the points raised in this section, the following are a series of examples relating to the instruction set for the M6800/6802 microprocessors. The 6802 has the same basic architecture and instruction set as the earlier 6800, but includes a built-in clock generator and a small amount on chip RAM. Figure 16.5 lists the order code for these processors and shows, for each instruction, the various addressing modes, the machines codes in hexadecimal notation, the instruction mnemonics and other details. At the right hand side of the figure, each condition code register flag is shown with an indication of the effect of the instruction upon the flag.

Details of the Motorola devices discussed are reproduced by kind permission of Motorola Semiconductors Ltd, to which company we express our gratitude.

Addressing mode examples

The 6800/6802 microprocessors have immediate, direct, index, extended, inherent and relative modes of addressing. The direct mode assumes that the first byte of the relevant address is 00, thus only the second byte of the address is to be specified. The extended mode is a form of direct mode where both bytes of the address have to be specified. The direct addressing mode thus provides a faster operation, since the instruction is only two bytes in length when compared with the three bytes of the extended mode.

Inherent mode example

Enter all zeroes into accumulator B.

Memory address	Machine code	Mnemonic
0100	8F	CLRB
0101	next instruction	

Inherent mode example

Reduce the contents of accumulator A by 1.

Memory address	Machine code	Mnemonic
0050	4A	DECA
0051	next instruction	

Immediate mode example

Load the data byte D3 (hexadecimal) into accumulator A.

Memory address	Machine code	Mnemonic
0030	86 D3	LDAA
0032	next instruction	

Because the accumulator A is an 8 bit register, the LDAA immediate mode instruction is a two byte instruction, one byte containing the operation code itself while the next byte contains the 8 bit binary data. The complete instruction, when entered into memory, must therefore require two memory locations.

Immediate mode example

Add to the contents of accumulator A the data 56 and then load the index register with the data ABCD:

Memory address	Machine code	Mnemonic
0020	8B 56	ADDA
0022	CE AB CD	LDX
0025	next instruction	

ADDRESSING MODES

ACCUMULATOR AND MEMORY OPERATIONS	MNEMONIC	IMMED OP	IMMED ~	IMMED #	DIRECT OP	DIRECT ~	DIRECT #	INDEX OP	INDEX ~	INDEX #	EXTND OP	EXTND ~	EXTND #	INHER OP	INHER ~	INHER #	BOOLEAN/ARITHMETIC OPERATION (All register labels refer to contents)	H (5)	I (4)	N (3)	Z (2)	V (1)	C (0)
Add	ADDA	8B	2	2	9B	3	2	AB	5	2	BB	4	3				A + M → A	↕	●	↕	↕	↕	↕
	ADDB	CB	2	2	DB	3	2	EB	5	2	FB	4	3				B + M → B	↕	●	↕	↕	↕	↕
Add Acmltrs	ABA													18	2	1	A + B → A	↕	●	↕	↕	↕	↕
Add with Carry	ADCA	89	2	2	99	3	2	A9	5	2	B9	4	3				A + M + C → A	↕	●	↕	↕	↕	↕
	ADCB	C9	2	2	D9	3	2	E9	5	2	F9	4	3				B + M + C → B	↕	●	↕	↕	↕	↕
And	ANDA	84	2	2	94	3	2	A4	5	2	B4	4	3				A · M → A	●	●	↕	↕	R	●
	ANDB	C4	2	2	D4	3	2	E4	5	2	F4	4	3				B · M → B	●	●	↕	↕	R	●
Bit Test	BITA	85	2	2	95	3	2	A5	5	2	B5	4	3				A · M	●	●	↕	↕	R	●
	BITB	C5	2	2	D5	3	2	E5	5	2	F5	4	3				B · M	●	●	↕	↕	R	●
Clear	CLR							6F	7	2	7F	6	3				00 → M	●	●	R	S	R	R
	CLRA													4F	2	1	00 → A	●	●	R	S	R	R
	CLRB													5F	2	1	00 → B	●	●	R	S	R	R
Compare	CMPA	81	2	2	91	3	2	A1	5	2	B1	4	3				A − M	●	●	↕	↕	↕	↕
	CMPB	C1	2	2	D1	3	2	E1	5	2	F1	4	3				B − M	●	●	↕	↕	↕	↕
Compare Acmltrs	CBA													11	2	1	A − B	●	●	↕	↕	↕	↕
Complement, 1s	COM							63	7	2	73	6	3				$\overline{M}$ → M	●	●	↕	↕	R	S
	COMA													43	2	1	$\overline{A}$ → A	●	●	↕	↕	R	S
	COMB													53	2	1	$\overline{B}$ → B	●	●	↕	↕	R	S
Complement, 2s (Negate)	NEG							60	7	2	70	6	3				00 − M → M	●	●	↕	↕	①	②
	NEGA													40	2	1	00 − A → A	●	●	↕	↕	①	②
	NEGB													50	2	1	00 − B → B	●	●	↕	↕	①	②
Decimal Adjust, A	DAA													19	2	1	Converts Binary Add. of BCD Characters into BCD Format	●	●	↕	↕	↕	③
Decrement	DEC							6A	7	2	7A	6	3				M − 1 → M	●	●	↕	↕	④	●
	DECA													4A	2	1	A − 1 → A	●	●	↕	↕	④	●
	DECB													5A	2	1	B − 1 → B	●	●	↕	↕	④	●
Exclusive OR	EORA	88	2	2	98	3	2	A8	5	2	B8	4	3				A ⊕ M → A	●	●	↕	↕	R	●
	EORB	C8	2	2	D8	3	2	E8	5	2	F8	4	3				B ⊕ M → B	●	●	↕	↕	R	●
Increment	INC							6C	7	2	7C	6	3				M + 1 → M	●	●	↕	↕	⑤	●
	INCA													4C	2	1	A + 1 → A	●	●	↕	↕	⑤	●
	INCB													5C	2	1	B + 1 → B	●	●	↕	↕	⑤	●

COND. CODE REG.

Operation	Mnemonic	IMMED OP	~	#	DIRECT OP	~	#	INDEX OP	~	#	EXTEND OP	~	#	INHER OP	~	#	Boolean/Arithmetic Operation
Load Acmltr	LDAA	86	2	2	96	3	2	A6	5	2	B6	4	3				M → A
	LDAB	C6	2	2	D6	3	2	E6	5	2	F6	4	3				M → B
Or, Inclusive	ORAA	8A	2	2	9A	3	2	AA	5	2	BA	4	3				A + M → A
	ORAB	CA	2	2	DA	3	2	EA	5	2	FA	4	3				B + M → B
Push Data	PSHA													36	4	1	A → M_{SP}, SP − 1 → SP
	PSHB													37	4	1	B → M_{SP}, SP − 1 → SP
Pull Data	PULA													32	4	1	SP + 1 → SP, M_{SP} → A
	PULB													33	4	1	SP + 1 → SP, M_{SP} → B
Rotate Left	ROL							69	7	2	79	6	3				M — rotate left through carry
	ROLA													49	2	1	A
	ROLB													59	2	1	B
Rotate Right	ROR							66	7	2	76	6	3				M — rotate right through carry
	RORA													46	2	1	A
	RORB													56	2	1	B
Shift Left, Arithmetic	ASL							68	7	2	78	6	3				M — shift left arithmetic
	ASLA													48	2	1	A
	ASLB													58	2	1	B
Shift Right, Arithmetic	ASR							67	7	2	77	6	3				M — shift right arithmetic
	ASRA													47	2	1	A
	ASRB													57	2	1	B
Shift Right, Logic	LSR							64	7	2	74	6	3				M — shift right logic
	LSRA													44	2	1	A
	LSRB													54	2	1	B
Store Acmltr	STAA				97	4	2	A7	6	2	B7	5	3				A → M
	STAB				D7	4	2	E7	6	2	F7	5	3				B → M
Subtract	SUBA	80	2	2	90	3	2	A0	5	2	B0	4	3				A − M → A
	SUBB	C0	2	2	D0	3	2	E0	5	2	F0	4	3				B − M → B
Subract Acmltrs	SBA													10	2	1	A − B → A
Subtr. with Carry	SBCA	82	2	2	92	3	2	A2	5	2	B2	4	3				A − M − C → A
	SBCB	C2	2	2	D2	3	2	E2	5	2	F2	4	3				B − M − C → B
Transfer Acmltrs	TAB													16	2	1	A → B
	TBA													17	2	1	B → A
Test, Zero or Minus	TST							6D	7	2	7D	6	3				M − 00
	TSTA													4D	2	1	A − 00
	TSTB													5D	2	1	B − 00

Fig. 16.5 M6800 order code.

INDEX REGISTER AND STACK POINTER OPERATIONS

POINTER OPERATIONS	MNEMONIC	IMMED OP	~	#	DIRECT OP	~	#	INDEX OP	~	#	EXTND OP	~	#	INHER OP	~	#	BOOLEAN/ARITHMETIC OPERATION	5 H	4 I	3 N	2 Z	1 V	0 C
Compare Index Reg	CPX	8C	3	3	9C	4	2	AC	6	2	BC	5	3				$(X_H/X_L) - (M/M+1)$	•	•	⑦	↕	⑧	•
Decrement Index Reg	DEX													09	4	1	$X - 1 \rightarrow X$	•	•	•	↕	•	•
Decrement Stack Pntr	DES													34	4	1	$SP - 1 \rightarrow SP$	•	•	•	•	•	•
Increment Index Reg	INX													08	4	1	$X + 1 \rightarrow X$	•	•	•	↕	•	•
Increment Stack Pntr	INS													31	4	1	$SP + 1 \rightarrow SP$	•	•	•	•	•	•
Load Index Reg	LDX	CE	3	3	DE	4	2	EE	6	2	FE	5	3				$M \rightarrow X_H, (M+1) \rightarrow X_L$	•	•	⑨	↕	R	•
Load Stack Pntr	LDS	8E	3	3	9E	4	2	AE	6	2	BE	5	3				$M \rightarrow SP_H, (M+1) \rightarrow SP_L$	•	•	⑨	↕	R	•
Store Index Reg	STX				DF	5	2	EF	7	2	FF	6	3				$X_H \rightarrow M, X_L \rightarrow (M+1)$	•	•	⑨	↕	R	•
Store Stack Pntr	STS				9F	5	2	AF	7	2	BF	6	3				$SP_H \rightarrow M, SP_L \rightarrow (M+1)$	•	•	⑨	↕	R	•
Indx Reg → Stack Pntr	TXS													35	4	1	$X - 1 \rightarrow SP$	•	•	•	•	•	•
Stack Pntr → Indx Reg	TSX													30	4	1	$SP + 1 \rightarrow X$	•	•	•	•	•	•

JUMP AND BRANCH

OPERATIONS	MNEMONIC	RELATIVE OP	~	#	INDEX OP	~	#	EXTND OP	~	#	INHER OP	~	#	BRANCH TEST	5 H	4 I	3 N	2 Z	1 V	0 C
Branch Always	BRA	20	4	2										None	•	•	•	•	•	•
Branch If Carry Clear	BCC	24	4	2										$C = 0$	•	•	•	•	•	•
Branch If Carry Set	BCS	25	4	2										$C = 1$	•	•	•	•	•	•
Branch If = Zero	BEQ	27	4	2										$Z = 1$	•	•	•	•	•	•
Branch If ≥ Zero	BGE	2C	4	2										$N \oplus V = 0$	•	•	•	•	•	•
Branch If > Zero	BGT	2E	4	2										$Z + (N \oplus V) = 0$	•	•	•	•	•	•
Branch If Higher	BHI	22	4	2										$C + Z = 0$	•	•	•	•	•	•
Branch If ≤ Zero	BLE	2F	4	2										$Z + (N \oplus V) = 1$	•	•	•	•	•	•
Branch If Lower Or Same	BLS	23	4	2										$C + Z = 1$	•	•	•	•	•	•
Branch If < Zero	BLT	2D	4	2										$N \oplus V = 1$	•	•	•	•	•	•
Branch If Minus	BMI	2B	4	2										$N = 1$	•	•	•	•	•	•
Branch If Not Equal Zero	BNE	26	4	2										$Z = 0$	•	•	•	•	•	•
Branch If Overflow Clear	BVC	28	4	2										$V = 0$	•	•	•	•	•	•
Branch If Overflow Set	BVS	29	4	2										$V = 1$	•	•	•	•	•	•
Branch If Plus	BPL	2A	4	2										$N = 0$	•	•	•	•	•	•
Branch To Subroutine	BSR	8D	8	2										} See Special Operations	•	•	•	•	•	•
Jump	JMP				6E	4	2	7E	3	3					•	•	•	•	•	•
Jump To Subroutine	JSR				AD	8	2	BD	9	3					•	•	•	•	•	•
No Operation	NOP										01	2	1	Advances Prog. Cntr. Only	•	•	•	•	•	•
Return From Interrupt	RTI										3B	10	1	} See Special Operations	•	•	⑩	•	•	•
Return From Subroutine	RTS										39	5	1		•	•	•	•	•	•
Software Interrupt	SWI										3F	12	1		•	•	•	•	•	•
Wait for Interrupt	WAI										3E	9	1		•	S ⑪	•	•	•	•

CONDITIONS CODE REGISTER OPERATIONS	MNEMONIC	INHER OP	~	#	BOOLEAN OPERATION	5 H	4 I	3 N	2 Z	1 V	0 C
Clear Carry	CLC	0C	2	1	0 → C	•	•	•	•	•	R
Clear Interrupt Mask	CLI	0E	2	1	0 → I	•	R	•	•	•	•
Clear Overflow	CLV	0A	2	1	0 → V	•	•	•	•	R	•
Set Carry	SEC	0D	2	1	1 → C	•	•	•	•	•	S
Set Interrupt Mask	SEI	0F	2	1	1 → I	•	S	•	•	•	•
Set Overflow	SEV	0B	2	1	1 → V	•	•	•	•	S	•
Acmltr A → CCR	TAP	06	2	1	A → CCR	⑫					
CCR → Acmltr A	TPA	07	2	1	CCR → A	•	•	•	•	•	•

CONDITION CODE REGISTER NOTES:
(Bit set if test is true and cleared otherwise)

① (Bit V) Test: Result = 10000000?
② (Bit C) Test: Result = 00000000?
③ (Bit C) Test: Decimal value of most significant BCD Character greater than nine?
  (Not cleared if previously set.)
④ (Bit V) Test: Operand = 10000000 prior to execution?
⑤ (Bit V) Test: Operand = 01111111 prior to execution?
⑥ (Bit V) Test: Set equal to result of N ⊕ C after shift has occurred.
⑦ (Bit N) Test: Sign bit of most significant (MS) byte of result = 1?
⑧ (Bit V) Test: 2's complement overflow from subtraction of LS bytes?
⑨ (Bit N) Test: Result less than zero? (Bit 15 = 1)
⑩ (All) Load Condition Code Register from Stack. (See Special Operations)
⑪ (Bit I) Set when interrupt occurs. If previously set, a Non-Maskable Interrupt is required to exit the wait state.
⑫ (ALL) Set according to the contents of Accumulator A.

LEGEND:
OP   Operation Code (Hexadecimal);
~   Number of MPU Cycles;
#   Number of Program Bytes;
+   Arithmetic Plus;
−   Arithmetic Minus;
.   Boolean AND;
M_{SP} Contents of memory location pointed to be Stack Pointer;
+   Boolean Inclusive OR;
⊕   Boolean Exclusive OR;
$\overline{M}$ Complement of M;
↑   Transfer Into;
0   Bit = Zero;

00   Byte = Zero;
H   Half-carry from bit 3;
I   Interrupt mask
N   Negative (sign bit)
Z   Zero (byte)
V   Overflow, 2's complement
C   Carry from bit 7
R   Reset Always
S   Set Always
↕   Test and set if true, cleared otherwise
•   Not Affected
CCR Condition Code Register
LS   Least Significant
MS   Most Significant

Fig. 16.5 *Continued*

The first instruction is, as in the previous example, a two byte instruction. However the second instruction is a three byte instruction, since it refers to the index register which is 16 bits in length. The LDX instruction must therefore specify the data to be loaded into all the 16 bits.

Note that, although it is convenient to write the instruction and the associated data on the same line as shown above, the memory is actually located as follows:

Memory address	Contents
0020	8B
0021	56
0022	CE
0023	AB
0024	CD
0025	next instruction

Extended mode example
Load accumulator A with the data contained in memory address 89 AB.

Memory address	Machine code	Mnemonic
0075	B6 89 AB	LDAA
0078	next instruction	

Extended mode example
Store the data contained in accumulator B into the memory location 1234.

Memory address	Machine code	Mnemonic
0060	F7 12 34	STAB
0063	next instruction	

Extended mode example
Rotate left the data contained in memory location 5000.

Memory address	Machine code	Mnemonic
0000	79 59 99	ROL
0003	next instruction	

Direct mode example
Load accumulator A with the data contained in memory address 00AB.

Memory address	Machine code	Mnemonic
2000	96 AB	LDAA
2002	next instruction	

Direct mode example
Store the data contained in accumulator B into the memory location 0034.

Memory address	Machine code	Mnemonic
1020	D7 34	STAB
1022	next instruction	

Relative mode example
Branch forwards five places if the result of the previous instruction is zero.

Memory index	Machine code	Mnemonic
1850	27 05	BEQ
1852	next instruction	
1853		
1854		
1855		
1856		
1857	next instruction after branching	

The correct value of the offset is obtained by subtracting the address of the instruction following the branch instruction from the address to which it is desired to branch. Thus $1857 - 1852 = 0005$. The offset is the least two significant hexadecimal characters of the result. Note that the maximum forward offset is given by 7F (+127) and the maximum backward offset by 80 (−128), see Section 14.2.3.

Relative mode example
From a branch instruction in address 2000, branch to address 1FD2.

Memory address	Machine code	Mnemonic
1FDO		
1FD1		
1FD2		
.		
.		
.		
1FFF		
2000	20 D0	BRA
2002		

Note that $1FD2 - 2002 = FFD0$. The required offset byte is the two least significant hexadecimal characters, i.e. D.

Indexed mode example
Load accumulator A with the contents of the memory location 09 plus the index register contents.

Memory address	Machine code	Mnemonic
0100	CE 01 00	LDX
0103	A6 09	LDAA
0105	next instruction	
0106		
0107		
0108		
0109	55	
010A		

The index register contents 0100 are added to the offset 09 to give the address 0109. The contents of address 0109, in this example an arbitrary 55, are loaded into accumulator A. The indexed mode is especially suitable for accessing data tables where successive entries in the table can be obtained by incrementing the index register.

Machine code programming examples

All programs intended for use in dedicated microprocessor systems, e.g. traffic light controller, automatic washing machine controller etc., would, when completely developed and debugged, be *burned* into ROM so that the program is permanent and available as soon as the system is switched on. In order that such a program will start automatically, as soon as the power is applied, the address of the first program instruction is entered into memory in two memory locations, FFFE and FFFF, known as the RESET vector. The M6800 processor, upon switch-on, will immediately examine the two locations to obtain the start address of the program, and will commence operating at that address. Thus if the program start address is 0500, then the byte 05 would be entered into location FFFE and the byte 00 would be entered into location FFFF. Programs under development, however, will normally reside in the system RAM and microprocessor development systems, which are intended for use in developing new programs, are often provided with extensive sections of RAM memory. Consequently the addresses in RAM where a new program under development may be placed are largely a matter of convenience and are rather arbitrary. However it is convenient to use the *first page* of memory when possible, i.e. addresses 0000 to 00FF, in order to make use of the efficient Direct mode of addressing. In the following examples, the chosen memory addresses are in the first page but other than that are arbitrary.

Example   Add together the contents of memory locations 0050 and 0051 and place the result in location 2000.

Memory	Machine code	Mnemonic	Mode
0000	96 50	LDAA	Direct
0002	9B 51	ADDA	Direct
0004	A7 20 00	STAA	Extended
0007	3E	WAI	Inherent

Note that when the programme segment is run by the processor, some instruction must be included at the end of the segment to effectively stop the processor. If this is not done, the processor will attempt to decode the data randomly held in the next memory locations and the desired program operation will not in general be obtained. As the M6800 microprocessor cannot be stopped merely by stopping the system clock, some programming device must be used to place the processor in a short closed programming loop which effectively inhibits the processor operation. One such method is to use the machine code instruction 3E (mnemonic-WAI-wait for interrupt). In practice this difficulty will not arise since most practical programs are non-ending. For example, on completing a washing program the washing machine controller will automatically perform a programming loop, examining the system inputs and waiting to be told the next washing sequence.

In all the examples in this section, the program is ended with the code 3E to perform the *stop execution* function.

Example   Add together the first ten hexadecimal numbers and place the result in location 0000.

Memory	Machine code	Mnemonic	Mode
0000			
0001	4F	CLRA	Inherent
0002	5F	CLRB	Inherent
0003	1B	ABA	Inherent
0004	5C	INCB	Immediate
0005	C1 11	CMPB	Immediate
0007	26 FA	BNE	Relative
0009	97 00	STAA	Direct
000B	3E		Inherent

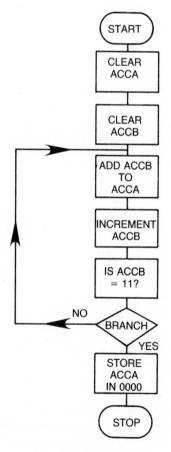

Fig. 16.6 Flow map.

The steps of this program will perhaps be more easily followed by referring to Fig. 16.6 which is a *flow map* for the example. In this example, the ADD and INCREMENT loop must continue until the ACCB has been incremented to 11; i.e. when the contents of ACCB are equal to 11, this is NOT added to the ACCA. To test for this condition, the COMPARE instruction is used. When the two quantities being compared are equal (i.e. ACCB = 11) then the zero flag Z is set to 1. This is so because the COMPARE instruction effectively performs the EXCLUSIVE OR function between the quantities being compared.

The Z flag can therefore be tested using the BNE branch instruction. The branch offset may be calculated as previously explained by substracting the source address of the branch from the desired address. Thus

$$
\begin{array}{r}
0003 \\
- \ 0009 \\
\hline
FFFA
\end{array}
$$

The branch offset is the lowest two hexadecimal characters of the result, i.e. FA.

For other than trivial problems, it is usually much easier if the program is designed in the form of a flow map before the machine code program is attempted.

Example Store the hexadecimal numbers 0 to 20 in the memory locations 0040 to 0060.

Figure 16.7 shows a flow map which represents a possible solution to this exercise. The solution uses the index register as a pointer to the memory location in which the data are to be stored. By incrementing the index register once per loop, the data are stored in successive locations in the memory.

Memory	Machine code	Mnemonic	Mode
0000	4F	CLRA	Inherent
0001	CE 00 40	LDX	Immediate
0004	A7 00	STAA	Indexed
0006	08	INX	Inherent
0007	4C	INCA	Inherent
0008	81 21	CMPA	Immediate
000A	26 F8	BNE	Relative
000C	3E	WAI	Inherent

Branch offset

$$
\begin{array}{r}
0004 \\
- \ 000C \\
\hline
FFF8
\end{array}
$$

Note: The same store location would have been obtained if the index register had been loaded with the base address 0000 and the offset used in the indexed store instruction had been 40.

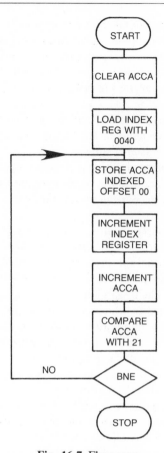

Fig. 16.7 Flow map.

16.6 Entering the program

The microcomputer system will consist of the microprocessor and memory unit, together with peripheral equipment to allow programs to be entered into the memory. The peripheral equipment will consist of either a teletypewriter (TTY) or a keyboard and visual display unit (VDU). The TTY consists of:

(1) A keyboard, similar to a typewriter keyboard, which produces individual ASCII coded alphanumerical characters at a rate of 10 per second for transmission serially to the CPU.
(2) A printer which accepts similarly coded alphanumerical characters from the CPU.
(3) Optionally, a 10 character per second paper tape reader.
(4) Optionally, a 10 character per second paper tape punch.

The microprocessor, in its unprogrammed state, cannot understand even the input signals from the TTY. In order to use the TTY as an input/output peripheral, a program must be written and entered into the memory, usually in the form of prepogrammed ROM devices, which will include the peripheral handler programs. Such a program is referred to as the system *monitor*, and it will also usually include routines which will allow the user to perform, via the TTY peripheral, several important functions:

(1) Examine the contents of any memory locations.
(2) Change the contents of any RAM memory locations.
(3) Examine a register contents.
(4) Load a program via the paper tape reader.
(5) Dump a program to the paper tape punch.
(6) Run a program starting in any address.

Using the memory examine and memory change routines, the machine code program, once developed, can be entered and run in the microprocessor system. It is unusual for any but trivial programs to run correctly first time; in general the development of a working program is a process of repeated trial and correction. The process of fault finding an incorrect program is helped if it can be determined at which point in the program the operation departs from the expected. For this reason, facilities are often incorporated in the system monitor to insert *breakpoint*, i.e. to insert stops temporarily in the program to allow small portions of the program to be run individually. Alternatively, *trace* facilities can allow the program to be run normally, but with each CPU register contents to be printed upon the system output consol after the completion of every program step. Using a TTY for this purpose can be tedious, due to the slow rate of printing and data interchange. The use of a separate keyboard and VDU, however, will allow much faster operation but, in turn, suffers the disadvantage of not providing a *hard copy* print-out, i.e. a print-out on paper. All fault finding, i.e. program debug, must therefore involve the use of the VDU screen.

16.7   Assembly language programs

The use of machine code for programming is tedious, if the program is not short, for several reasons:

(1) For a program with several loops, the calculation of branch offsets becomes repetitive.
(2) Any error which, when corrected, requires the insertion of extra program bytes will require changes to the branch offsets.
(3) The machine code instructions are not easily recognizable; the machine code program is not easily readable.

For the majority of programming applications, therefore, a higher level language is chosen, and in most cases the language used is the *assembly* language. This is a language based upon the mnemonics already introduced. It is thus more akin to an English language program and includes facilities for introducing explanatory comments, etc., to make the purpose of each program segment clear.

A program written in Assembly language is termed a source program. The process does, however, require the use of another computer program, which may be included in the system monitor, which will allow the microcomputer to accept the source program via the system input keyboard and from it produce the machine code or object program which is always the program which finally runs in the machine.

The *source program* consists of a sequence of *statements*, successive statements being separated by the carriage return (CR) character. A source statement contains from one to four fields. These are

LABEL   OPERATOR   OPERAND   COMMENT

The successive fields in a statement are separated by either one or more space (SP) characters or a horizontal (TAB).

An entry in the *label field* is used as a name by which to refer to a particular address in the memory, or occasionally refer to a particular value of data. The name or label can, for example, be used as the branch address of a branch instruction and the actual branch offset byte will be calculated automatically during the assembly process, i.e. when the source program is being *assembled* into machine code.

The following rules apply to labels:

- A label consists of from one to six alphanumeric characters, the first character must be alphabetic. The label must begin in the first character position of a statement.
- A label must not consist of a single character A, B or X as these reserved names refer to the accumulators or the index register.
- All labels within a program must be unique.
- Two special rules should be noted:
  A space (SP) in the first character position indicates that no label is present in the label field.
  An asterisk (*) in the first character position indicates that the entire statement is a *comment*, i.e. is a comment included to make the purpose of the program more clear. A comment will be ignored by the assembler during the assembly process.

The *operator field* consists of the mnemonic found in the order code, e.g. LDAA, CMPA, STAB, etc., in Fig. 16.5.

The kind of information placed in the *operand field* depends on the particular mnemonic operator and upon the addressing mode required. The assembler will recognize numbers, symbols and expressions in the operand field. Thus, for example, for the M6800/6802:

Numbers are accepted by the assembler in the following formats:

Operand	Assembler assumes	Example
Number	Decimal	584
$ Number	Hexadecimal	$8E
Number H	Hexadecimal	4FH
@ Number	Octal	@37
Number 0	Octal	370
Number Q	Octal	37Q
% Number	Binary	%1101
Number B	Binary	1101B

where Number is a positive integer.

Symbols are accepted by the assembler with the same rules as for labels.

An *expression* is a combination of symbols and/or numbers separated by one of the arithmetic operators +, −, *, or /. The assembler evaluates the expressions algebraically from left to right. The rules of precedence, for example, of multiply over addition, which apply in normal arithmetic, do not apply. The addressing mode chosen by the assembler for any instruction depends upon the entry in the operand field, according to the following table:

Mode	*Operand field*
Immediate	# Number
	# Symbol
	# Expression
Relative	Label
Indexed	X
	Number, X
	Symbol, X
	Expression, X
Direct or	Number
Extended	Symbol
	Expression

In choosing between the direct and extended modes, the assembler will select the direct mode if the address falls within the memory *first page*, i.e. between 0000 and 00FF.

The fourth field, the *comment field*, is optional and allows the programmer to include any comments which may make the program more understandable to a reader. The comment will be ignored by the assembler during the production of the machine code object program, but would be listed as part of the source program.

In addition to the statements which will be assembled to produce machine code, facilities exist for including lines which give information to the assembler as to how the machine code object program is to be produced; these are called *assembler directives*. Assembler directives, which are placed in the operator field are used in the organization of the format of the assembler output, for equating numbers to labels, and for reserving blocks of memory, for example, for use in data tables. Some of the allowable directives are:

ORG – Origin. Defines the numerical memory address for the program steps following the ORG directive.

EQU – Equates a symbol to a numerical value, another symbol or to an expression.

FCB – Form constant byte. Locates constants in particular memory locations.

FCC – Form constant characters. Translates strings of characters into ASCII code (see Table 16.3).

FDB – Form double constant byte. Stores the 16 bit binary number from the operand in two successive memory locations.

RMB – Reserve memory bytes. Reserves a block of memory for future use.

END – Indicates the end of the source program.

NAM – Name. Gives the program name. This must be the first statement in the source program.

SOURCE PROGRAM

```
NAM STORE HEX NUMBERS
ORG 0
BEGIN CLRA
LDX #0 INITIALISE INDEX REGISTER
MORE STAA $40,X
INX INCREMENT INDEX REGISTER
INCA INCREMENT ACCA
CMPA #$21 ARE ALL 20 NUMBERS STORED?
BNE MORE STORE NEXT NUMBER
WAI STOP PROCESSING
END
```

ASSEMBLER OUTPUT

Line No.	Addr-ess.	Machine Code	Label	Oper-ator.	Oper-and.	Comment
00001				NAM	STORE	HEX NUMBERS
00002	0000			ORG	0	
00003	0000	4F	BEGIN	CLR A		
00004	0001	CE 0000		LDX	#0	INITIALISE INDEX REGISTER
00005	0004	A7 40	MORE	STA A	$40,X	
00006	0006	08		INX		INCREMENT INDEX REGISTER
00007	0007	4C		INC A		INCREMENT ACCA
00008	0008	81 21		CMP A	#$21	ARE ALL 20 NUMBERS STORED?
00009	000A	26 F8		BNE	MORE	STORE NEXT NUMBER
00010	0000	3E		WAI		STOP PROCESSING
00011		0000		END		

TOTAL ERRORS 00000

Fig. 16.8 Example.

Example   Repeat, using assembler language, the program to store the hexadecimal numbers 0 to 20 in the memory locations 0040 to 0060. The source program and the assembled program are shown in Fig. 16.8.

16.8   Input/output programming

The programming exercises used so far involve the manipulation of data completely within the microprocessor system, i.e. no input signals are required (other than the program itself) during the running of the program and no output signals are produced. The microprocessor system is being used in a manner analogous to a desk top calculator in the sense that no interaction with the outside world, other than with the operator, is involved. The main application of microprocessors is in the real time control of external devices or processes and this requires the ability at various states of the program to output control signals to external equipment or to read into the microprocessor control signals generated by external equipment.

Such input/output data transfers are arranged via an additional LSI device which can be added to the microprocessor system and which shares the same data, address and control buses with the microprocessor. A popular device, marketed by Motorola Semiconductors Ltd, for use with the 6800 and other microprocessors is known as a peripheral interface adaptor (PIA) (MC6820 or MC6821). The PIA is a general purpose input/output device and provides 16 input/output connections, configured as two 8 bit input/output ports. Each individual connection of the two ports may be used separately as an input or as an output. Together with the two ports, four other connections are provided which are intended for use in controlling the input/output data transfers.

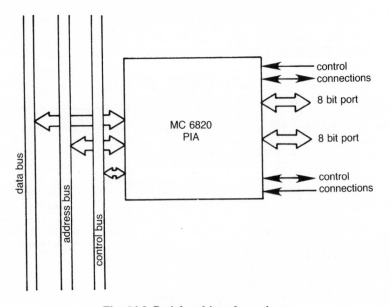

Fig. 16.9 Peripheral interface adapter.

The PIA is packaged as a 40 pin DIL chip. All inputs and outputs are designed to be TTL device compatible which makes interfacing the microprocessor to external logic relatively easy. The PIA is represented in Fig. 16.9. From the point of view of the external connections, the PIA divides into two almost identical parts, each containing one input/output port and two control connections. The two sections can be differentiated by the letters A and B. Thus the A and B input/output ports are basically similar in operation; the slight differences which do exist are in their electrical characteristics and in the operation of the control connections. For a first understanding therefore, the two sections A and B may be considered identical, and the following discussion of the A section applies equally well to the B section.

16.8.1 Data transfers via the ports A and B

Associated with each port there are three 8 bit registers internal to the PIA. These are illustrated for port A in Fig. 16.10. The equivalent port B registers are similar.

The PIA is intended to be used as a memory-mapped device, i.e. the chip appears to the microprocessor as though it was four memory locations. The microprocessor, in fact, has no way of knowing that anything other than memory locations exists. Two of memory locations are assigned to the input/output registers to which the external connections are made. These registers are normally referred to as output register A – (mnemonic (ORA)) and output register B – (mnemonic (ORB)) even though they can be used for data input as well as data output. The names and mnemonics normally used for the remaining registers are control register A (CRA), control register B (CRB), data direction register A (DDRA) and data direction register B (DDRB).

By suitable arrangement of the connection of the address bus to the PIA, any of the 64K allowable addresses may be assigned to the PIA registers. A typical memory assignment is as follows:

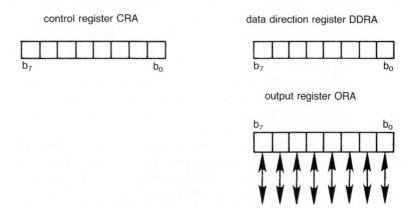

Fig. 16.10 PIA port A registers.

Register		Assigned address
Output register A	ORA	$8008
Data direction register A	DDRA	$8008
Control register A	CRA	$8009
Output register	ORB	$800A
Data direction register B	DDRB	$800A
Control register B	CRB	$800B

The output registers, i.e. the data input/output connections, can be used directly for data transfers, as follows:

To input data to the microprocessor, use the program step

    LDAA   ORA

Either accumulator, of course, may be used and the data transfer is made effectively by the microprocessor reading the contents of the ORA memory location.

To output data from the microprocessor, use the program step

    LDAA   #DATA
    STAA   ORA

Again either accumulator can be used. The data are loaded first into the accumulator and the accumulator contents are stored in the ORA memory location. The data transfers are thus very easily arranged.

One small complication however is introduced by the need to *initialize* the PIA prior to using it in this way. This is necessary because the individual register bits must be set up first as either inputs or outputs. In other words, the system must be told which output register bits are to be inputs and which outputs, prior to being used. Any bit in either output register may be used as an input or as an output.

The specification of an output register bit as an input or output is arranged by placing a pattern of bits in the data direction register. Thus if an 0 is placed in a bit in the DDRA, then the equivalent bit of the ORA is configured as an input. Conversely, if a 1 is placed in a bit in the DDRA, the equivalent bit of the ORA is configured as an output. As an example, the pattern hexadecimal 0F ($0F) placed in the DDRA would configure bits 0 to 3 of the ORA as outputs and bits 4 to 7 as inputs.

Refer again to the table of register assigned addresses. You will note that the data direction register and the output register in each section of the PIA share the same address. How can two registers be used separately at the same address? Obviously they cannot without some extra switch to connect the desired register to the address connections. The switch used is bit 2 of the control register. With bit 2 of the CRA set to 0, the address $8008 connects to the data direction register DDRA. With bit 2 of the CRA set to 1, the address $8008 connects to the output register ORA. The remaining bits of the control register are used in interrupt programs, which will not be dealt with here. These bits should therefore be set always to 0.

An initialization segment, to configure ORA for four inputs (bits 4 to 7) and four outputs (bits 0 to 3) is as follows:

```
* INITIALIZATION SEGMENT
        CLR     CRA              Swtiches address to DDRA
        LDAA    #$0F
        STAA    DDRA             Configures 4 inputs, 4 outputs
        LDAA    #$04
        STAA    CRA              Switches address to ORA
* END OF INITIALIZATION SEGMENT
```

Example   Write a program to generate a square waveform at each bit of ORA. Estimate the period of the square waveform.

A square wave voltage can be generated by alternately outputting from the ORA logic 0 levels and logic 1 levels. A delay would be required between the changes of output level in order to determine the square wave periods. The solution can initially be represented as a flow map as, in Fig. 16.11.

One circuit of the main program loop produces one half period of the waveform, with the data output from ACCA being changed in successive half

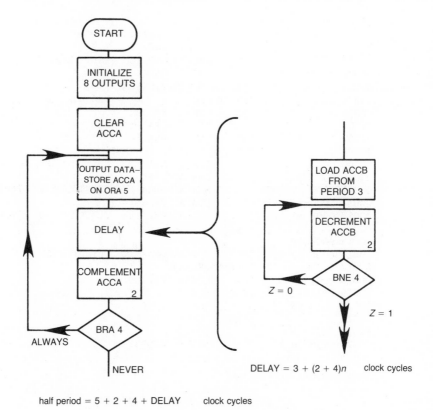

Fig. 16.11 Flow map for square wave generator.

SOURCE PROGRAM
```
 NAM SQUARE WAVE GENERATOR
 ORG O
*SYSTEM DEFINITIONS
PERIOD RMB 1 RESERVE ONE BYTE AS BUFFER
ORA EQU $8008
DDRA EQU ORA
CRA EQU ORA+1
*INITIALISATION SEGMENT
 CLR CRA SWITCH $8008 TO DDRA
 LDAA #$FF
 STAA DDRA CONFIGURE 8 OUTPUTS
 LDAA #$04
 STAA CRA SWITCH $8008 TO ORA
*MAIN PROGRAM
 CLRA
STORE STAA ORA OUTPUT DATA
*DELAY SEGMENT
 LDAB PERIOD
LOOP DECB
 BNE LOOP
*END OF DELAY SEGMENT
 COMA  COMPLEMENT ACCA CONTENTS
 BRA STORE
 END
```

ASSEMBLER OUTPUT

Line No.	Addr-ess.	Machine Code	Label	Oper-ator.	Oper-and.	Comment
00001				NAM	SQUARE	WAVE GENERATOR
00002	0000			ORG	O	
00003				*SYSTEM DEFINITIONS		
00004	0000	0001	PERIOD	RMB	1	RESERVE ONE BYTE AS BUFFER
00005		8008	ORA	EQU	$8008	
00006		8008	DDRA	EQU	ORA	
00007		8009	CRA	EQU	ORA+1	
00008				*INITIALISATION SEGMENT		
00009	0001	7F 8009		CLR	CRA	SWITCH $8008 TO DDRA
00010	0004	86 FF		LDA A	#$FF	
00011	0006	B7 8008		STA A	DDRA	CONFIGURE 8 OUTPUTS
00012	0009	86 04		LDA A	#$04	
00013	000B	B7 8009		STA A	CRA	SWITCH $8008 TO ORA
00014				*MAIN PROGRAM		
00015	000E	4F		CLR A		
00016	000F	B7 8008	STORE	STA A	ORA	OUTPUT DATA
00017				*DELAY SEGMENT		
00018	0012	D6 00		LDA B	PERIOD	
00019	0014	5A	LOOP	DEC B		
00020	0015	26 FD		BNE	LOOP	
00021				*END OF DELAY SEGMENT		
00022	0017	43		COM A		COMPLEMENT ACCA CONTENTS
00023	0018	20 F5		BRA	STORE	
00024	0000			END		

TOTAL ERRORS 00000

Fig. 16.12 Example.

periods by the complement ACCA instruction. The box marked DELAY is itself a small segment of program which is shown separately in the figure. The delay is obtained by decrementing ACCB down to zero, each decrementing loop taking six clock cycles. Note that the time of operation of each code is given in the order code table in Fig. 16.5 in the columns headed ~. As most M6800 systems operate with a 1 MHz system clock, then one clock cycle is of 1 µs duration.

The delay counter, ACCB, is set to an initial value by loading the accumulator from the RAM buffer location assigned the label PERIOD. The square wave period can thus be changed by changing the contents of the buffer location. Note also that the maximum value of delay will be obtained when the value inserted in PERIOD is 00. You should verify that the maximum half period is then equal to 1.55 ms.

The source program and the assembly listing for this example are shown in Fig. 16.12.

16.9 Interrupt sources

In most practical systems, arrangements are made whereby the microprocessor can perform not only one, but perhaps several different control tasks at the same time. At any instant in time, however, the microprocessor will be concerned in performing only one of its available operation codes and must therefore, at that instant, be involved in dealing only with the program which includes that operation code; in other words the microprocessor can only deal with one particular program at any one instant in time. Arrangements can be made, however, whereby the microprocessor operation of a particular program may be interrupted to allow a section of a different program to be performed. In this way, a single microprocessor may be able to perform satisfactorily several perhaps quite unrelated control tasks at the same time.

Consider an industrial application in which 8 bit parallel data samples are to be read into a microprocessor system from a data transducer. The data samples could perhaps relate to the speed of a particular process. Data samples are required at fixed time intervals, say every one second. The transducer would be connected to the input port of a PIA via suitable signal processing circuits to produce the correct TTL switching voltage levels. Each data sample is read from the output register of the PIA and stored in a memory location. A flow map for a suitable program is shown in Fig. 16.13.

Each circuit of the loop shown will result in one data sample being read into storage in memory. The total loop execution time must therefore be one second in this application. The time of execution of the loop codes, as shown in the figure, totals 24 µs, leaving a total of 999 976 µs required in the padding delay to ensure an overall loop time of one second. The delay would be generated, as in the previous example, by loading and decrementing to zero the necessary number of registers. The microprocessor, used in this manner, obviously spends the majority of its time in a waiting state, waiting for the delay to end.

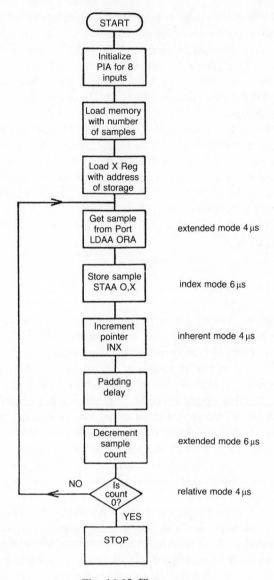

Fig. 16.13 Flow map.

The use of interrupt programming will allow the time wasted in the delay to be put to use in some other way. For example, it may be that analysis of the data samples is required, to produce the results of the process test. This can be arranged by installing an external timer circuit, which is set to produce a timing pulse every one second, and which can be connected so as to interrupt the microprocessor and transfer program execution from its current task to an *interrupt program*, which can be written to read in one data sample. At the conclusion of the interrupt program, execution is transferred back to the *background program*. The background program could then be the data analysis program. Obviously the background task must be, by its very nature, a program which is not time critical, since its execution will be suspended every time an interrupt program is called.

Further tasks, which could be included in this example and which could also be incorporated by using interrupt methods, would be

(1) Data transfer to a printer to output results as hard copy.
(2) Continuous checking of the power supply to the process, so that a stand-by power supply could be switched on, in case of main supply failure.

An interrupt input is, therefore, essentially an asynchronous input signal, which causes the microprocessor to suspend operation of its current program in order to execute a separate interrupt program. The input is asynchronous because, as the main and interrupt programs in general concern completely unrelated tasks, it is not possible to predict when the interrupt signal will occur. After completion of the interrupt program, the microprocessor will return to its original program and continue its execution, at the exact point in that program at which it was interrupted.

In the example above, greatest interrupt priority would be given to the stand-by power system, followed presumably by the data sampling system, with the printer control system having least interrupt priority. The power stand-by system would be allowed to interrupt the data sampling system, but not vice versa. Both of these systems would be able to interrupt the printer control and all three would be allowed to interrupt the background data analysis program. As can be seen by this example, a practical system must include provision for establishing the priority of the various interrupt input sources.

16.9.1   *Motorola M6800 interrupt handling*

Three of the pin connections of the 40 pin, M6800 microprocessor chip are used to allow interrupt signals to cause a system interrupt. These are the pins labelled $\overline{\text{IRQ}}$, $\overline{\text{NMI}}$ and $\overline{\text{RES}}$. These are shown in Fig. 16.17 and, in line with common practice, the overlining bar on top of the letters indicates that the input is an active low level input. By this it is indicated that, provided each pin is held at its logic HIGH level state (nominally +5 V), the pin will not be activated. However, a logic LOW level input to any pin will cause the input to become active, and an interrupt will be caused. Consequently, in normal use, such input pins are always connected to the +5 V power supply, via a resistor, to ensure that, unless actually driven to the LOW state by the signal line, the input is held in the HIGH state and thus a spurious interrupt may not be caused by an interference source in the vicinity.

The pin labels are as follows:

(1) $\overline{\text{IRQ}}$ This stands for INTERRUPT REQUEST and this pin is the general purpose interrupt request input.
(2) $\overline{\text{NMI}}$ This stands for NON MASKABLE INTERRUPT REQUEST and is

a further interrupt request input which has a higher priority than the IRQ interrupt.

(3) $\overline{\text{RES}}$ This stands for RESET and is a special purpose interrupt input which is used to bring the microprocessor under control when it is first switched on or, alternatively, after its current program execution is aborted.

When any of these three signal input pins is driven to the active LOW state, an interrupt will be requested, and the sequence of events which will then occur is fixed by the design of the M6800 microprocessor device itself. The sequences can best be described by considering first the general purpose input, $\overline{\text{IRQ}}$. The resulting interrupt sequence can be shown in the form of a flow map, Fig. 16.14.

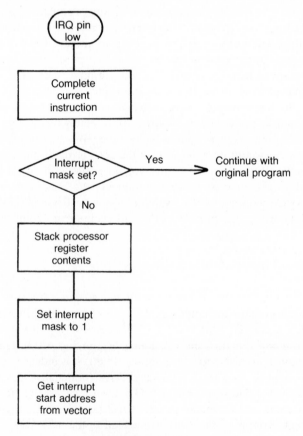

Fig. 16.14 $\overline{\text{IRQ}}$ interrupt sequence.

It must be emphasized that this flow map does not represent a program to be written by the user, but represents a sequence which is inherent in the design of the microprocessor chip itself, and this sequence is brought about solely by driving the $\overline{\text{IRQ}}$ pin to the LOW state. The sequence may be described as follows:

(1) The $\overline{IRQ}$ pin is driven LOW.

(2) The interrupt request may occur at any time and, in general, the processor will be only part of the way through the execution of its current operation code. The current code instruction is always fully completed before the interrupt request is serviced, as represented by the first box in the flow map. Consequently there will be a delay in the servicing of the request which may be several microseconds in duration, depending upon which operation code is current.

(3) When the current instruction has been completed, the interrupt mask (I) is examined. This special purpose flag was mentioned in Section 16.5.3 and is bit 4 of the condition code register. If this flag is SET to 1, then further servicing of the interrupt request will be prevented and the microprocessor will return to the next operation code in the program it was currently operating. If, however, the interrupt mask (I) is CLEAR (i.e. I = 0) then servicing of the interrupt is allowed. The mask I may be SET or CLEARED under program control by the programmer using the instructions SEI (set interrupt mask) or CLI (clear interrupt mask). At certain times the mask I may be SET or CLEARED by the microprocessor itself.

(4) If the mask (I) is CLEAR, then the interrupt request will be accepted. The processor will cease operation of its original program and will transfer to an interrupt program. However, in order that operation of the original program can be recommenced when the interrupt program has been completed, all data held in the microprocessor internal registers regarding the state of the program when the interrupt occurred must be stored for later use. In the M6800 microprocessor, this temporary data storage is accomplished by transferring the current values of the internal registers into an area of the main system memory. This area of memory is termed the *stack*, and the position of the stack is determined by entering the highest address of the chosen stack area into a special register, the *stack pointer*. Obviously the stack must be located in a section of memory in which RAM read/write memory devices have been configured. The stack pointer, therefore, always holds the address of the highest unused memory location in the stack area; data storage is accomplished by transferring the data byte to the memory location pointed to by the stack pointer, following which the stack pointer is automatically decremented to point to the next lower stack address, which then automatically becomes the highest unused address.

Figure 16.15 shows a section of memory used as the stack, both before and after stacking all the internal register values. The order of stacking is always the same, commencing with the program counter (LOW BYTE) and finishing with the condition code register. When required, the original values may be returned to the processor internal registers by reversing the stacking process; each register is restored by incrementing the stack pointer and transferring the contents of the memory location pointed to into the register.

A complete stack of all register values uses seven stack area locations.

HIGH ADDRESS

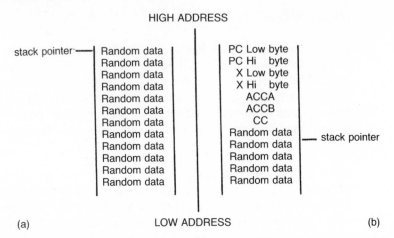

(a)   LOW ADDRESS   (b)

Fig. 16.15 Section of main memory used as a stack, (a) before stacking, (b) after stacking.

(5) Following the stacking process, the interrupt mask (bit 4 of the condition code register) is set to 1. This protects the interrupt process against further interrupts, although these may be allowed if the programmer wishes by subsequently clearing the interrupt mask to 0 again in the interrupt program, using the CLI (clear interrupt mask) instruction.

(6) All that remains in the interrupt sequence is to load the program counter with the start address of the desired interrupt program, when execution of the interrupt program will commence. The required start address is obtained from two memory locations, known as the *interrupt vector*, where it must be entered by the programmer when the interrupt program is written. The $\overline{IRQ}$ interrupt vector addresses are $FFF8 and $FFF9. The start address is thus obtained by automatically transferring the contents of location $FFF8 into the high byte of the program counter and the contents of $FFF9 into the low byte of the program counter.

Execution of the interrupt program will now commence and any of the processor registers may be used, secure in the knowledge that their previous contents are safely stored in the system stack.

Operation of the interrupt program will end when the operation code RTI (return from interrupt) is reached. At this point the processor will automatically return the data from the seven locations in the stack, in sequence, into the correct internal registers, so that the processor is then in exactly the same state as it was when the interrupt was requested. Execution of the original program will then recommence.

It should be noted that if the interrupt mask is cleared by the programmer while the interrupt program is being executed, then it may itself be interrupted by a further interrupt request. Again the registers will be saved on the system stack; each RTI instruction as it is encountered will then cause a transfer of seven bytes of data from the stack to the registers. The system is thus fully automatic in operation and will cope fully with interrupt requests which are *nested* in this way. Each level of stacking will, however, require seven bytes of memory in the stack area.

The sequence which occurs when the interrupt pin, $\overline{\text{NMI}}$, is driven LOW is very similar. This interrupt input is intended as a high priority input. It differs from the sequence already described only in that the interrupt mask, I, is not examined and an interrupt will be accepted via this input pin irrespective of the state of the I bit in the condition code register. The vector addresses for the $\overline{\text{NMI}}$ interrupt input are also different, being \$FFFC and \$FFFD. This allows a completely separate interrupt program to be accessed.

The third interrupt input pin, $\overline{\text{RES}}$, is used to bring the processor under program control after the power is switched on to the microprocessor or, alternatively, to bring the processor under control after aborting the operation of a faulty program. Consequently, after accepting an $\overline{\text{RES}}$ reset input, the register values are not saved on the system stack since, by the very nature of this input, no data of any value will be held there. Further, the $\overline{\text{RES}}$ reset vector addresses, \$FFFE and \$FFFF, will contain the start address of the main program, the background program in the example we are considering.

There is one further way in which the processor can be forced to go through an interrupt sequence. This is by including in a program the program instruction SWI (software interrupt) which is the machine code \$3F. As we have defined an interrupt input as an asynchronous input signal from an external source which occurs at an unpredictable time, then an instruction which is included in a program to cause an interrupt sequence can, in no way, really be an interrupt. However, it is useful, in some circumstances, to be able to force an interrupt sequence to occur, and the inclusion of the SWI instruction will cause a non-maskable interrupt program to be brought from the vector addresses \$FFFA and \$FFFB. Perhaps the most common use of this instruction is when its interrupt program causes the processor to print, on the system VDU, the current data values of the processor registers and then to wait for a further instruction to proceed from the system keyboard. The substitution of the machine code \$3F for any other machine code in a program will thus cause the state of the processor registers to be printed on the system VDU when it is

FFFF FFFE	$\overline{\text{RES}}$	Interrupt Program start address
FFFD FFFC	$\overline{\text{NMI}}$	Interrupt Program start address
FFFB FFFA	SWI	Interrupt Program start address
FFF9 FFF8	$\overline{\text{IRQ}}$	Interrupt Program start address

Fig. 16.16 Interrupt vector addresses.

encountered; this procedure can be of great value when attempting to debug a faulty program, i.e. to trace program faults. It can be seen that the interrupt vector addresses occupy the eight highest memory locations in the system memory. The addresses are summarized in Fig. 16.16.

16.9.2 Connecting interrupt signals via the PIA

In Motorola microprocessor systems, signals from external interrupting sources are commonly connected via an M6821 peripheral interface adapter (PIA). Figure 16.17 shows a 6800 microprocessor system with, as described in Section 16.8, a PIA providing two 8 bit input/output ports, together with four control connections CA1, CA2, CB1 and CB2. These four connections are used to control interrupt operation of the microprocessor. It can be seen that, in addition to the data, address and control buses, two connections are brought out of the PIA, $\overline{\text{IRQA}}$ and $\overline{\text{IRQB}}$ which pass interrupt requests to the processor, both being connected to the IRQ pin, so that an interrupt request from CA1, CA2 or CB1, CB2 will effectively drive the $\overline{\text{IRQ}}$ pin of the processor to the active LOW state.

When considering the initialization of a PIA in Section 16.8.1, only bit 2 of the control register (CRA or CRB) was used, this being the software switch which enabled data transfers, either to the output register or the data direction register. All other bits in the control register are used in interrupt programming, and were cleared to 0 in order to disable interrupt inputs. In order, therefore, to use interrupt programs, the PIA initialization segment must be altered; in effect, the final initialization step which stores $04 data into the

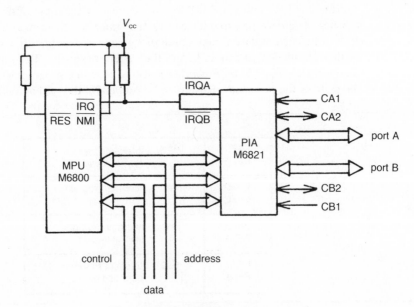

Fig. 16.17 PIA with interrupt connections.

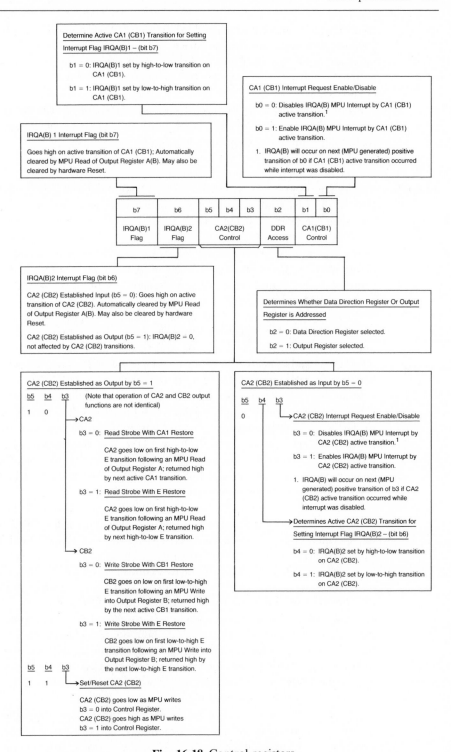

Fig. 16.18 Control registers.

control register must be changed. Since the PIA is intended to be a general purpose peripheral interfacing device, it is arranged so that it can cope with interrupt input signals from a variety of sources. Its input signals from a peripheral must be of TTL levels, but may be of positive or negative going polarity.

Figure 16.18 describes the use of the various bits of the control registers in configuring correctly the peripheral interface. The use of the interface is perhaps best understood by consideration of specific examples.

Example   Modity the program described in Fig. 16.12, so that a TTL level pulse input to CA1 of the PIA would interrupt the square wave generator and cause a step increase in the period of the square wave.

In the square wave generator of Fig. 16.12, the period of the square wave was determined by the contents of a memory location labelled PERIOD, the contents being transferred into accumulator B before being decremented to zero, to form the delay. As required in this example, the period may be increased by adding a fixed quantity to the contents of the buffer location PERIOD. Consequently, it should be arranged that on receipt of an interrupting pulse at input CA1, the square wave generator program should be interrupted and execution transferred to the interrupt program which would add the required incremental value to the contents of PERIOD.

From Fig. 16.18, we can determine that the input CA1 of the PIA is configured using bits 0, 1 and 7 of the control register CRA. The control inputs to the PIA are edge sensitive, i.e. the interrupt can be activated by either a LOW to HIGH transition or a HIGH to LOW transition between TTL voltage levels as determined by the state of bit 1 of the control register. Thus, if the interrupting source produces a LOW to HIGH edge, then bit 1 of the CRA must be SET during the initialization to 1.

In order to remember that an interrupt has been requested, bit 7 of the control register is used as a latch or interrupt flag. This will be SET to 1 when the specified edge occurs at the CA1 input. Bit 7 is a READ ONLY bit and cannot be SET to 1 in any other manner.

A further facility which is provided is controlled by bit 0 of the CRA. This bit controls an enable, use of which will allow the interrupt call to be held up (bit 0 = 0) or passed immediately to the processor $\overline{\text{IRQ}}$ pin (bit 0 = 1). Bits 3, 4, 5 and 6 of the CRA, whose use has not yet been described, should be left in the CLEARED state. Consideration of these points will show that, to initialize the PIA to accept interrupt calls immediately via the input CA1, from a LOW to HIGH interrupting edge, then the data to be loaded into the control register during the initialization segment will change from the data value $04 (Fig. 16.12) to $07 for this new example.

Figure 16.19(a) shows the source program and Fig. 16.19(b) the resulting assembler output program for this example. It can be seen that the interrupt program has been included at the end of the original program immediately following the instruction

    BRA   STORE

As the main program will always branch to the address STORE, the interrupt program cannot be entered other than by causing an interrupt input as described. Then, following the normal interrupt procedure, the interrupt program will be started via the start address which has been entered into the vector $FFF8 and $FFF9. The interrupt program, for this example, consists merely of the addition of an arbitrary constant $10 to the value held in the buffer PERIOD which is used to set the length of the delay. However, also included in the interrupt program is an instruction

    LDAA   ORA

i.e. a read to the accumulator of the output register ORA. This is explained as follows.

```
        NAM SQUARE WAVE GENERATOR WITH INTERRUPT
        ORG 0
       *SYSTEM DEFINITIONS
        PERIOD RMB I RESERVE ONE BYTE AS BUFFER
        ORA EQU $8008
        DDRA EQU ORA
        CRA EQU ORA+I
       *INITIALISATION SEGMENT
        CLR CRA SWITCH $8008 TO DDRA
        LDAA #$FF
        STAA DDRA CONFIGURE 8 OUTPUTS
        LDAA #$07
        STAA CRA SWITCH $8008 TO ORA, ALLOW CAI INTS
       *MAIN PROGRAM
        CLI CLEAR INTERRUPT MASK
        CLRA   INITIAL DATA
        STORE STAA ORA OUTPUT DATA
       *DELAY SEGMENT
        LDAB PERIOD INITIALISE DELAY COUNTER
        LOOP DECB
        BNE LOOP
       *END OF DELAY SEGMENT
        COMA COMPLEMENT DATA
        BRA STORE
       *INTERRUPT PROGRAM
        INT LDAA PERIOD GET DELAY VALUE
        ADDA #$10 ADD AN ARBITRARY INCREMENT
        STAA PERIOD RETURN VALUE TO BUFFER
        LDAA ORA DUMMY READ TO CLEAR LATCH
        RTI
       *END OF INTERRUPT PROGRAM
        ORG $FFFE
        FDB INT
        END
   (a)
```

Fig. 16.19 (a) Source code, (b) assembler output (overleaf).

```
                        NAM              SQUARE WAVE GENERATOR WITH INTERRUPT
          0000          ORG   O
                   *SYSTEM DEFINITIONS
0000 0001     PERIOD RMB   1             RESERVE ONE BYTE AS BUFFER
     8008     ORA    EQU   $8008
     8008     DDRA   EQU   ORA
     8009     CRA    EQU   ORA+1
                   *INITIALISATION SEGMENT
0001 7F8009          CLR   CRA           SWITCH $8008 TO DDRA
0004 86FF            LDAA  #$FF
0006 B78008          STAA  DDRA          CONFIGURE 8 OUTPUTS
0009 8607            LDAA  #$07
000B B78009          STAA  CRA           SWITCH $8008 TO ORA, ALLOW CA1 INTS
                   *MAIN PROGRAM
000E 0E              CLI                 CLEAR INTERRUPT MASK
000F 4F              CLRA                INITIAL DATA
0010 B78008   STORE  STAA  ORA           OUTPUT DATA
                   *DELAY  SEGMENT
0013 D600            LDAB  PERIOD  INITIALISE DELAY COUNTER
0015 5A       LOOP   DECB
0016 26FD            BNE   LOOP
                   *END OF DELAY SEGMENT
0018 43              COMA                COMPLEMENT DATA
0019 20F5            BRA   STORE
                   *INTERRUPT PROGRAM
001B 9600     INT    LDAA  PERIOD  GET DELAY VALUE
001D 8B10            ADDA  #$10    ADD AN ARBITRARY INCREMENT
001F 9700            STAA  PERIOD  RETURN VALUE TO BUFFER
0021 B68008          LDAA  ORA     DUMMY READ TO CLEAR LATCH
0024 3B              RTI
                   *END OF INTERRUPT PROGRAM
     FFFE            ORG   $FFFE
FFFE 001B            FDB   INT
                     END

***** TOTAL ERRORS    O
```

(b)

Fig. 16.19. *Continued*

When the interrupt input occurs, bit 7 of the control register CRA is SET to 1 as a latch. The interrupt request is then passed to the processor and the interrupt sequence will be operated. At the end of the interrupt program, when the instruction RTI is encountered, operation will be returned to the main program. However, unless the interrupt latch, b of the control register, has been cleared to 0, the interrupt request will be started again and the processor will never return correctly to the main program. The latch bit is a READ ONLY bit and cannot be SET or CLEARED by a STORE instruction. The bit will be SET only by an interrupt input and can only be CLEARED under program control by a read of the output register. Consequently, a dummy read is included in the interrupt program, merely to reset the interrupt latch.

A label INT has been included to mark the start of the interrupt program.

The interrupt vector can be loaded quite easily with the interrupt start address, as shown in Fig. 16.19, by including an assembler directive

    ORG $FFF8

to move the assembler address to the vector, followed by the directive

    FDB   INT

which places the address INT in the next two memory bytes, $FFF8 and $FFF9, i.e. the assembler will Form a Double Byte of the value of INT in the two memory locations.

Example   Modify the program of Fig. 16.19 so that TTL level pulse inputs to CA1 and CA2 of the PIA would interrupt the square wave generator and respectively cause a step increase or a step decrease in the period of the square wave.

This example requires the handling of interrupt inputs from two sources, the inputs CA1 and CA2. As can be seen from Fig. 16.17, the PIA connection CA2 can be used as an input or as a control signal output, the choice being determined by b5 of the control register. In this example, when CA2 is required as an input, b5 must be initialized as 0. With b5 = 0, the operation of CA2 is identical to that of CA1, with the control register bits b3, b4 and b6 performing the equivalent function to b0, b1 and b7. Thus b4 will set the polarity of the active edge of the interrupt input, b6 will act as a READ ONLY latch and b3 is the enable/disable control. These functions are summarized fully in Fig. 16.18.

We can thus provide two interrupting input sources connected to CA1 and CA2 by initializing the control register as follows:

    b0 = 1   to enable interrupts via CA1
    b1 = 1   to set the CA1 input active edge to LOW to HIGH
    b2 = 1   to switch address $8008 to the output register
    b3 = 1   to enable interrupts via CA2
    b4 = 1   to set the CA2 input active edge to LOW to HIGH
    b5 = 0   to configure CA2 on an input
    b6 and b7 are READ ONLY latches and can be thought of as 0

The required data value for the PIA initialization segment is thus $1F.

Figure 16.20(a) shows the source code and Fig. 16.20(b) the resulting assembler output.

It should be realized that when either interrupt input CA1 or CA2 is driven, the interrupt request is passed to the processor via the $\overline{\text{IRQ}}$ pin. Consequently there is no way for the processor to know which interrupting source has requested service. The interrupt program must therefore examine the interrupt latches in the PIA control registers to determine which latch is SET, and consequently which interrupt has requested service. The interrupt program will then be of branched form so that a different response can occur for different inputs. Figure 16.21 shows a flow map which describes the operation of the interrupt program. The interrupt latches are examined by reading the control register contents into the accumulator. If the interrupting source had been CA1, requiring an increase in the square wave period, then the latch b7 would be set to 1. Consequently the programming step

LDAA   CRA

will cause the CRA contents to be read into accumulator A and b7 being set will cause the negative flag N of the condition code register to be SET. The branching step

BMI   INTCA1

```
        NAM SQUARE WAVE GENERATOR WITH INTERRUPTS
        ORG 0
*SYSTEM DEFINITIONS
PERIOD RMB I RESERVE ONE BYTE AS BUFFER
ORA EQU $8008
DDRA EQU ORA
CRA EQU ORA+I
*INITIALISATION SEGMENT
 CLR CRA SWITCH $8008 TO DDRA
 LDAA #$FF
 STAA DDRA CONFIGURE 8 OUTPUTS
 LDAA #$IF
 STAA CRA SWITCH $8008, ALLOW CAI CA2 INTS
*MAIN PROGRAM
 CLI  CLEAR INTERRUPT MASK
 CLRA  INITIAL DATA
STORE STAA ORA OUTPUT DATA
*DELAY SEGMENT
 LDAB PERIOD INITIALISE DELAY COUNTER
LOOP DECB
 BNE LOOP
*END OF DELAY SEGMENT
 COMA  COMPLEMENT DATA
 BRA STORE
*INTERRUPT PROGRAM
INT LDAA CRA DETERMINE SOURCE OF INTERRUPT
 BMI INTCAI
 LDAA PERIOD GET DELAY VALUE
 SUBA #$IO SUBTRACT ARBITRARY DECREMENT
 STAA PERIOD RETURN VALUE TO BUFFER
 BRA DONE
INTCAI LDAA PERIOD GET DELAY VALUE
 ADDA #$IO ADD AN ARBITRARY INCREMENT
 STAA PERIOD RETURN VALUE TO BUFFER
DONE LDAA ORA DUMMY READ TO CLEAR LATCH
 RTI
*END OF INTERRUPT PROGRAM
 ORG $FFFE
 FDB INT
 END
```

(a)

Fig. 16.20 (a) Source code, (b) assembler output.

```
                            NAM              SQUARE WAVE GENERATOR WITH INTERRUPTS
            0000            ORG     O
                         *SYSTEM DEFINITIONS
   0000    0001  PERIOD   RMB     I           RESERVE ONE BYTE AS BUFFER
           8008  ORA      EQU     $8008
           8008  DDRA     EQU     ORA
           8009  CRA      EQU     ORA+I
                         *INITIALISATION SEGMENT
   0001    7F8009         CLR     CRA         SWITCH $8008 TO DDRA
   0004    86FF           LDAA    #$FF
   0006    B78008         STAA    DDRA        CONFIGURE 8 OUTPUTS
   0009    863F           LDAA    #$IF
   000B    B78009         STAA    CRA         SWITCH $8008, ALLOW CAI CA2 INTS
                         *MAIN PROGRAM
   000E    OE             CLI                 CLEAR INTERRUPT MASK
   000F    4F             CLRA                INITIAL DATA
   0010    B78008 STORE   STAA    ORA         OUTPUT DATA
                         *DELAY SEGMENT
   0013    D600           LDAB    PERIOD   INITIALISE DELAY COUNTER
   0015    5A      LOOP   DECB
   0016    26FD           BNE     LOOP
                         *END OF DELAY SEGMENT
   0018    43             COMA                COMPLEMENT DATA
   0019    20F5           BRA     STORE
                         *INTERRUPT PROGRAM
   001B    B68009 INT     LDAA    CRA         DETERMINE SOURCE OF INTERRUPT
   001E    2B08           BMI     INTCAI
   0020    9600           LDAA    PERIOD   GET DELAY VALUE
   0022    8010           SUBA    #$IO        SUBTRACT ARBITRARY DECREMENT
   0024    9700           STAA    PERIOD   RETURN VALUE TO BUFFER
   0026    2006           BRA     DONE
   0028    9600   INTCAI  LDAA    PERIOD   GET DELAY VALUE
   002A    8BIO           ADDA    #$IO        ADD AN ARBITRARY INCREMENT
   002C    9700           STAA    PERIOD   RETURN VALUE TO BUFFER
   002E    B68008 DONE    LDAA    ORA         DUMMY READ TO CLEAR LATCH
   0031    3B             RTI
                         *END OF INTERRUPT PROGRAM
           FFFE           ORG     $FFFE
   FFFE    001B           FDB     INT
                          END
```

***** TOTAL ERRORS   O

(b)

Fig. 16.20. *Continued*

is controlled by the state of the N flag and will thus branch if the interrupt is via CA1. Otherwise, if the interrupting source was the CA2 input, then latch b6 of the control register would be SET and the BMI branch instruction would not result in a branch in the program. A different result would therefore occur for interrupting inputs from the two sources. In each case, however, a dummy read of the output register must be included to clear the interrupt latches to zero before returning to the main program.

Note that, if further interrupting sources are used, then the inputs CB1 and CB2 can be configured in exactly the same way by using the control register B.

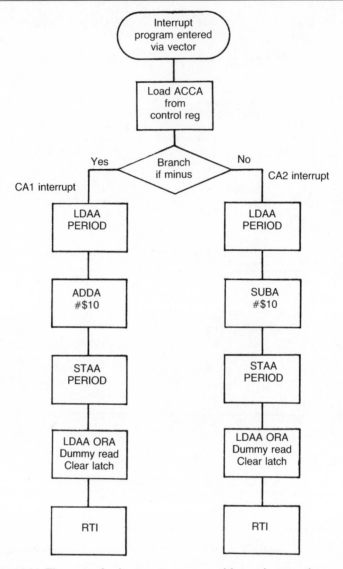

Fig. 16.21 Flow map for interrupt program with two interrupting sources.

Similarly, the interrupting source will be determined by reading successively the four interrupt latches.

16.9.3 Asynchonous handshake data interchange via peripherals

The PIA has been designed to facilitate the interchange of parallel bytes of data between the processor system and a peripheral, such as a printer. A printer will accept bytes of data on eight parallel signal lines, each data byte representing an alphanumeric character for printing, probably coded according

Table 16.3 ASCII code.

		ASCII character set (7-bit code)							
	MS char	0 000	1 001	2 010	3 011	4 100	5 101	6 110	7 111
LS char									
0	0000	NUL	DLE	SP	0	@	P	`	p
1	0001	SOH	DC1	!	1	A	Q	a	q
2	0010	STX	DC2	"	2	B	R	b	r
3	0011	ETX	DC3	#	3	C	S	c	s
4	0100	EOT	DC4	$	4	D	T	d	t
5	0101	ENQ	NAK	%	5	E	U	e	u
6	0110	ACK	SYN	&	6	F	V	f	v
7	0111	BEL	ETB	'	7	G	W	g	w
8	1000	BS	CAN	(	8	H	X	h	x
9	1001	HT	EM	)	9	I	Y	i	y
A	1010	LF	SUB	*	:	J	Z	j	z
B	1011	VT	ESC	+	;	K	[	k	{
C	1100	FF	FS	,	<	L	\	l	|
D	1101	CR	GS	−	=	M	]	m	}
E	1110	SO	RS	•	>	N	↵	n	~
F	1111	SI	US	/	?	O	or	o	DEL

to the ASCII code described in Table 16.3. The speeds of operation of printers vary greatly, depending upon their mode of operation, but in all cases will be very slow when compared with the clock frequency of the processor itself. Consequently, following the transfer of one data byte to the printer, many clock cycles will occur before the printer is ready to receive the next byte. Such peripherals are termed asychronous, since the rate of data transfer is determined not by the system clock, but by the peripheral itself.

The 6821 PIA is designed, as shown in Fig. 16.17, with two 8 bit parallel ports together with the control connections. Although the ports are very similar in operation, there are minor differences which make port A more suitable for parallel data input to the processor, while port B is more suitable for data output. An asychronous data transfer, as described, would be controlled by the two control connections, configured as shown in Fig. 16.22.

The figure shows a printer with an 8 bit parallel data connection to port B of the PIA. Two control connections are also used: control CB1 is used to output a pulse to the printer, while control CB2 is configured as an input to accept a pulse output from the printer.

The mode of operation is as follows. When the processor has a data byte, representing an alphanumeric character, ready for printing, the byte is made available to the printer from the port B output register. The availability of the byte is signalled to the printer by a pulse output from CB1. This informs the printer that the character can be printed and when this is completed the printer

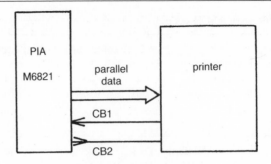

Fig. 16.22 Printer connection.

informs the processor by a pulse output to the processor via control CB2. The CB2 signal can be used to interrupt the current processor program, so causing a new character byte to be output to the printer; a further CB1 pulse will cause the printer to print the new character. In this manner, the relatively slow data transfer to the printer will take place at a rate determined by the speed of the printer itself, with the transfer of each character controlled by the two signals CB1 and CB2. Such a data transfer is said to be asychronous, because the transfer is not timed by the system clock, and the two control signals are termed *asynchronous handshake* signals.

The processor program can be written so that the printer data interchange is handled as an interrupt program. In this way the processor system can be performing a completely different background program and is interrupted by the printer CB2 signal to obtain a new character. The interrupt program itself would then output the new character byte and output the control signal CB1 before returning from interrupt to its background program. The printer routine would then cause minimal interference with other programs. All information required to produce the interrupt operated printer handler in this way is included in Fig. 16.18, but will not be considered in greater detail here.

16.10 Systems software

In Section 16.6 a basic microprocessor system was briefly described, consisting of the processor *hardware*, i.e. microprocessor, memory devices, input/output ports, VDU and power supply, together with the processor *software*, the system monitor program. When the processor system is switched ON, it is arranged that a RESET interrupt is performed automatically, i.e. the processor RES input pin is held LOW for a short time. This causes a RESET interrupt to occur and the start address of the program to be entered is obtained from the RESET interrupt vector addresses $FFFE and $FFFF. The address entered into the vector would be the start address of the monitor program, which would exist as a binary machine code program held in permanent memory, ROM. Consequently, the effect of switching ON the processor power would be to run directly the monitor program which, in effect, would bring the processor under control and would normally print a PROMPT character on the system

VDU and would then stand in a waiting program loop, examining the VDU keyboard and waiting for an input instruction to be keyed in.

If the system is required to accept a program written in machine code, then the system described contains all that is necessary to enter the program, run the program and perhaps to debug the program to remove any faults found. Such a monitor program would be between 1 and 3 kilobytes in size and would physically exist in one or two ROM or EPROM memory chips, configured so that they are in the highest memory addresses. For example, if the monitor program was small enough to fit into a 2 kilobyte chip, then it could be addressed in memory locations $F800 to $FFFF; the start address of the monitor program would be $F800 and this address would be entered into the RESET vector, i.e. addresses $FFFE and $FFFF. As these addresses also lie within the same 2 kilobyte memory space, the required start address is also conveniently and permanently held in the ROM. Such an arrangement is necessary if the microprocessor system is to be self-starting on power up.

If the microprocessor system is to be used for source code programs, then further system software is commonly required. An assembler program, depending upon its sophistication, may occupy an address space of several kilobytes and could also be accommodated in ROM memory chips, together with some read/write RAM memory space for temporary data storage. However, such a minimal system would be of limited application, and commonly the system hardware is extended by the addition of a single or double magnetic disk drive to provide a relatively large back-up memory system. With this addition, large system software programs can be held on disk, being downloaded into memory only when required. The relatively small system main memory can thus be used very efficiently and the availability of large back-up memory space allows the addition of further system software.

The storage and retrieval of data on the magnetic disks is under the control of a further large system program – the disk operating system (DOS), which is itself stored on the disk memory. The disks are brought into operation by using the monitor to load into memory a small loader program which, in turn, loads the DOS program down into main memory space. The DOS contains routines which will allow the storage of data as *files* on the disks and which can be retrieved and returned to main memory using simple commands and file names. A system *editor* will be included, which will allow the creation of new files and alterations to old files.

The Assembler program would also exist in this way as a separate file on the disk. In operation, the assembler language source file would be entered into the system using the editor and stored on the disk. Then, in answer to an

assemble filename

command, the assembler program would be downloaded into memory, the source code file would be downloaded into memory, the assembly of the source code into machine code would occur and the resultant machine code would be stored on disk as yet another file, which when required can be downloaded into the system memory and run. Other system software may include programs which can assist in tracing errors or compiler programs which would allow the compilation into machine code of files written in one of the high level languages such as BASIC, FORTRAN or PASCAL.

Facilities such as these are available in specially designed manufactures' development systems (MDS) which are produced by the microprocessor manufacturers and by others to provide all the facilities required for efficient hardware and software development. Such large scale facilities can of course be very expensive. It must be emphasized that the final aim is always to produce a machine code program which will run in a microprocessor system to perform a task. In general the task will be a specific one, e.g. to generate a square wave, or to control a central heating system or to operate an accounting system for a petrol station. Whatever the specific task, the overall hardware and software should consist of only those facilities, i.e. memory size, input/output port devices, software routines, etc., which are necessary for that task. Consequently many microprocessor installations consist only of a microprocessor chip together with one or two other chips to provide the other functions.

16.11   The use of microprocessors in digital system design

As was stated at the beginning of this chapter, the advent of microprocessors has changed the philosophy of digital system design. Before the introduction of microprocessors, digital hardware was mainly TTL device based and the production of a working system followed a well-defined path.

(1) The logic problem is defined.
(2) The required logic hardware is designed.
(3) The logic devices are obtained – from stock if possible, but special devices from suppliers.
(4) A printed circuit layout is drawn.
(5) The printed circuit is manufactured.
(6) The circuit is assembled.
(7) The circuit is tested.
(8) The complete system is assembled and tested.

Any mistakes which occur in any step necessitate very difficult changes in expensive hardware, and unfortunately the easiest place to make a mistake is in step (1) in the problem definition. The equivalent steps for a microprocessor-based design are as follows.

(1) The problem is defined.
(2) An algorithm to solve the problem is worked out.
(3) The software (program) is written.
(4) The software is checked and run in a microprocessor development system.
(5) The complete system is assembled and tested with the program running in the development systems.
(6) The complete system is assembled with the development system replaced

by standard microprocessor circuit cards obtained from the manufacturer, and with the tested program *burned* into ROM.

The important points are these:

(1) The microprocessor *hardware*, i.e. the circuit cards containing the microprocessor, the memory and the input/output modules, are identical for each and every problem, apart from interface electronics needed to interface the input/output modules to the external components, e.g. transducers, stepping motors, etc. The microprocessor hardware, being standard, can be bought ready and tested from suppliers and in fact is not needed until the complete design has been tested.
(2) Errors in any step in the design will necessitate changes only in the system software, i.e. the program, and will not affect the microprocessor hardware. Such changes are usually easy to make and are relatively cheap.

Digital system design has thus become very much a software oriented subject and electronic design engineers have made a remarkable change in professional expertise over a very short period.

Problems

The problems have, in general, been designed for solution using an 8 bit microprocessor, such as the Motorola 6800/6802. The answers are given for these microprocessors. However, they can be tackled for other microprocessors.

1. Convert the following binary numbers to hexadecimal notion:

    0010 0110   1001 0001   0000 1111
    0001 0000 1101 0110   1010 0000 1111 0100

2. For the Motorola 6800/6802 microprocessor:
    (a) How many numbers of bytes are needed to specify a direct addressing operation?
    (b) What form of addressing is used by branch instructions?
    (c) What is the operation code (in hex) required to specify the instruction of add to accumulator A in direct addressing mode?
    (d) In how many addressing modes can the ADD A instruction be used?
    (e) What does the CLR operation do?
    (f) What does the COM B operation do?

3. Write a machine code program starting at address 0020 which loads accumulator A with the data in memory location 0050, subtracts it from the data in memory 0060 and stores the result in memory location E155. Terminate the program with a wait for interrupt instruction.

4. Write a machine code program starting at address 0020 which loads accumulator A with the data in memory location 0050, adds to it the data in memory

0051 and stores the result in memory location 0060. Terminate the program with a software interrupt instruction.

5. Write a machine code program starting at address 0020 which exchanges data in memory locations 0060 and 0080. Terminate the program with a software interrupt instruction.

6. Write a machine code program starting at address 0020 which multiplies the number at memory location 0050 by 5 and stores the product at memory location 0060. The number is between 00 and 40. Terminate the program with a software interrupt instruction.

7. Write a program in machine code and assembly language to subtract a number in memory location 0060 from a number in memory location 0050 and store the difference in memory location 0080. Terminate the program with a software interrupt instruction.

8. Write a program in machine code and assembly language to add a number in memory location 0060 to a number in memory location 0050 and store the result in memory location 0080. Terminate the program with a software interrupt instruction.

9. Write a program to convert a hexadecimal number, in the range 00 to 0F and in memory location 0050, into its decimal equivalent and store it in memory location 0060.

10. Write a program to initialize a 6821 PIA so that the B-side connections are all inputs and the A-side connections are all outputs. Use memory addresses 8008 to 800B for the PIA registers.

11. What values should be stored in the control registers of a 6821 PIA to obtain the following mode of operation?

CA1: Disabled
CB1: Enabled interrupt input, set by a low-to-high transition
CA2: Enabled and used as a set/reset output
CB2: Enabled and used as a write strobe output with E Restore.

Chapter 17
Electrical Transducers

17.1 Basic function of transducers

The basic function of an electrical transducer is to convert some variable, such as temperature or pressure, into an electrical signal which may subsequently be processed. A thermocouple is an example of a transducer, giving an output of an e.m.f. for an input of temperature. Transducers may be divided into two basic types: passive and active. Passive transducers, such as resistive, capacitive and inductive elements, require an external power supply for their operation, whilst active ones, such as electromagnetic and thermoelectric elements, do not. These two types can be further divided into those transducers providing an analogue output signal and those producing a digital output.

17.1.1 Static properties of transducers

Before considering the properties of individual types of transducer, certain properties of transducers in general may be defined.

Range
The range of a transducer may be defined in terms of the range of the input variables over which the transducer will operate with acceptable characteristics. Thus a linear displacement transducer may give acceptable readings over an input range of, say, 0.01 to 4.0 mm.

Linearity
The input/output relationship for a transducer system may be in the general form

$$y = a_0 + a_1x + a_2x^2 + a_3x^3 + \ldots + a_nx^n \tag{17.1}$$

where y is the output variable and x is the input variable. The coefficients $a_0 \ldots a_n$ are the calibration factors, whose values determine the form of the relationship between y and x. If a_2 and all higher order coefficients are zero, then the relationship reduces to

$$y = a_0 + a_1x \tag{17.2}$$

which represents a linear relationship between the variables. Further, when a_0 is also zero, then the relationship

$$y = a_1x \tag{17.3}$$

represents a linear relationship which passes through the origin, giving a zero output signal when the input quantity is zero. This ideal linear relationship is rarely achieved in practice, except over a very limited range of input variation.

Figure 17.1 illustrates the form of a linear input–output relationship, and also of two non-linear relationships, one containing only even order powers and one only odd powers of the variable x beyond the linear term. An arrangement of two identical transducers, differentially connected (i.e. with one deflected in a positive sense and the other deflected in a negative sense) can sometimes be used, and can be shown to give cancellation of the even order powers to give the odd power only characteristic shown in the figure.

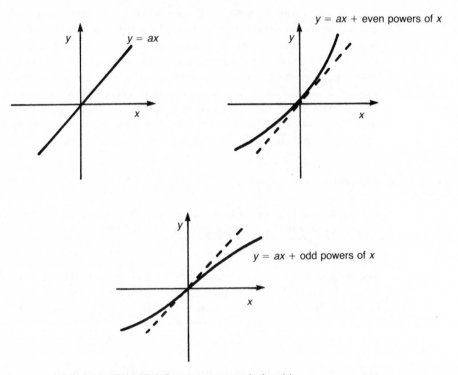

Fig. 17.1 Input–output relationships.

Sensitivity
The sensitivity of a system specifies the magnitude of the output variable change caused by a unit change of the input variable. In the equation

$$y = a_1 x \qquad (17.3)$$

the sensitivity is specified by the magnitude of the constant a_1. For example, if the output quantity y is measured in V, and the input quantity x is a temperature, measured in °C, then the unit of the constant a_1 is

$$a_1 = \frac{y}{x} \, \text{V/°C}$$

Resolution

The resolution of a measurement system is the smallest change in the input variable to which the measuring system will respond.

Accuracy

The accuracy of a measurement system is a measure of the difference between a measured value and the so called true value of the quantity. Bear in mind, however, that our knowledge of the true value of any quantity is only known as a result of one or more measurements of the value, even though none may, in fact, be stated to be the true value. The true value can only be inferred by a statistical examination of a series of measurements. The accuracy of a system will also, in general, vary throughout the range of the system, and may be specified over the range by means of a calibration curve.

Drift

The drift of a system describes the gradual change in the characteristics of the system, with age, temperature, or other external property. Such an external change may affect the system in either or both of two ways. Zero drift: as shown in Fig. 17.2(a), a drift in the zero level of a system results in a shift in the whole characteristic. Sensitivity drift: illustrated in Fig. 17.2(b), results in a change in the value of the constant a_1 and hence in the slope of the characteristic.

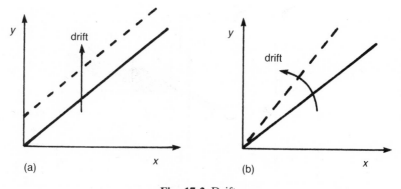

Fig. 17.2 Drift.

Repeatability

This property describes the degree to which a measurement of a given input value may be repeated.

The characteristics briefly mentioned in this section can be measured by the application of steady or perhaps slowly varying signals. Where more than one external quantity may affect the input–output relationship, only one variable is allowed to change while a characteristic is measured. In this way a family of static characteristics covering the operation of a measuring system may be obtained.

17.2 Dynamic properties of transducers

Under dynamic conditions, when input variables may change rapidly, the system characteristics may be quite different from the static characteristics. In practice, the variation of an input quantity will not, in general, follow a simple mathematical relationship with time; a study of the dynamic behaviour of systems can, therefore, be complex. However, a knowledge of the response of a system to simple time varying inputs can allow a reasonable prediction of the response of the system to more complex time variations. In particular, a knowledge of the step response of a system, i.e. the response of the system to a sudden step change in the value of the input variable, can be very useful.

17.2.1 A zero order transducer

The simplest possible form of transducer system is one in which no time delaying elements of any sort occur and is called a *zero order system*.

Figure 17.3 shows a simple potentiometer which is to be used to give an output voltage proportional to the displacement of the slider from the zero position. For a zero order system the equation relating the output y and input x must be

$$y = a_1x \tag{17.3}$$

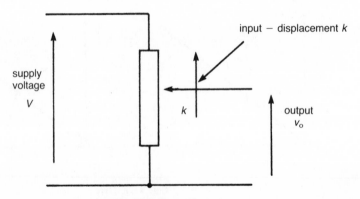

Fig. 17.3 Potentiometer.

The potentiometer output voltage may be written

$$V_0 = Vk \tag{17.4}$$

where k is the fractional displacement of the slider. Thus when $k = 0$, the output voltage is zero, and when $k = 1$, the output voltage is V. The equation is a zero order equation.

This is only true so long as there are no elements in the construction of the potentiometer which will prevent the output voltage following instantly the movement of the slider. We must ensure that the potentiometer winding has no inductance; that there is no stray capacitance between the winding or the output terminal and any other part of the circuit. These requirements are obviously impossible to fulfil. Further, the system must be considered as a whole, from both the electrical and the mechanical point of view. It must be ensured, therefore, that the moving parts must have no mass, so that no inertia time delays occur, and also that the moving parts are perfectly rigid, so that no elastic bending effects can occur. The zero order instrument is obviously an unrealizable objective, but may be approximated in some conditions for very slowly varying inputs. A consideration of this basic instrument, however, leads to a very important conclusion. In considering a transducer, the complete system must be examined, i.e. the mechanical and the electrical properties of the transducer must be examined together before its operation can be fully understood.

If a truly zero order instrument could be achieved, it would respond to any input change without delay, and produce an output signal of the same form as the input signal.

17.2.2   *A first order transducer*

A *first order system* is one which contains only one time delay element. Thus a mechanical system could contain either mass or stiffness in addition to friction, while an electrical system could contain either inductance or capacitance in addition to resistance.

Figure 17.4 shows two simple systems which are of first order. The mechanical system consists of a moving plate, which has negligible mass, connected to a frame by a spring. The spring has a stiffness of k, where stiffness is defined as the force required to displace the spring by unit distance. The system has also mechanical friction represented by the dashpot damper, producing a resistive force equal to $r\,dx/dt$, where dx/dt is the velocity of the plate. The electrical first order system consists of a capacitor C in series with an electrical resistor R across a voltage source. The inputs of the two systems are a time varying force

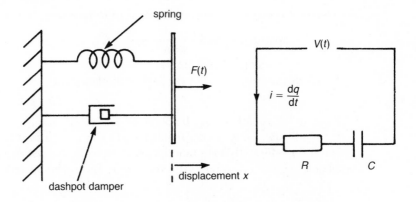

Fig. 17.4 Mechanical and electrical first order systems.

$F(t)$ in the mechanical case, and a time varying voltage $V(t)$ in the electrical case. Each system may be analysed as follows.

For the *mechanical system*, equating the forces on the plate

$$F(t) = r\frac{dx}{dt} + kx$$

or

$$\frac{1}{k}F(t) = \frac{r}{k}\frac{dx}{dt} + x \qquad (17.5)$$

The term r/k is known as the *time constant* τ of the system. The term $1/k$ is known as the *static sensitivity*. The general form of the equation is

$$\frac{1}{k}F(t) = \tau\frac{dx}{dt} + x \qquad (17.6)$$

For the *electrical system* summing the voltages round the circuit

$$V(t) = R\frac{dq}{dt} + \frac{1}{C}q$$

where q is the charge and dq/dt is the circuit current

$$CV(t) = CR\frac{dq}{dt} + q \qquad (17.7)$$

The product CR is known as the *time constant*.

The term C also becomes the *static sensitivity*. Again, the general form of the equation is

$$CV(t) = \tau\frac{dq}{dt} + q \qquad (17.8)$$

It can be seen that Equations (17.6) and (17.8) are of the same form. If corresponding terms are taken from the equations for the two systems, it can seen that a correspondence may be drawn between

electrical resistance	and	mechanical resistance
capacitance	and	$\dfrac{1}{\text{stiffness}}$
voltage	and	force
electrical charge	and	mechanical displacement
current	and	velocity

Solutions to the equations may be obtained for various inputs.

The *steady state solution* is obtained when the force or voltage has been applied for a sufficiently long time for all changes in the system to have been completed. All derivative terms are then zero. The equations reduce to

$$\frac{1}{k}F(t) = t \qquad (17.9)$$

and

$$CV(t) = q \qquad (17.10)$$

The equations have reduced to the ideal static equations of Section 17.1. If the voltage or the force is taken as the input quantity of Equation (17.3), then

$$x = \frac{1}{k}F(t) \qquad (17.9)$$

and

$$q = CV(t) \qquad (17.10)$$

A *step function input* is produced when the input quantity makes an abrupt change of magnitude. Consider a force input which is zero up to time $t = 0$ when it becomes instantly a force of F. The eventual steady state value of the displacement, x_s, is given by the steady state equation (17.9)

$$x_s = \frac{1}{k}F$$

The response of the first order mechanical system to such a step input is

$$x = x_s(1 - e^{-t/\tau}) \qquad (17.11)$$

Similarly the response of the first order electrical system to a step voltage input which rises instantly from zero to a value V at time t is given by

$$q = q_s(1 - e^{-t/\tau}) \qquad (17.12)$$

where, again, q_s is the steady state solution given by the steady state equation (17.10)

$$q_s = CV$$

These responses are illustrated in Fig. 17.5. The responses show that the first order mechanical and electrical systems are completely similar in operation and

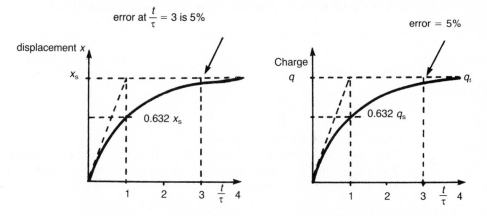

Fig. 17.5 Step responses of first order mechanical and electrical systems.

can be considered as analogous systems. A useful dynamic characteristic can be derived from the first order responses. The *settling time* of a system is defined as the time which elapses after the application of a step input for the output of the system to reach and stay within a given tolerance band about its final value. A system or instrument with a small settling time has, therefore, a fast response. For a first order instrument, a 5% tolerance settling time is equivalent to approximately three time constants.

17.2.3   A second order transducer

If the mass of the moving system is not negligible, or if the electrical system has inductance included, the two systems become second order. These systems are illustrated in Fig. 17.6.

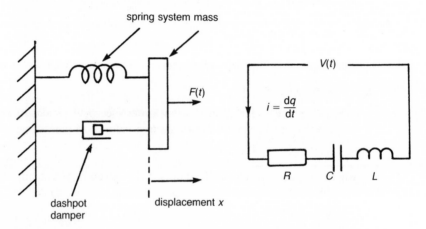

Fig. 17.6 Mechanical and electrical second order systems.

The equations to the two systems are: for the *mechanical system*

$$F(t) = m\frac{d^2x}{dt^2} + r\frac{dx}{dt} + Kx \tag{17.13}$$

where m is the mass of the moving parts, and for the *electrical system*

$$V(t) = L\frac{d^2q}{dt^2} + R\frac{dq}{dt} + \frac{1}{C}q \tag{17.14}$$

where L is the added series inductance.

The response of the second order system to a *step input* is shown in Fig. 17.7. The steady state response, when all the transients have ended, is the same as in the previous examples, i.e. Equations (17.9) and (17.10),

$$x_s = \frac{1}{k}F$$

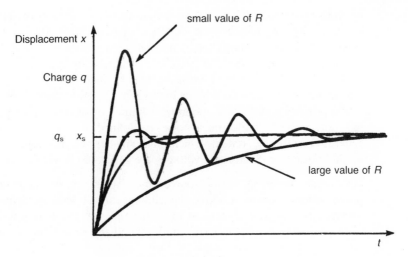

Fig. 17.7 Second order system response.

or

$$q_s = CV$$

With a second order system, the response becomes oscillatory if the value of the mechanical or electrical resistance is reduced below a critical value. The choice of the correct constants, i.e. the correct values for the mass, stiffness and the frictional resistance of the mechanical circuit, is important if the system output is to reach its true value quickly, and yet with minimum overshoot.

In general, a transducer system will consist of both mechanical and electrical components, and its dynamic response is determined by the combined effects of all of its physical constants. The analysis of such systems, to determine their dynamic properties, is a difficult process, and is outside the scope of this book. It must be remembered, however, that a simple analysis of a transducer, which may give quite accurate results for operation of the system at low frequencies, will usually become useless as the operating frequency is increased.

17.3   Resistance transducers

In this section, transducers which depend for their operation upon a changing resistance value will be considered.

17.3.1   *Potentiometer transducers*

Variable potentiometers are in wide use in measurement systems, mainly because of the ease with which they can be adapted to give an indication of linear or angular displacement, but also because of their relatively high electrical output. Potentiometers for measurement purposes are usually made by winding a resistance wire upon a rigid former, the variable sliding contact being a pressure contact upon the wire surface. Alternatively, the resistive element may be a carbon film, deposited upon a former. The output voltage per unit deflection of the sliding contact is dependent upon the supply voltage to the potentiometer; within limits imposed by the heating of the element, the sensitivity may be increased by increasing the supply voltage.

The *resolution* of the potentiometer is limited by electrical noise generated in the potentiometer. Electrical noise is defined as small random voltages generated within the system, which mask smaller voltages generated correctly by the transducer. Such noise voltages are generated in several ways.

(1) *Contact noise*. A major source of noise is the sliding contact. If the potentiometer is wire wound, the sliding contact will, in general, contact several adjacent wound turns at the same time. If this were not so, then the slide would necessarily be narrow and pointed and would, therefore, penetrate into the valleys between adjacent turns, causing rapid wear. As the contact slides over the turns, it alternately short-circuits several adjacent turns causing small but definite steps to occur in the output voltage characteristic, which cannot be distinguished from similar small steps in the output voltage due to small displacements of the slider. Another source of contact noise is wear or dirt upon the track.

(2) *Thermal* or *Johnson noise*. Thermal noise is characterized by randomness of amplitude and frequency distribution. The noise signal is developed because of the random motion of the current carriers in the electrical conductor. Noise of this type is spoken of as *white noise*, its frequency distribution being constant over the whole spectrum. Consequently, the total noise power developed is proportional to the frequency bandwidth of the signal processing system. The root mean square noise voltage generated in a resistor is given by

$$E_{noise} = \sqrt{(4kTR\,\delta f)}\ \text{V} \tag{17.15}$$

where T is the absolute temperature, in K, R is the resistor value, in Ω, δf is the frequency bandwidth of the signal processing system, in Hz, k is the Boltzmann constant, of value equal to 1.38×10^{-23} J/K.

The *linearity* of a potentiometer device is affected by the loading applied to the transducer output. Figure 17.8 shows a potentiometer of total resistance R, loaded at its output by a resistor R_L. The linearity is determined by the ratio

$$m = \frac{\text{total potentiometer resistance}}{\text{load resistance}}$$

$$= \frac{R}{R_L} \tag{17.16}$$

If the load resistor R_L is disconnected, the output voltage V_{L_0} is given by

$$\frac{V_{L_0}}{V} = k \tag{17.17}$$

With the load resistor reconnected, the output voltage is

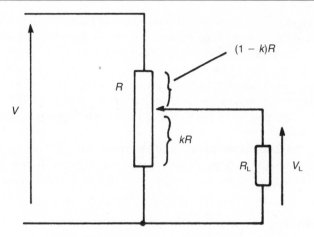

Fig. 17.8 Loading of a potentiometer.

$$\frac{V_L}{V} = \frac{R_L k R/(R_L + kR)}{(1 - k)R + R_L kR/(R_L + kR)}$$

or

$$\frac{V_L}{V} = \frac{1}{1 + mk(1 - k)} \tag{17.18}$$

The difference between the unloaded output voltage and the loaded voltage, expressed as a fraction of the unloaded voltage, is

$$\frac{V_{L_0} - V_L}{V_{L_0}} = \frac{mk(1 - k)}{1 + mk(1 - k)} \tag{17.19}$$

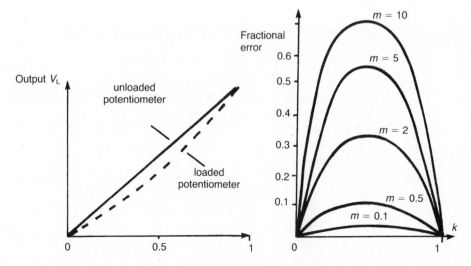

Fig. 17.9 Effect of loading.

Figure 17.9 shows typical unloaded and loaded output voltage characteristics, and also a plot of the fractional error for different values of the ratio m.

The following conclusions may be drawn.

(1) The maximum error occurs at mid deflection, i.e. when $k = 0.5$.
(2) For a maximum error of 1%, the load resistor value must not be lower than about 25 times that of the complete potentiometer.

Variable potentiometers of the types described are most useful for measuring relatively large displacements. Resolutions of the order of 0.5 mm are obtainable. A linearity of the order of 0.01% of full scale is typical.

17.3.2 Strain gauges

For small deflections, the use of strain gauges is very common. Strain is defined as the change in length per unit length of the body. A wire resistance strain gauge consists of a length of fine wire, usually folded lengthwise, so that although small in overall dimensions, its effective length in the direction it is to be strained is large.

The resistance of a length of uniform wire is

$$R = \rho \frac{L}{A} \, \Omega \tag{17.20}$$

where L and A are its length and cross-sectional area, and ρ is the material resistivity. When the wire is strained, the resistance of the wire changes, due to a change in all of the parameters. The sensitivity of a gauge is usually expressed in the form

$$\text{strain sensitivity} = \frac{\delta R / R_0}{\delta L / L_0} = S, \text{ the } \textit{gauge factor} \tag{17.21}$$

For most wire strain gauges, the value of S is about 2, so that

fractional change in resistance $= 2 \times$ fractional change in length.

Commercially available wire strain gauges are formed by bonding the folded wire grid to an insulating former. The gauges are commonly constructed from copper–nickel foil upon a plastic backing strip. Special adhesives have been developed by which the gauge may be bonded to the element being strained. For maximum sensitivity, the length of the gauge must be mounted in the direction of the strain. Gauges are available in single or multiple element rosettes, the multiple elements allowing the resolving of strains of unknown direction. Gauges are available with a wide range of resistance values, with $120 \, \Omega$, $350 \, \Omega$, $600 \, \Omega$ and $1000 \, \Omega$ as standard preferred values.

A typical gauge specification is

$$R = 120 \, \Omega \pm 2\% \qquad \text{gauge factor } S = 2.1 \pm 1\%$$

Sensitivity is limited by heating, normal current levels being of the order of 10

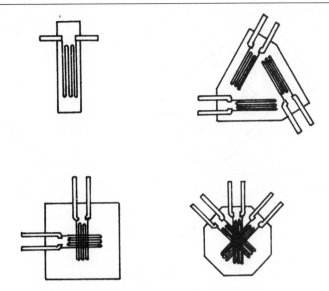

Fig. 17.10 Strain gauges.

to 30 mA. Figure 17.10 shows several different styles of construction which are commercially available.

In the early 1960s, semiconductor strain gauges became available. These strain gauges are normally produced from single crystal silicon. A distortion of the crystal lattice, due to externally applied stress, produces large changes in the electrical resistivity of the material. This effect is known as the piezo-resistance effect. In bulk form silicon is a brittle material, but in a very thin strip it is sufficiently flexible to be bent round a radius of as little as 1 mm. Silicon strain gauges are very small, typical dimensions being 1 mm overall length, 0.02 mm width and 0.002 mm thick. The great advantage of the semi-conductor strain gauge is in its sensitivity; the gauge factor S may exceed 200. Gauges are also available with both positive and negative gauge factors. A maximum strain of the order of 3000 μ strain (i.e. fractional change in length of 3000×10^{-6}) is normally quoted. Strain gauge techniques are applicable down to strains of the order of 10^{-6} or 10^{-7}.

The Wheatstone bridge provides a convenient method for the measurement of gauge resistance. The bridge will normally be balanced while the gauge is unstrained; the bridge output as the gauge is then strained can provide a sensitive and reasonably linear indication over a wide range of strain. Figure 17.11 shows a Wheatstone bridge with one arm consisting of a strain gauge. The gauge has an unstrained resistance of R Ω; the bridge can, therefore, be balanced by including resistors of equal value in the other arms.

When strained, the gauge resistance changes by δR Ω. If the bridge detector is of high impedance, so that the detector current may be neglected, then the bridge output voltage is

$$\frac{\delta V}{V} = \frac{\delta R}{R} \times \frac{R}{2(2R + \delta R)} \tag{17.22}$$

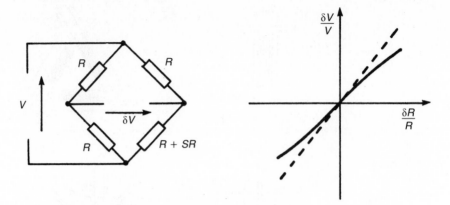

Fig. 17.11 Wheatstone bridge.

The voltage/defection characteristic is, therefore, non-linear, but for small deflections, i.e. if $\delta R \ll 2R$, the equation reduces to

$$\frac{\delta V}{V} = \frac{\delta R}{R} \times \frac{1}{4} \tag{17.23}$$

The characteristic may then also be written as

$$\frac{\delta V}{V} = \frac{S}{4} \times \text{strain} \tag{17.24}$$

If the bridge is arranged to have two active arms, i.e. if gauges in adjacent arms are strained in the opposite sense, or if gauges in opposite arms are strained in the same sense, then the sensitivity is doubled, giving

$$\frac{\delta V}{V} = \frac{S}{2} \times \text{strain} \tag{17.25}$$

With four active arms, with gauges strained with the appropriate sense,

$$\frac{\delta V}{V} = S \times \text{strain} \tag{17.26}$$

The voltage supply to the bridge may be either direct or alternating current. If, however, an a.c. supply is used, the effect of stray capacitance may make a null balance difficult to achieve. A small trimmer capacitance across one or other arms of the bridge may be adjusted to give a better balance. It must be remembered, however, that an a.c. bridge can only be balanced at one frequency, i.e. the balance is frequency sensitive. Any harmonic distortion components existing in the bridge supply will tend to mask the balance null, since the bridge will be unbalanced at these frequencies.

An increase in temperature will affect the gauges in the following ways:

(1) The resistance of the gauge increases.
(2) The gauge will change in length due to the effect of the temperature upon the test piece to which it is attached.
(3) The gauge factor changes.

Temperature compensation can be achieved by circuit design in a variety of ways. An identical but unstrained gauge may be connected into an adjacent arm of the bridge. If this inactive gauge is mounted close to the active gauge on the test piece, but in such a direction as to be unstrained, it will compensate for temperature effects in the active gauge.

A method which is useful when equal positive and negative strains are available, for example on opposite surfaces of a bending beam, is to use four identical gauges, one in each arm of the measuring bridge. Adjacent arms of the bridge should include gauges which are subjected to strains of opposite sign.

17.3.3 *Temperature sensitive resistors*

A general equation giving the variation of resistance of a material with temperature is

$$R = R_o(1 + \alpha_o t + \beta_o t^2 + \gamma_o t^3 \ldots) \tag{17.27}$$

For many materials, the values of the constants α_o, β_o, etc. are very small with the result that the equation

$$R = R_o(1 + \alpha_o t) \tag{17.28}$$

is a fair approximation to the resistance characteristic over a restricted range of temperature variation. The constant α_o is the *resistance temperature coefficient* of the material, at the base temperature from which the temperature change t is calculated. R_o is the resistance of the piece of material at the same base temperature. The metals in particular show reasonably linear resistance/temperature characteristics over a restricted temperature range. For temperature sensing applications, however, the value of the temperature coefficient is rather small. Table 17.1 lists the values of α_o for the common metals, and also for carbon, which exhibits a negative temperature coefficient. The most

Table 17.1 Resistance temperature coefficients.

Material	Resistance temperature coefficient near room temperature/°C
Aluminium	0.0045
Copper	0.0043
Gold	0.0039
Mercury	0.00099
Silver	0.00414
Platinum	0.00391
Lead	0.0041
Tungsten	0.00486
Manganese	±0.00002
Carbon	−0.00070

commonly used material for resistance thermometers is pure platinum. Platinum is linear within ±0.3% over the temperature range −20°C to +120°C.

If linearity of the resistance–temperature characteristic is not of prime importance, much greater sensitivity may be obtained by using a semiconductor sensing element instead of a metal sensor. As explained in Section 6.1, the resistivity of semiconductor materials decreases with an increase in temperature. Many semiconductor temperature sensors are now commercially available in the form of beads, open or glass encapsulated, discs or rods, or washers. The materials are various mixtures of manganese, nickel, cobalt, copper and other metals, formed with binders into the various shapes and sintered at very high temperatures (1100 to 1400°C). They become a hard ceramic material. The sensors are referred to as *thermistors*.

Figure 17.12 shows the resistance–temperature characteristic for a typical thermistor. Because of the wide resistance variation, the resistance axis scale is plotted in logarithmic form. The general form of the thermistor resistance characteristic is

$$R_{\mathrm{Th}} = R_{\mathrm{o}} e^{[(B/T)-(B/T_0)]} \tag{17.29}$$

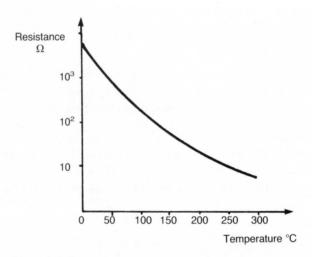

Fig. 17.12 Thermistor resistance/temperature characteristic.

where R_{Th} is the thermistor resistance, R_0 is the thermistor resistance at a base temperature of T_{o} K and B is the thermistor constant, which is a characteristic of the particular material. Typical values for the constant B lie in the range 2000 to 4500.

The effective temperature coefficient may be derived using a modified version of Equation (17.28), where $\alpha = (R - R_{\mathrm{o}})/R_{\mathrm{o}}t$, or

$$\alpha = \frac{1}{R_{\mathrm{Th}}} \frac{dR_{\mathrm{Th}}}{dT}$$

giving

$$\alpha = \frac{1}{R_o e^{[(B/T)-(B/T_0)]}} R_o e^{[(B/T)-(B/T_0)]} \left(-\frac{B}{T^2}\right)$$

or

$$\alpha = -\frac{B}{T^2} \qquad (17.30)$$

For a typical value of B of 4000, $\alpha = -0.047\,\Omega/\Omega/°C$ at 20°C.

In temperature sensing applications, thermistors are commonly used in a Wheatstone bridge circuit. The basic bridge techniques are similar to those described for use with strain gauges. Figure 17.13 shows a simple Wheatstone bridge with the thermistor sensor forming one arm of the bridge. With a high impedance detector, the bridge output is (Equation (17.22))

$$\frac{\delta V}{V} = \frac{\delta R}{R} \frac{R}{4R + 2\delta R}$$

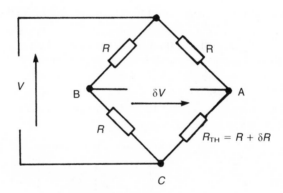

Fig. 17.13 Wheatstone bridge.

The rate of change of the bridge output voltage may be estimated as follows. When the variation is due to the $R - R_{Th}$ side of the bridge only, with respect to point C, the potential of point A is

$$V_A = V\frac{R_{Th}}{R + R_{Th}} \qquad (17.31)$$

whence

$$\frac{dV_A}{dT} = V\frac{R}{(R + R_{Th})^2}\frac{dR_{Th}}{dT} \qquad (17.32)$$

For use around a given temperature T, it is useful to be able to maximize the rate of change of output voltage. At the temperature T, the thermistor resistance R_{Th} is fixed. The bridge sensitivity may, therefore, be maximized by choosing the best value for R; this is the value of R which makes the factor $R/(R + R_{Th})^2$ a maximum. Thus, putting

$$\frac{\mathrm{d}}{\mathrm{d}T}\frac{R}{(R + R_{\mathrm{Th}})^2} = 0$$

we obtain the optimum relationship

$$R = R_{\mathrm{Th}} \tag{17.33}$$

Errors may occur in thermistor bridge circuits due to self heating of the thermistor or due to temperature or noise effects in long leads which may be used to connect a remote sensor to the bridge. Self heating errors are caused by heating of the thermistor above its ambient temperature because of the power dissipation in the thermistor due to the bridge currents. This effect can obviously be reduced by decreasing the bridge supply voltage, but this also reduces the sensitivity. If the temperature being measured is relatively constant, or only changing slowly, several measurements may be made at different transducer currents. A graph of measured temperature against transducer current may be extrapolated backwards to give a deduced value at zero transducer current. Another possibility is the excitation of the bridge circuit by short duration, high amplitude voltage pulses. This method relies on the measurement being completed during the excitation pulse before the temperature of the sensor has time to change appreciably.

Errors due to long sensor leads may be minimized as shown in Fig. 17.14, which shows the inclusion of dummy leads in an adjacent bridge arm, the dummy leads being physically adjacent to the sensor leads. Because similar effects may be expected to occur in both the active and dummy leads, their effect on the bridge is negligible.

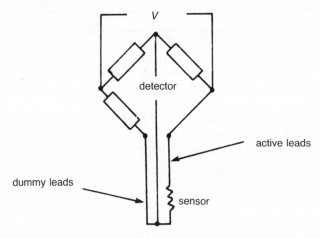

Fig. 17.14 Three-lead sensor.

17.3.4 Hot wire measurements on fluids

Measurements of fluid velocity may be made by inserting a temperature sensitive resistor in the fluid and passing an energizing current through it. The power dissipated in the resistor will cause its temperature to rise above that of the fluid. Analysis of the heat transfer between the resistor and the fluid is not simple, because in practice the heat transfer coefficient is complex in its dependence upon surface conditions. In the steady state, however, the resistor will attain a steady temperature in a time relatively long when compared to the thermal time constant of the system. If the fluid is in motion the temperature attained by the resistor is a function of the velocity of the fluid. Measurements of the resistance of the transducer can, by calibration, be used to give velocity measurements.

A suitable sensor material will have a high temperature coefficient of resistance, a convenient resistivity to enable sensors of reasonable dimensions to be produced, robustness to withstand turbulent flow conditions and high oxidation resistance. Tungsten, platinum or platinum–iridium alloy wire sensors are commonly used in the form shown in Fig. 17.15. Such sensors can respond over a frequency range up to about 500 Hz. For applications which do not require such a rapid response, thermistors have found wide use, giving a frequency response up to about 2 Hz. Hot film sensors are available, consisting of a conducting film of perhaps platinum held on a ceramic substrate, and having a better frequency response than wires.

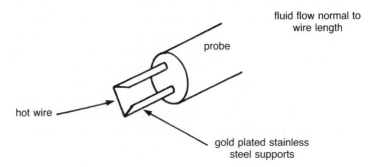

Fig. 17.15 Fluid velocity sensor.

Effects due to a variation in the temperature of the fluid itself may be compensated by the inclusion of a further sensor in the fluid, shielded from flow effects, whose resistance is then a function of the fluid temperature only. The combination of the two signals, to give an output dependent upon the flow rate only, is, however, not a simple problem.

There are two basically different methods by which transducer systems such as the hot wire sensor system may be operated.

(1) The constant current method. A constant known current is passed through the resistance sensor. The equilibrium temperature is reached when the power dissipated in the resistor is balanced by the heat loss from the sensor to the fluid. The resistance of the sensor at the balance temperature is measured by determining the voltage across it.

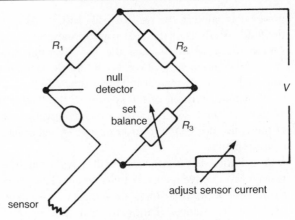

Fig. 17.16 Bridge circuit.

(2) The constant temperature method. Figure 17.16 shows a bridge circuit which includes the hot wire sensor as one of its arms. The bridge can be balanced using the adjustable arm R_3. Once balanced, the bridge will unbalance again if the sensor resistance changes, i.e. if the sensor temperature changes. At any other fluid flow rate, therefore, the temperature may be brought back to its balance value by adjusting the bridge supply voltage, so altering the current in the sensor. The sensor current, which may be measured by using an ammeter, is then a function of the fluid flow rate. Arrangements may conveniently be made for such a system to be self-balancing, providing a continuous record of flow rate changes.

Notice that flow rate measuring systems as described are not absolute measuring systems, but require calibration in fluids moving with known velocity. A calibration curve will be required to cover the full range of the device, because such systems are not linear in operation.

17.3.5 *Other resistance transducers*

Many other forms of variable resistance transducer have been adapted to suit different physical variables. For example, the silicon *photodiode*, operated with reverse voltage bias, functions in effect as a light sensitive resistor. The diode consists basically of a light sensitive pn junction. When irradiated with light, electron–hole pairs are generated on both sides of the junction. Extra minority carriers are then available and the effective resistance of the diode in the reverse biased sense is reduced. The cell resistance is, therefore, varied by the cell illumination, the resistance falling as the illumination intensity increases. Figure 17.17 shows the characteristic of such a photodiode, together with a load line for the resistor load R_L. The reverse resistance of the diode is also temperature sensitive, reducing with increasing temperature.

Acoustic variable resistance transducers have been in common use for many years in the form of the *carbon granule microphone*, which is the

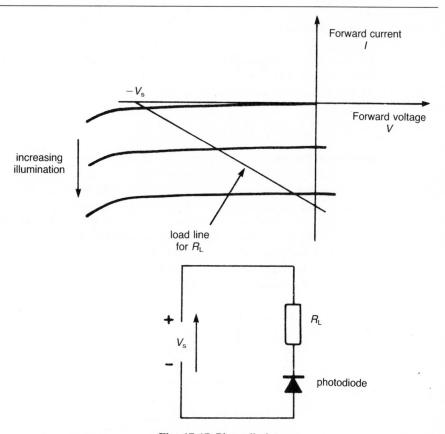

Fig. 17.17 Photodiode.

standard microphone used in telephone systems. Microphones are of two basic types: pressure microphones, which operate because of the acoustic pressure upon one side of the microphone diagram, and velocity or pressure gradient microphones, which depend for their operation upon the acoustic pressure difference between the two sides of the diaphragm. The carbon granule microphone is a pressure device and makes use of the fact that the resistance of a mass of carbon granules varies with the pressure applied to the granules. A direct current is passed through the granules, which are mounted so that the diaphragm can exert pressure upon them. Sound waves striking the diaphragm vary the pressure exerted upon the granules, so varying the resistance of the microphone. The current thus varies in accordance with the sound pressure applied to the diaphragm.

17.4 Reactance transducers

Various transducers can be constructed which operate by means of the production of a variable reactance, which may be measured by means of alternating current techniques.

17.4.1   *Capacitance transducers*

The capacitance of a parallel plate capacitor, which consists of a pair of parallel plates, each of area A m^2, spaced a distance d m and separated by a material of relative permittivity ε_r is given by

$$C = \frac{\varepsilon_o \varepsilon_r A}{d} \qquad\qquad (17.34)$$

where ε_o is the absolute permittivity of free space.

A variable transducer can be produced by any system by which the input variable can alter the value of the capacitor variables ε_r, A or d. The most common capacitor transducers are those in which the spacing d is varied, giving a direct reading of distance. The area A may be varied, perhaps by altering the overlap between two plates. The dielectric constant may be varied by moving in or out of the space between the two plates a wedge shaped dielectric. These effects are illustrated in Fig. 17.18, which shows three different methods of producing a displacement transducer.

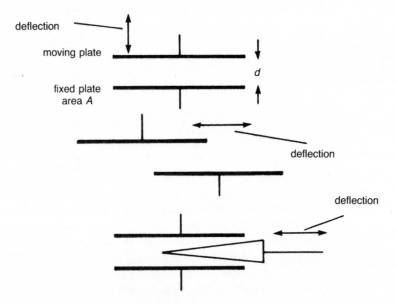

Fig. 17.18 Displacement transducers.

A major difference between variable resistor and variable capacitor transducers is the impedance level of the sensor. Resistive sensors can be produced, in general, with a resistance value of any order which is convenient to use. Capacitor devices are, on the other hand, normally of extremely high reactance. Consider a parallel plate capacitor consisting of circular plates of diameter 1 cm, spaced by 1 mm. The capacitance of such a component is

$$C = \frac{0.7}{10^3}\text{pF}$$

if the dielectric is air. At a frequency of 1 kHz, the reactance of the capacitor is of the order of 230 MΩ.

The reactance of the capacitor may be reduced by using a higher measuring frequency. At higher frequencies, however, the capacitive sensor will cease to behave as a pure capacitor and would be represented by the equivalent circuit of Fig. 17.19. In the figure, R_s and L_s represent the series resistance and inductance of the capacitor leads respectively, while R_p represents the effective leakage in the capacitor dielectric.

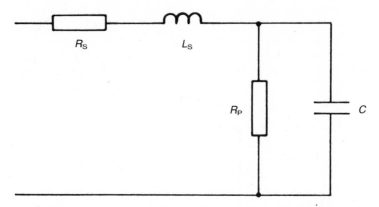

Fig. 17.19 High frequency equivalent circuit of a capacitor.

The linearity of the variable spacing capacitor transducer is given, for very small variations in spacing, by differentiating Equation (17.34), with respect to the spacing d. Thus

$$\frac{dC}{dd} = -\frac{\varepsilon_o\varepsilon_r A}{d^2} = -\frac{C}{d}$$

This may be expressed as

$$\frac{\delta C}{C} = -\frac{\delta d}{d} \qquad (17.35)$$

For larger changes in spacing, the relationship may be shown to be

$$\frac{\delta C}{C} = -\frac{\delta d}{d + \delta d}$$

and the relationship between displacement and variation in capacitance is no longer linear. The relationship is illustrated in Fig. 17.20.

Alternating current bridge techniques can be applied satisfactorily to capacitive transducer measuring systems. Figure 17.21 shows a relatively simple a.c. bridge. When the bridge is balanced

$$i_1 R_1 = i_2 \frac{1}{j\omega C_1}$$

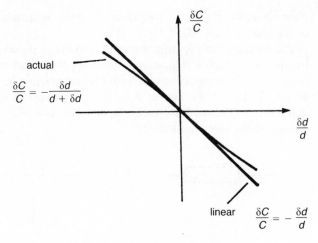

Fig. 17.20 Linearity.

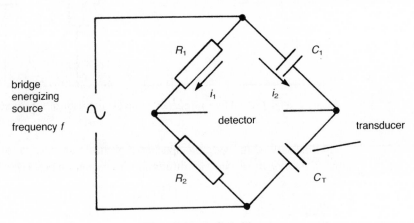

Fig. 17.21 A.C. bridge.

and

$$i_1 R_2 = i_2 \frac{1}{j\omega C_T}$$

whence

$$C_T = C_1 \frac{R_1}{R_2} \qquad (17.36)$$

The precautions outlined previously for a.c. bridges are equally applicable here; the energizing source should be reasonably free from harmonic distortion. A problem commonly met when using a.c. bridges to measure small reactance changes is the effect of stray capacitances between various leads and earth, or between various leads. Readings taken with such a bridge can be checked by reversing connections, for example to the detector, or from the bridge source or to the ratio arms R_1 and R_2. If stray capacitance effects are negligibly small, the same measurement results will be obtained. For maximum bridge sensitivity, the two capacitors should be equal in value and the resistor values should equal that of the capacitive reactance at the measuring frequency. A good balance will not be achieved unless both capacitors are equally loss-free.

The linearity of the capacitor sensor may be improved by using a differentially connected transducer, as shown in Fig. 17.22, which also shows how such a transducer could be connected in two adjacent arms of the a.c. bridge. Displacement of the central plate will increase one capacitor and decrease the other. The change in capacitance of the two capacitors may be made additive by including them in adjacent arms of the measuring bridge. Figure 17.23 shows schematically a differential capacitor pressure transducer.

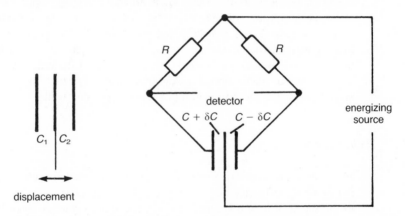

Fig. 17.22 Differentially connected transducer.

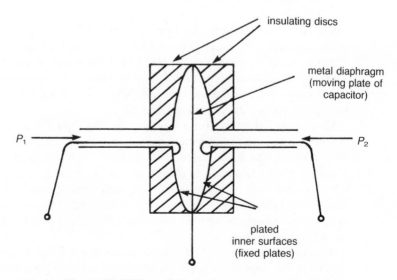

Fig. 17.23 Differential capacitor pressure transducer.

An alternative measuring system is used with the capacitor microphone which, because of its stable characteristics and linearity, forms a standard piece of equipment in acoustic laboratories. The microphone, which is a pressure actuated device, consists of a very thin diaphragm, stretched to raise its resonant frequency. The capacitance between the diaphragm and a back plate, which forms the fixed plate of the transducer, is varied by the pressure variations exerted upon the diaphragm by the sound pressure waves. The capacitor microphone is polarized by a direct voltage of several hundred volts via a resistor.

The capacitor charges to the potential of the polarizing voltage. Rapid variations in the capacitor value produce voltage changes across the capacitor, because the time constant of the capacitor–resistor circuit is too long for the charge stored in the capacitor to change. A slow change in capacitor value, however, produces only a very small capacitor voltage, and this method produces zero output voltage for static deflections of the transducer. Such a frequency response is perfectly adequate for acoustic purposes. To prevent loading effects, the circuit connected across the capacitor transducer must be of very high impedance, usually of the order of $10\,M\Omega$. For this purpose, many capacitor microphones are equipped with very high impedance buffering amplifiers at the microphone end of the microphone cable.

17.4.2 Inductance transducers

Many different types of transducer can be classified loosely under the heading of variable inductance transducer. The inductance of a uniformly wound toroidal coil of length l m and cross-sectional area A m^2 is given by

$$L = \mu_o \mu_r \frac{AN^2}{l}\,H \tag{17.37}$$

where N is the total number of turns, μ_o is the absolute permeability of free space and μ_r is the relative permeability of the magnetic core. The inductance of the coil may be varied by changing any of the equation variables. However, changing the physical dimensions of an inductor is not normally a convenient process, especially as the coils are normally wound upon a ferromagnetic core in order to reduce the overall physical size.

Various methods are in use whereby the inductance may be varied. If the magnetic core path is completed, via a ferromagnetic armature which may be moved relative to the remainder of the core, a change in the air gap in the magnetic circuit produces a change in the inductance. Figure 17.24 shows schematically both single ended and differential forms of this type of transducer. In the differential form, the inductance of the two coils will change in opposite senses; the coils may conveniently be included in adjacent arms of the measuring bridge.

A variation of this form of differential transducer, which is suitable for use in displacement measuring systems, is illustrated in Fig. 17.25. As in the previous case, the inductance, and hence the impedance, is varied by changing the

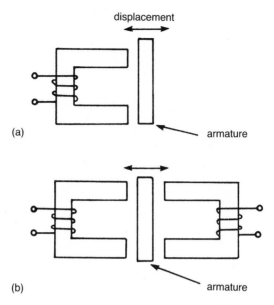

Fig. 17.24 Ferromagnetic armature form of transducer, (a) single ended armature type, (b) ferromagnetic armature type.

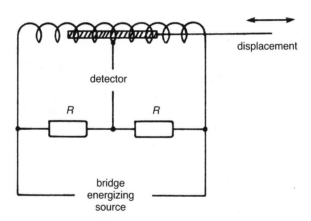

Fig. 17.25 Displacement measurement transducer.

properties of the magnetic circuit. This is achieved by making the core of the form of a ferromagnetic plunger which may be moved axially along the centre tapped inductor. The figure shows the transducer connected in an a.c. bridge with the two halves of the device in adjacent bridge arms. If the two halves are identical and if the plunger is centrally placed in the inductor then the bridge will be balanced. A deviation in the position of the plunger will unbalance the bridge. The use of a differential system in this way produces a relatively linear output characteristic for a limited range of deviation of the plunger.

One of the most commonly used inductive type transducers is the *linear variable differential transformer*, or LVDT as it is widely known. This device is

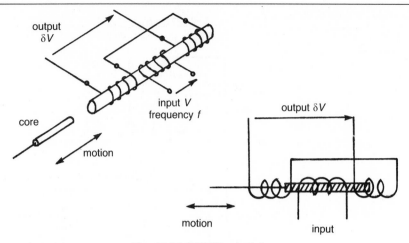

Fig. 17.26 LVDT principle.

also basically a displacement measuring transducer. It is based upon a variable transformer effect, and is made in differential form to improve the linearity. Figure 17.26 shows two identical secondary windings which are connected in series but in opposing directions. The primary winding is symmetrically placed with respect to the secondary windings. With the sliding magnetic core in the central position, the whole device is symmetrical; the e.m.f.s generated in the secondary windings are then equal and the output voltage is, therefore, zero. Deviation of the core from the central position causes one secondary e.m.f. to increase and the other to decrease; the output voltage, being the difference between the two, thus increases for a deviation of the core on either side of the null position. With correct design the magnitude of the output voltage is very nearly proportional to the deviation of the core from the null. With a sinusoidal input voltage the two secondary voltages are also sinusoidal. Because of the differential connection, the phase of the output voltage changes by 180° as the core moves through the null position. If the detector used with the measuring system is phase sensitive, it is possible to discriminate between deflections on different sides of the null. Figure 17.27 shows a phase-sensitive detector which uses as its phase reference the phase of the a.c. source with energizes the primary of the LDVT.

The detector circuit itself is basically a bridge circuit, which, from the point of view of the reference signal, is balanced. With zero input to the detector, therefore, the output voltage V_o is also zero. The input to the detector should be either in phase with or 180° out of phase with the reference signal. It then adds to the reference signal in one half of the bridge, and subtracts from the reference signal in the other half of the bridge, resulting in the output characteristic also shown in Fig. 17.27. In order to achieve the correct phasing of the detector input signal, some phase correction will normally have to be applied to the output of the LDVT. If the transducer is to be used only on one side of

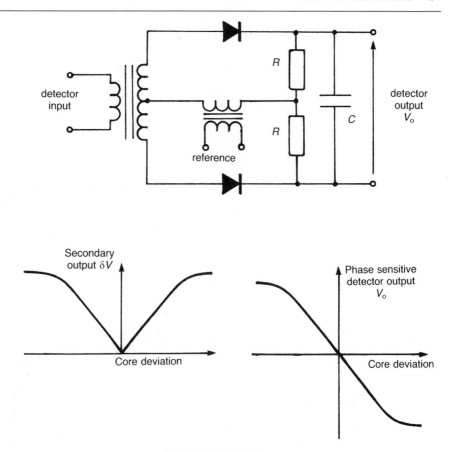

Fig. 17.27 LVDT.

the null position, i.e. accepting a smaller linear deflection range, then of course phase sensitive detection is not necessary.

The output voltage from the LDVT at the balance point will not, in general, be zero. The effect of harmonic distortion in the input voltage and also stray capacitance effects leads to a small minimum voltage which obscures the null point. LDVTs are available with ranges of up to several centimetres, with a linearity of about $\frac{1}{2}\%$ of full scale and sensitivity of the order of 2 volts per volt per cm of deflection. With a normal energizing voltage of, say, 10 V rms this will give an output voltage of 2 V per mm deflection. It is possible, using these transducers, to resolve to about 10^{-6} mm.

17.5   Electrodynamic transducers

Direct velocity sensing transducers may be based upon the relationship

$$e = Blv \text{ volts} \tag{17.38}$$

where e is the e.m.f. induced in a conductor, length l metres, moving with velocity v metres per second in a magnetic flux density B teslas (Wb/m^2).

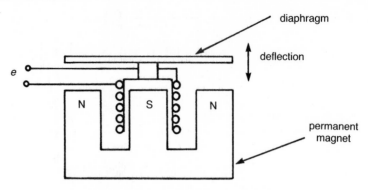

Fig. 17.28 Electromagnetic transducer.

An example of such a transducer is illustrated in Fig. 17.28, which shows a moving diaphragm or table which is supported by a spring mount so that a moving coil which is attached to it moves axially in the field produced by a permanent magnet. For a sinusoidal velocity of the coil in the magnetic field the induced e.m.f. e will also be sinusoidal in form. The device is reversible in that the application of a sinusoidal e.m.f. to the coil will result in a sinusoidal velocity in the coil and diaphragm. Good examples of such transducers are the moving coil loudspeaker, which is almost universally used for audio applications (electrostatic variable capacitor loudspeakers are also used) and the moving coil microphone. These two devices, based upon the same principle, are good examples of overall system design in that their construction must be carefully related to the conditions under which they are used.

The *moving coil loudspeaker* is designed to accept relatively high power alternating current, and to convert this power into acoustic power in space in the form of a radiated sound wave. The design requires a large moving diaphragm or cone in order to couple efficiently into free space. Particular attention is paid to the flexible mount in order to produce a very free moving system with linear characteristics. In order to respond efficiently at very low frequencies (down to 10 Hz) the resonant frequency of the moving system must be very low. The cone must be light and yet rigid to prevent the *break up* of the cone movement into complex oscillations at different driving frequencies. It is impossible to achieve these design aims in high quality units over a wide frequency range. Consequently, wide range systems will, in general, use two or three different transducers with the input audio signal split between them using filters.

In the *microphone* the design problem is to couple with relatively small sound pressure variations and to produce a moving coil velocity proportional to the varying sound pressure. This is achieved by making the resonant frequency of the system in the range 500 to 2000 Hz, and by correcting the upper and lower frequency response by using resonant air chambers coupled acoustically to the diaphragm. It is possible to produce a relatively flat response over a frequency range of 40 to 13 000 kHz.

17.6   Piezoelectric transducers

The operation of piezoelectric devices is based upon the fact that in certain crystalline materials, mechanical deformation of the crystal lattice results in a relative displacement of positive and negative charges within the material. This results in the generation of an e.m.f. between opposite faces of the crystal which is proportional to the degree of distortion of the material. The effect is reversible in that the application of an e.m.f. results in a physical deformation of the crystal.

Materials commonly used for such devices include quartz, Rochelle salt and barium titanate ceramics. The basic device is in effect a capacitor with the piezoelectric material as the dielectric between two electrodes. The charge Q produced by the deformation produces, across the device, a voltage V whose magnitude is dependent upon the capacitance of the system. A static deformation of the crystal produces a static charge and hence a voltage across the device, but the voltage will only remain as long as the charge does not leak away via the measuring system connected across the transducer. The detector must, therefore, have a very high input impedance and specially developed circuits are used if a very low frequency response is required, with an input impedance which may reach $10^{14}\,\Omega$. A true zero frequency response is, however, not possible.

17.7   Thermocouples

Another transducer which produces directly an output e.m.f. is the thermocouple, which is widely used as a temperature measuring device, or more accurately a temperature difference measuring device.

If two wires of different materials are joined together, a small e.m.f. is developed across the junction which is dependent upon its temperature. Figure 17.29(a) shows a simple circuit loop consisting of two dissimilar metals. There are obviously two junctions, J_A and J_B, associated with the circuit. If the two junctions are at the same temperature the e.m.f.s produced at the two junctions are equal and opposite, and hence no current flows in the circuit.

Figure 17.29(b) shows an ammeter connected in one branch of the circuit. This does not alter the current flowing in the circuit (other than by increasing the total resistance of the circuit loop). Although two extra junctions J_C and J_D have been introduced into the loop, and in general the ammeter will be constructed of a third type of metal, the two new junction effects cancel out if the junctions are at the same temperature, which would normally be the case. This is also true if the ammeter is moved in the circuit so that junctions J_A and J_C are coincident, as shown in Fig. 17.29(c). Although there are now only three junctions in the loop, and all three are between different materials, the ammeter material does not alter the loop current as long as the junctions J_A and J_D are at the same temperature. In use, one junction is used as the active probe, measuring the difference in temperature with reference to the temperature of the other junction, the reference junction. The accuracy of

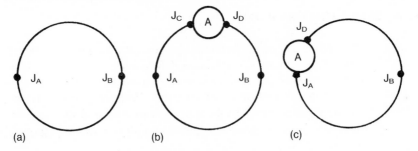

Fig. 17.29 Thermocouple.

the measurement can be no better than the accuracy to which the temperature of the reference junction is known.

The thermocouple materials in common use are copper–constantin, iron–constantin, chromel–alumel and platinum–rhodium. The sensitivity varies from a maximum of about 60 µV/°C for copper–constantin down to less than 10 µV/°C for platinum–rhodium. The usable temperature range varies for different couples, being from −200°C to +380°C for copper–constantin, and from 0 to 1500°C for platinum–rhodium.

Problems

1. A resistor bridge is initially balanced with all four arms equal in value. It is energized by a supply of voltage V. If one arm is deflected from its initial value of $R\ \Omega$ by a small amount δR, show that the out of balance voltage is given by

$$\frac{\delta V}{V} = \frac{\delta R}{R} \times \frac{R}{2(2R + \delta R)}$$

The detector impedance may be assumed to be infinite.

2. If two arms of the bridge in Problem 1 are equally deflected an amount δR, show that the out of balance voltage is given by

$$\frac{\delta V}{V} = \frac{\delta R}{R} \frac{R}{(2R + \delta R)}$$

3. If the bridge of Problem 1 is used with a detector of zero resistance, show that the detector current is given by

$$\frac{\delta I}{V} = \frac{\delta R}{R} \frac{1}{(4R + 3\delta R)}$$

4. If the bridge of Problem 1 is used with a detector of resistance $R_D\ \Omega$, show that the out of balance voltage is given by

$$\frac{\delta V}{V} = \frac{\delta R}{R} \times \frac{1}{4 + 4\dfrac{R}{R_D} + R\left(\dfrac{3}{R_D} + \dfrac{2}{\delta R}\right)}$$

5. A first order low pass instrument is to respond to a frequency of 80 Hz with an accuracy of not less than 2%. What must be its minimum cut-off frequency?

6. A copper resistor has a resistance of 90 Ω at 20°C and is connected to a d.c. supply. By what percentage must the supply voltage be increased in order to maintain the resistor current constant if the temperature of the resistor rises to 60°C? The temperature coefficient of resistance of copper may be taken as 0.00428 at 0°C.

7. A temperature measuring device consists of a thermistor and a 10 kΩ resistor connected in series across a 5 V direct supply. The voltage across the 10 kΩ resistor is measured by means of a very high resistance voltmeter. The thermistor has a resistance of 2000 Ω at 25°C and a B value of 3000 at 25°C. Determine the rate of change of output voltage with temperature (a) at 25°C, (b) at 50°C.

8. A capacitor displacement transducer consists of two parallel plates of area A spaced a distance d in the air. Show that, if the spacing is altered by a small amount δd, then the fractional change in capacitance is given by

$$\frac{\delta C}{C} = -\frac{\delta d}{d + \delta d}$$

9. A capacitive displacement transducer consists of two square, parallel, metal plates each of side length 50 mm and separated by 1 mm. A square sheet of dielectric 1 mm thick and of side length 50 mm can be slid between the plates, replacing the air, relative permittivity 1, by the sheet, relative permittivity 2.5. The displacement of this sheet between the plates is the input variable for the transducer. What will be the capacitance of the transducer when the displacement is (a) 0 (no sheet between the plates), (b) 25 mm, (c) 50 mm (sheet completely filling the space between the plates)?

10. An iron–constantin thermocouple, when used to measure temperatures between 0°C and 400°C, gives an e.m.f. of 5.268 mV at 100°C and 21.846 mV at 400°C. What would be the error at 100°C as a percentage of the full-scale reading if a linear relationship was assumed?

11. A platinum resistance coil is used as a temperature sensor between 0 and 200°C. At 0°C the coil has a resistance of 100.0 Ω, at 100°C it is 138.5 Ω and at 200°C it is 175.8 Ω. What would be the error at 100°C as a percentage of the full-scale reading if a linear relationship was assumed?

12. A variable inductance transducer is of the form shown in Fig. 17.24(a), consisting of a coil of N turns wound on a constant cross-section core. Show that the inductance of the coil is related to the distance d between the ferromagnetic armature and the core by an equation of the form

$$L = \frac{L_0}{1 + kd}$$

where $L_0 = N^2/S_0$ and $k = 2/\mu_0 A S_0$, with S_0 being the reluctance when $d = 0$ and A the constant cross-sectional area of the core.

13. Use the equation derived in Problem 11 to obtain the inductances of the two coils used with the inductance transducer described by Fig. 17.24(b). The two cores and coils are identical. Take the distance between the cores to be constant at $2d$ and the displacement x of the armature, of negligible thickness, to be measured from its central position between the cores.

14. Show that, for the differentially connected capacitive transducer used in the a.c. bridge shown in Fig. 17.22, the output voltage from the bridge is related to the displacement x of the central plate by

$$V = \frac{E_s x}{2d}$$

where E_s is the voltage of the bridge energizing source and $2d$ is the separation between the two outer plates. The two bridge resistors are equal.

15. A cantilever has four identical strain gauges attached to it. When subject to a force the cantilever bends, subjecting two of the strain gauges to identical compressive strain and the other two strain gauges to the same magnitude strain but tensile. The four gauges are used as the four active arms of a Wheatstone bridge. The resistance of each strain gauge is $120\,\Omega$ and they each have the same gauge factor of 2.1. What will be the output voltage from the bridge when the strain on each gauge has the magnitude of 0.001 and the bridge has a supply voltage of 5 V?

Chapter 18
Signal Processing

18.1 Noise

We have already considered in Section 17.3.1 the effect of noise generated internally in a transducer. Noise, defined as any unwanted signal, is, in fact, introduced at all parts of any signal system and, as previously discussed, the amplitude of the total noise signal determines the smallest detectable signal level in the system.

The noise level in a system is conveniently quoted as a signal to noise ratio, defined as the ratio of the magnitude of the wanted signal to that of the total noise signal, and is normally expressed in decibels, so that

$$\text{signal to noise ratio} = 20 \lg \frac{v_s}{v_n} \qquad (18.1)$$

Although it is possible to recover signals which are hidden by noise, even with a signal to noise ratio which is less than unity, for normal instrumentation processes the minimum signal to noise level must be very much better; great care must, therefore, be taken in the design and physical layout of circuits where the signal is at its lowest level. For high quality audio systems, for example, signal to noise ratios of the order of 60 to 80 dB are expected. Bearing in mind that a decibel ratio of 60 dB corresponds to a ratio of 1000 to 1, the achievement of such noise levels requires extreme care to be taken with the low signal level amplifier stages, and in particular with the amplifier input stage.

Wideband (white) noise will be generated, as shown in Section 17.3.1, in the input components of the amplifier, and also in the transducer, which may be, for example, a tape recorder replay head, which is providing the signal for the amplifier input. In addition, white noise will also be introduced by the active devices in the amplifier circuit. Any semiconductor device or saturated vacuum device is a source of a further form of wideband (white) noise, known as *shot noise*, caused by the random generation of current carriers. Both shot noise and Johnson, or thermal, noise may be classed as internal noise, being generated by physical processes within the transducer and other input devices themselves. The effects of such internal noise may be minimized by choosing only good quality, low noise components for use in low signal level circuits, and by operating the components under the optimum low noise level conditions specified by the component manufacturers. For some communication systems, a low internal noise level has been achieved by operating the low signal level amplifying devices at a low temperature.

There are two types of noise: random noise, which is generated internally within components and which has been discussed above, and interference, which is due to the interaction between external signals and fields and the system. Interference can occur in three ways: multiple earths, magnetic coupling or capacitive coupling.

If the system has more than one connection to earth then if the earth points are not precisely at the same potential an interference current can be produced, as illustrated in Fig. 18.1(a). The resulting circuit is known as a ground loop. Such effects can be avoided by having only a single earthing point, as in Fig. 18.1(b).

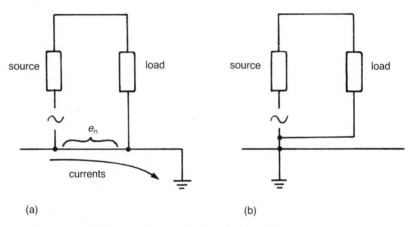

(a)   (b)

Fig. 18.1 Noise coupling due to ground loops.

A changing magnetic field is set up by an alternating or varying direct current. Thus any part of the system which has a circuit loop which is linked by a portion of the varying field produced by external circuits will have an e.m.f. induced in it of magnitude equal to the rate of change of flux linkages in the circuit loop. This magnetic coupling effect may be increased severely by the presence nearby of ferromagnetic material, which can given local increases in the magnetic flux density. Common sources of such interference are the currents in nearby power cables and cables in the system itself, and abruptly changing currents that occur in the operation of relays and motors. Ferromagnetic material may be used to screen a circuit from magnetic coupling effects, it needing to surround the circuit completely and in a continuous manner (without air gaps). A useful rule of thumb is that ferromagnetic material may encircle signal circuits but signal circuits should never encircle ferromagnetic material. The use of twisted pairs of wires for the out and return currents not only reduces the area of the circuit loop for flux linkage, but the magnetic flux linked by successive circuit loops in the twisted wire arrangement will give rise to induced e.m.f.s in opposite directions and so lead to a self cancelling effect.

Capacitive coupling arises because the earth and nearby items such as power cables are separated by just a dielectric, generally air, from conductors in

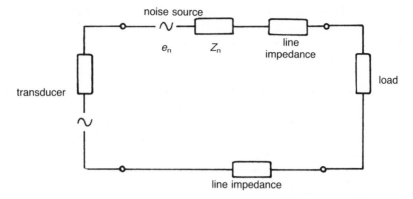

Fig. 18.2 Normal mode noise.

the signal system. Such arrangements act as capacitors. A change in voltage applied to one of the plates of a capacitor affects the voltage on the other. The capacitors thus couple the signal system to other systems and so signals in the other systems can give rise to interference signals. Capacitance coupling can be eliminated by completely enclosing circuits by earthed metal screens. The use of screened cable is thus recommended for low level signal circuits.

Noise may arise within the signal source, and is referred to as *normal mode noise*, or between the earth terminal of the system and its lower potential terminal, and is referred to as *common mode noise*. Figure 18.2 shows the equivalent circuit for a system with normal mode noise. The normal mode noise is represented by a voltage generator in series with the signal being measured. The figure shows a transducer connected to a load, which might for instance be the input terminals of a high gain amplifier. The noise generator, of voltage e_n and internal impedance Z_n is shown in series with one input lead. The noise voltage produced across the load terminals is dependent upon the magnitude of e_n and also upon the relative impedances of the transducer, the

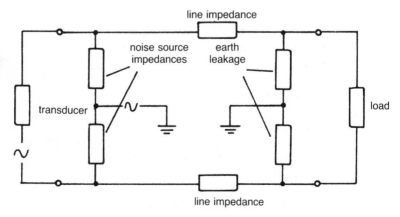

Fig. 18.3 Common mode noise.

load and the noise source. Figure 18.3 shows the equivalent circuit for a system with common mode noise. It results from ground loop currents or magnetic coupling inducing currents in earth loops.

18.2 Analogue/digital conversions

When studying analogue or digital techniques it must be borne in mind that the circuits considered are not necessarily complete in themselves and often form part of larger systems. The overall system may not be entirely analogue or entirely digital in nature. Thus a digital system may be controlled by input signals which are the amplified analogue outputs, perhaps of some measuring transducers. Similarly, a digital system output may be required to control the measured analogue system via analogue control valves. Interfacing is thus required between the analogue and the digital subsystems, and it is necessary to be able to convert an analogue signal into a digital equivalent signal, and vice versa. In Section 14.1 the basic analogue to digital conversion process was considered and it was shown that in a digital system, because of the time required to form each coded data value, it is only possible to produce digitally coded values of sample values of the analogue signal.

An analogue signal, by which we mean a continuously variable signal, cannot be represented exactly by a digital signal. The analogue signal must be sampled at intervals, the amplitude sample then being converted into a digital form. Samples must be taken at sufficiently frequent intervals for all the relevant information in the analogue signal variations to be retained. Sampling theory shows that in order to retain information which is contained in a signal, then at least two samples must be obtained per period of the highest frequency component. If the highest signal frequency component is f_s, then the period of the sampling signal is given by

$$T \leqslant \frac{1}{2f_s} \tag{18.2}$$

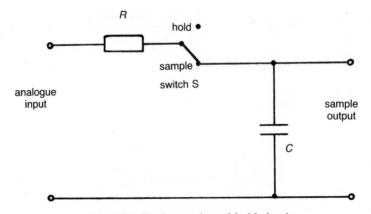

Fig. 18.4 Basic sample and hold circuit.

Figure 18.4 shows a basic *sample and hold circuit*. The capacitor C is used as a store or memory to hold the value of the sample. It is connected to the analogue signal input via the resistor R, which in practice will include the output impedance of the analogue signal source. The time constant CR is chosen to be sufficiently short so that the capacitor voltage can follow the required analogue signal variations. At the instant that the sample is to be taken, switch S is changed into the hold position, and the sample voltage is available to the succeeding analogue to digital convertor. The disadvantage of this simple sample and hold circuit lies in the drift which occurs in the capacitor voltage during the hold period. This is mainly due to the load placed upon the capacitor by the following circuitry. The use of as large a capacitor as is allowed by the time constant design will minimize the voltage drift, but this simple circuit is still not satisfactory for accurate use. An improvement may be made by including a high impedance buffer amplifier between the memory capacitor and its applied load, but for high accuracy systems, sample and hold circuits include some error correction arrangements. Whatever circuit is used, however, a certain time is necessary in the sample position for the memory device to store an accurate sample value. This is known as the *aperture time*, and corresponds to the sample time, t_s, shown in Fig. 14.6. An ideal sample and hold circuit would have a zero aperture time and an infinite hold time.

18.2.1 *Digital to analogue converters*

The most common digital to analogue converters are designed to operate with a digital input in a weighted code form, usually in simple binary coded form.

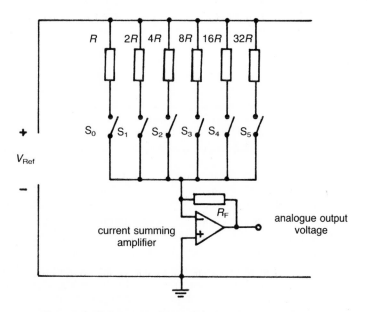

Fig. 18.5 Weighted resistor digital to analogue converter.

For digital inputs which are coded in some other form, for example in Gray binary code, it is necessary normally for the input signal to be converted into a weighted code form prior to the digital to analogue conversion.

Figure 18.5 shows a weighted resistor digital to analogue converter, which is arranged for digital input words of six bits. The six switches S_0 to S_5 are driven by the digital input, being closed when the relevant input bit has the logical value 1, and open when the bit has the value 0. The digital input must, therefore, be available in parallel form. The resistors are weighted so that each allows a current to flow into the summing amplifier of value proportional to the weighting of the bit; for the simple binary coded form shown the resistor values are graded according to the powers of two. The analogue output voltage is then proportional to the value of the digital input, having its maximum value when all switches are closed.

A modification of the circuit, which will accept binary coded decimal inputs, is shown in Fig. 18.6. Four graded resistors are used per decade, with a ratio of 10 between the values of adjacent decades. The disadvantage of converters of this type is the need for a large number of accurately graded resistors.

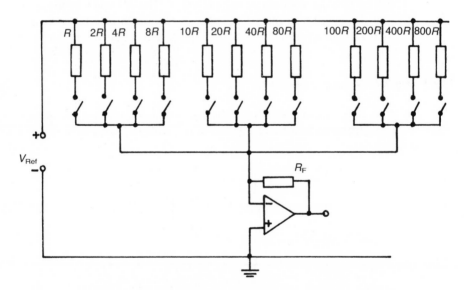

Fig. 18.6 Binary coded decimal digital to analogue converter.

Figure 18.7 shows an alternative type of converter for simple binary coded digital signals, and which is known as a resistor ladder converter. The converter is shown for a four bit parallel input, but the method is easily extended to any number of bits. The four bit inputs are again arranged to operate switch contacts; for each bit, a change over contact set is provided with the bottom end of the resistor being connected to the reference voltage V_{Ref} if the bit value is logical 1 and to earth if the bit value is logical 0. Switch S_0 is driven by the most significant bit of the digital input. In the figure, the switches are shown in the positions corresponding to the input word

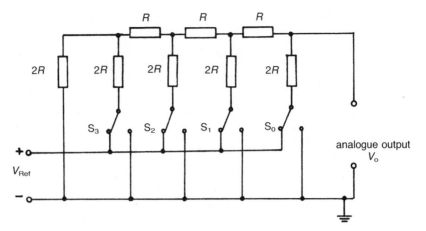

Fig. 18.7 Ladder resistor digital to analogue converter.

$F = 1111$

In Fig. 18.8 the circuit is redrawn with the switches omitted, but with the connections equivalent to this input word. Analysis of the circuit shows that the analogue output voltage V_0 corresponding to this input word is

$$V_0 = \frac{15}{16} V_{\text{Ref}}$$

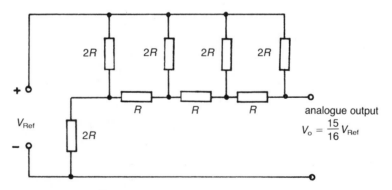

Fig. 18.8 Equivalent circuit of resistor ladder converter for i.

The decimal equivalent value of the binary word 1111 is 15. Analysis of the circuit for any input word shows that the output voltage V_0 is given by

$$V_0 = \frac{V_{\text{Ref}}}{16} \times \text{decimal equivalent value of input word} \qquad (18.3)$$

For a ladder network of this type, the actual value of the resistors does not affect the value of the analogue output voltage as long as the ratio $R:2R$ is correct, and as long as the output is not loaded significantly.

In the two types of converter described, the switches have been shown as contact switches; in modern fast systems these switches would normally be electronic switches. Both the converters require the digital input in parallel form – a serial digital input must therefore be converted into parallel form. This could be arranged by clocking the serial input into a parallel output shift register.

18.2.2   Analogue to digital converters

Analogue to digital converters are commonly based upon a comparison process in which a locally produced digital word is converted into its analogue equivalent and is compared with the input analogue sample. The locally produced digital word is varied until agreement with the sample is reached.

Figure 18.9 shows an analogue to digital converter known as a *continuous balance* converter. Clock pulses from the generator are clocked into a counter, whose parallel digital output is converted into its analogue equivalent, the voltage V_c in the figure. Voltage V_c is compared with the analogue input sample voltage V_a in a comparator, whose output controls the gating of the clock pulses to the counter. The system forms a continuous control loop which maintains the word stored in the counter as the digital equivalent to the analogue sample input.

The full range of the counter would be equivalent to the full voltage range of the input sample. If, therefore, the allowed analogue range is from 0 to 10 V, a

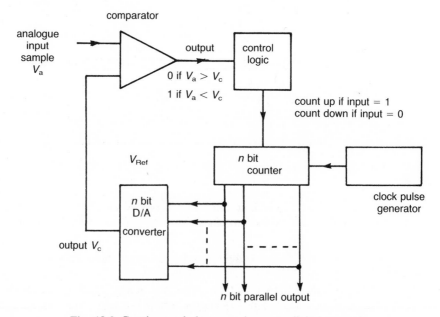

Fig. 18.9 Continuous balance analogue to digital converter.

10 bit binary counter would have a resolution equivalent to 1 least significant bit value of

$$v = \frac{10.0}{2^{10}} \text{V}$$

$$= \frac{10.0}{1024} \approx 10 \text{mV}$$

This is roughly a resolution of 1% of full scale.

The time taken for an analogue to digital converter to produce the digital output in response to a new analogue input sample is called the *conversion time*. The conversion time of the continuous balance type of converter is dependent upon the difference between the new sample and the previous sample. If, therefore, the previous sample was zero and the new sample is the full range voltage, then all bits of the counter will require setting, requiring $2^n - 1$ clock pulses where n is the number of bits in the counter. The maximum conversion time is thus

$$t = \frac{2^n - 1}{f_c} \text{ seconds} \tag{18.4}$$

where f_c is the clock pulse frequency (pulses per second).

An alternative form of converter is shown in Fig. 18.10. The block diagram is identical to that of the continuous balance converter, but it differs in the operation of the control logic. In this converter, the counter is emptied by resetting all bits to zero before a conversion is started. When the new analogue sample is present, the control logic starts the count, i.e. clock pulses are fed

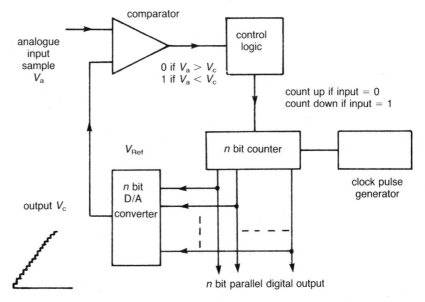

Fig. 18.10 Staircase ramp analogue to digital converter.

into the counter. The counter digital output thus increases bit by bit, at the clock frequency. The output from the digital to analogue convertor is a linear ramp, made up of equal incremental steps. The count continues until the generated staircase ramp exceeds the value of the analogue sample voltage, when the comparator output goes to logic 1 and stops the count. The counter output is, at this time, the digital equivalent of the analogue sample. The resolution and the conversion time are again decided by the number of bits in the counter, the range of the analogue input voltage and the clock frequency; the comparator must, of course, be capable of resolving sufficiently small changes of input voltages.

The conversion time of any converter using a counter is relatively long, if the counter is filled in the normal manner, with each input pulse equivalent in value to the least significant bit. The *successive approximation* type of converter differs from the previously described converters in the method of filling the counter. Its block diagram is again identical to that shown in Fig. 18.9, but it differs in the method of operation of the control logic. The counter is filled starting at the most significant bit (MSB) rather than at the least significant bit (LSB). This enables the output to approximate to the analogue sample in a much shorter time; this is, therefore, one of the fastest of such converters. The conversion starts by placing a 1 in the MSB of the counter; this produces an output from the digital to analogue convertor of one half of the full range voltage. The control logic thus indicates whether the analogue sample lies in the lower or upper half of the allowable voltage range. The 1 in the MSB is then removed if the sample lies in the lower half or retained if the sample lies in the upper half, and the process repeated for the next most significant bit. As each bit is tested, the resolution of the conversion is doubled until the required accuracy has been obtained.

18.3 Fourier representation of signals

In the previous sections of this book, signals have been considered without detailed reference to their frequency spectrum. A continuous sinusoid contains no information other than the presence of the sinusoid; all real information-carrying signals are, therefore, non-sinusoidal in nature and consist of a voltage or current, some property of which is varied as an analogue of the information. The use of the *Fourier series* allows any repetitive signal waveform to be represented as a d.c. component together with an infinite series of sinusoidal components; the lowest sinusoidal term, the fundamental, being of the same frequency as the original signal waveform, and the remaining sinusoidal terms being multiples or harmonics of the fundamental.

Expressed mathematically, any complex signal may be written

$$v = V_0 + V_1 \sin(\omega t + \phi_1) + V_2 \sin(2\omega t + \phi_2) + \ldots + V_n \sin(n\omega t + \phi_n)$$
$$(18.5)$$

where V_0 is the d.c. component, $V_1 \ldots V_n$ are the peak values of the sinusoidal terms and $\phi_1 \ldots \phi_n$ are the phases of these waves at time $t = 0$.

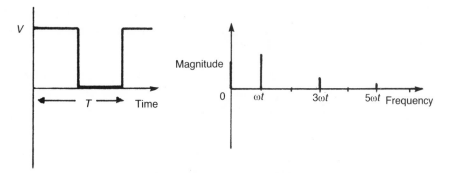

Fig. 18.11 A square wave and its amplitude spectrum.

Of great interest, especially in dealing with digital systems, is the square pulse waveform shown in Fig. 18.11, which shows a square wave of peak value V and period $T = 2\pi/\omega$ seconds. The d.c. component may be obtained by inspection, the average value of the wave being $0.5\,V$. A Fourier expansion gives the wave as

$$v = \frac{V}{2} + \frac{2V}{\pi}(\sin \omega t + \frac{1}{3}\sin 3\omega t + \frac{1}{5}\sin 5\omega t \ldots) \tag{18.6}$$

The component frequencies of a signal are conveniently displayed as an amplitude spectrum, also shown in the figure, in which the length of the line is proportional to the magnitude of the frequency component. The wave may be reconstituted by adding together the d.c. component and the alternating components in the correct magnitude and phase; all components, with frequencies extending up to infinity, must be included to obtain the original ideal square wave shape. Because, of course, all practical systems have a limited bandwidth, it follows that the realization of an ideal square wave in practice is impossible, and that limits are imposed by the system on the maximum rate of change of signal level that can be achieved.

It is interesting to consider the synthesis of a square wave using only a very limited number of harmonics. Figure 18.12 shows the synthesis of a square wave by adding together a d.c. component and fundamental and third harmonic terms, with amplitudes and phase according to Equation (18.6).

In Fig. 18.13 the same synthesis is shown with the phase of the third harmonic shifted by $\frac{1}{2}\pi$ radians. It can be seen that the resultant wave is very different from that shown in Fig. 18.12. It is important that when signals are processed by a system, if the waveform is not to be altered then all the signal components are processed in the same way, their relative magnitudes and phases being maintained. By maintaining the same relative phase, we in fact mean that each signal component is subject to the same time delay in the system. Bearing in mind that one degree of phase at the fundamental frequency is equivalent in time to three degrees of phase at the third harmonic frequency, we can say that the phase response of a system should be linearly proportional to frequency.

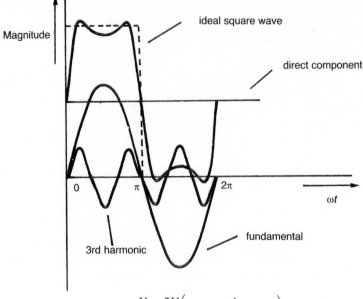

Fig. 18.12 $\dfrac{V}{2} + \dfrac{2V}{\pi}\left(\sin \omega t + \dfrac{1}{3}\sin 3\omega t\right).$

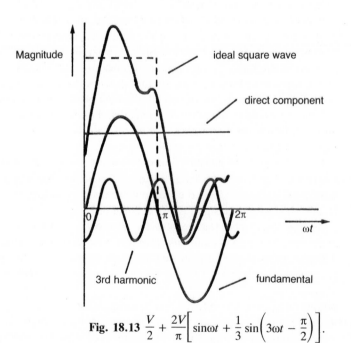

Fig. 18.13 $\dfrac{V}{2} + \dfrac{2V}{\pi}\left[\sin\omega t + \dfrac{1}{3}\sin\left(3\omega t - \dfrac{\pi}{2}\right)\right].$

In practice, pulse waveforms are usually specified in terms of their rise and fall times, which are defined as the times required for the signal voltage or current to change between 10% and 90% of its final value.

The frequency bandwidth required to process signals obtained from transducers obviously depends greatly upon the original physical variable. For a high quality audio signal, the required bandwidth is approximately from 30 Hz to 15 kHz. For commercial telephony, it is found that the normal speech bandwidth can be restricted to the range 300 Hz to 3400 Hz without loss of intelligibility. Signals originating from mechanical vibration measurements would normally lie in the frequency range up to 3 kHz, and may be required to respond to static deflections, which would require the system to have a response extending down to 0 Hz.

18.4 Data transmission

Data transmission between a signal source and its load may be via connecting wires or via a radiated link. If a d.c. voltage is applied to the source ends of a pair of conductors, the voltage received at the load end is equal to the source voltage minus the voltage drop across the resistance of the conductors. When an alternating signal voltage is applied to the source end of the line, the reactive properties of the line also affect the value of the voltage at the load. Capacitance between the two conductors and the series inductance of each conductor cause the line to take on different transmission properties. At low frequencies the reactive effects will be small, and are usually negligible. At high frequencies, however, the connecting wires can have considerable effect upon the wave shape received at the line termination. For systems in which the signal is in pulse form, high frequency techniques are necessary even at relatively low data rates, because the need to retain the pulse wave shape requires, as discussed in the previous section, a bandwidth of up to ten times the basic pulse repetition frequency.

As a rough guide, low frequency signals may be transmitted over very long wires, a good example being the commercial telephone system. For the TTL and other logic devices described in Chapter 14, frequency components in excess of 100 MHz are not uncommon, and connecting wires of lengths of several centimetres must be treated as transmission lines. In general, if the transmission length forms a significant fraction of the wavelength of the signal, then care must be taken. The velocity of propagation of the signal on a line is dependent upon the reactive properties of the line. The actual velocity is somewhat less than the velocity of light (3×10^8 m/s), giving a propagation time along a cable of the order of 5 ns/m.

18.4.1 Transmission lines

The term *transmission line* is used for conductors along which electrical signals and energy can be sent. This might, for instance, be a coaxial cable, a pair of parallel wires in a circuit, metallic strips on a printed circuit board or perhaps the power lines used to transmit electrical power over thousands of kilometres. Transmission lines have resistance, inductance, capacitance and what is termed conductance. This conductance represents the imperfection of the insulation between conductors which allows some current to leak from one to the other.

Figure 18.14(a) shows an equivalent circuit which, when used to represent successive lengths of a transmission line, gives similar transmission properties. The resistance of a section of the transmission line is represented by R, its inductance by L, its capacitance by C and its conductance by the shunt resistance R_s. The characteristic impedance Z_0 of a single symmetrical T-section is given by (see Section 13.2 for a discussion of the term characteristic impedance and Equation (13.6))

$$Z_0 = \sqrt{(\tfrac{1}{4}Z_1^2 + Z_1 Z_2)} \qquad (18.7)$$

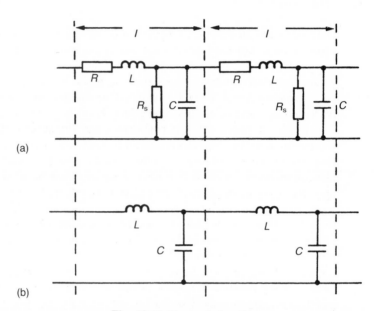

(a)

(b)

Fig. 18.14 Transmission line.

where $\tfrac{1}{2}Z_1$ is the impedance of series parts of the T and Z_2 that of the shunt impedance. In this case, Z_1 is the impedance due to R in series with L and Z_2 is that due to C in parallel with R_s. If R is the resistance per unit length, L the inductance per unit length, C the capacitance per unit length and R_s the shunt resistance per unit length then we can write, for a length l,

$$Z_1 = Rl + j\omega Ll$$

$$Z_2 = \frac{1}{Gl + j\omega Cl}$$

where G is the conductance per unit length, i.e. $G = 1/R_s$. Thus Equation (18.7) gives

$$Z_0 = \sqrt{\left[\frac{(R + j\omega L)^2 l^2}{4} + \frac{R + j\omega L}{G + j\omega C}\right]}$$

Since the length of each T-section is chosen to be small in comparison with the total length of the transmission line, then the first term in the above equation can be generally neglected to give

$$Z_0 = \sqrt{\left[\frac{R + j\omega L}{G + j\omega C}\right]} \tag{18.8}$$

For good quality cables or at high frequencies, the series resistance and the conductance components can generally be neglected in comparison with the reactive components (Fig. 18.14(b)). Such a line is referred to as a *loss-free line* and then Equation (18.8) approximates to

$$Z_0 = \sqrt{\left[\frac{L}{C}\right]} \tag{18.9}$$

Such an impedance is purely resistive. Because the impedance Z_0 is determined by the inductance and capacitance per unit length of line, the value is fixed by the physical construction of the cable, i.e. by the dimensions and the permittivity and permeability of the insulation. Typical values are $600\,\Omega$ for twin parallel open wires and $75\,\Omega$ for small coaxial cable similar to that used with domestic television aerials.

A correctly terminated transmission line has a terminating impedance which is the characteristic impedance of the line. When this occurs, all the energy delivered along the line by a signal is absorbed by the load. Thus voltage and current waves travel along the line and are absorbed by the load. If, however, the line is not correctly terminated then all the energy is not absorbed and some reflection of the voltage and current waves occurs. The reflected waves travel back towards the source end of the line. The current and voltage at any point along the transmission line is then the phasor sum of the incident and reflected waves at the point, the impedance at the point being the ratio of the phasor sum of the voltages to the phasor sum of the currents. The current and voltage along the line is then a stationary standing wave resulting from the combination of the forward and reflected travelling waves.

The ratio of the maximum value of the standing wave voltage to the minimum value is known as the *voltage standing wave ratio* and is used as a measure of the impedance mismatch in a transmission system. When the line is correctly terminated and no reflection occurs then the standing wave ratio has the value 1. When it is not correctly terminated and all the energy is reflected, the standing wave ratio is infinity. Standing wave ratios thus range from 1 to infinity, with the ideal value being 1.

In any transmission system, therefore, it is desirable that all connecting sections or devices should be matched in impedance so that reflections are avoided and all the energy transmitted along the line is delivered to the load.

Example   What is the characteristic impedance of a loss-free line having an inductance of 1.4 mH/km and a capacitance of 0.02 µF/km?

Using Equation (18.9)

$$Z_0 = \sqrt{\left[\frac{L}{C}\right]} = \sqrt{\left[\frac{1.4 \times 10^{-3}}{0.02 \times 10^{-6}}\right]} = 265\,\Omega$$

18.4.2 Impedance matching

In order to achieve impedance matching between devices or systems of different impedance, impedance transforming circuits must be used. These may take the form of an electronic circuit, such as an emitter follower transistor amplifier, or perhaps the simplest form of a matching transformer.

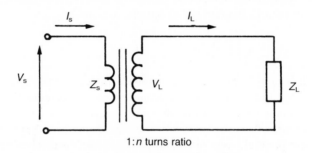

Fig. 18.15 Impedance matching.

Figure 18.15 shows a load impedance, Z_L, supplied via an ideal transformer from a voltage supply V_s. The transformer has a primary to secondary turns ratio of $1:n$, and because it is ideal, this results in a secondary load voltage of

$$V_L = nV_s$$

The secondary current is given by

$$I_L = \frac{nV_s}{Z_L}$$

and the primary current by

$$I_s = n \times \frac{nV_s}{Z_L}$$

or

$$I_s = n^2\frac{V_s}{Z_L}$$

The effective impedance presented to the voltage supply is therefore

$$Z_s = \frac{V_s}{I_s} = \frac{Z_L}{n^2} \tag{18.10}$$

The effect of the transformer is to transform the impedance of the load, Z_L, by a factor equal to the square of the transformer turns ratio.

Example A transducer of output resistance $3\,\Omega$ is to be connected to an amplifier of input resistance $5\,\text{k}\Omega$ by a cable of characteristic impedance $600\,\Omega$. Estimate the turn ratios of the required matching transformers.

Between the cable and the amplifier

$$\text{turns ratio} = \sqrt{\left(\frac{5 \times 10^3}{6 \times 10^2}\right)} = 2.89$$

Between the transducer and the cable

$$\text{turns ratio} = \sqrt{\left(\frac{6 \times 10^2}{3}\right)} = 14.1$$

In both cases, the secondary winding requires more turns than the primary winding.

18.4.3 Radio transmission

Data transmission via a radiated link may be used where the transmission path requires such a link because of the distance involved, or in some transducer applications because of the physical impossibility of making satisfactory connections via wires. Examples are the telemetry of medical data from the human body via a radio link from a swallowed transmitter enclosed in a small *pill* or the telemetry of temperature data from sensors embedded in the rotating armature of an electric motor. The use of radio links can enable instrumentation to be applied to a live subject without his or her knowledge; this is sometimes necessary in behavioural investigations where the attachment of cabling may alter the subject's responses.

The main problem in such data telemetry applications is again one of impedance matching in that energy must be radiated into space in the form of a radio wave. In order to couple efficiently into space the transmitting aerial should have dimensions which are comparable with the wavelength of the radiation; and if directional aerial characteristics are required, the aerial array will need dimensions considerably greater than the wavelength. For this reason, telemetry radio links require frequencies in excess of $10\,\text{MHz}$ where the wavelength is $30\,\text{m}$, with physically small systems using frequencies as high as 80 to $90\,\text{MHz}$, giving a wavelength of 3 to $4\,\text{m}$. Even so it is often necessary to operate with inefficient aerials which are too small, relying upon short transmission paths for adequate signal to noise ratio.

18.5 Modulation systems

The process of varying a property of a voltage or current in accordance with a data signal is termed *modulation*. The data are modulated on to a carrier signal, whose frequency is chose to suit the characteristics of the transmission link to be used. The carrier waveform may in general be of any wave shape, but is usually of sinusoidal or pulse form.

18.5.1 Amplitude modulation

Amplitude modulation (AM) is perhaps the form of modulation most frequently met, many transducer systems inherently producing a signal which is amplitude modulated. The basic process consists of multiplying the data signal

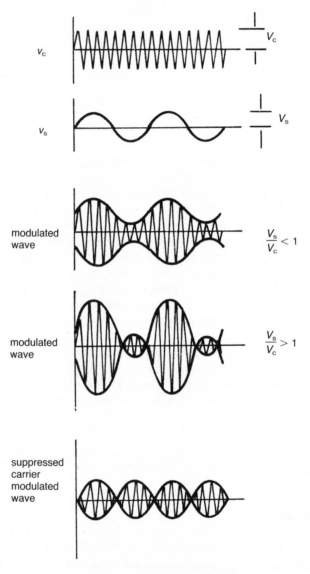

Fig. 18.16 Amplitude modulation.

by a constant amplitude, constant frequency carrier signal. If we represent the carier signal in sinusoidal form by

$$v_c = V_c \sin \omega_c t$$

and the data signal, in a simple form, by

$$v_s = V_s \sin \omega_s t$$

then the amplitude modulated carrier is presented by

$$v = (V_c + V_s \sin \omega_s t)\sin \omega_c t \tag{18.11}$$

The form of the amplitude modulated wave of Equation (18.11) is shown in Fig. 18.16.

The ratio

$$m = \frac{V_s}{V_c} \tag{18.12}$$

is known as the *modulation factor* or *modulation index* and is usually expressed as a percentage.

The modulated wave equation can be expanded using the identity

$$\sin \alpha \sin \beta = \tfrac{1}{2}\cos(\alpha - \beta) - \tfrac{1}{2}\cos(\alpha + \beta)$$

giving the result

$$v = V_c \sin \omega_c t + \frac{mV_c}{2}\cos(\omega_c - \omega_s)t - \frac{mV_c}{2}\cos(\omega_c + \omega_s)t \tag{18.13}$$

The modulated wave can be seen to consist of three terms: a term $V_c \sin \omega_c t$ which is the unchanged carrier signal a term $\tfrac{1}{2}mV_c \cos(\omega_c - \omega_s)t$ which is a cosinusoidal term at a frequency given by the difference between the carrier frequency and the modulating frequency and a term $\tfrac{1}{2}mV_c \cos(\omega_c + \omega_s)t$ which is a cosinusoidal term at a frequency given by the sum of the carrier and the modulating frequencies.

In Fig. 18.17(a) the frequency spectrum is shown for a sinusoidal carrier, angular frequency ω_c, modulated by a sinusoidal signal of angular frequency ω_s. The two sum and difference frequencies $\omega_c + \omega_s$ and $\omega_c - \omega_s$ are known as the side frequencies. The modulating signal, being of steady sinusoidal wave shape, contains no real information. A real signal will consist of a band of frequencies, from, say, ω_1 to ω_2 as shown in the spectrum of Fig. 18.17(b). The side frequencies thus become sidebands of frequencies, and are spoken of as the upper and lower sidebands. The lower sideband will extend from $\omega_c - \omega_1$ to $\omega_c - \omega_2$ while the upper sideband will extend from $\omega_c + \omega_1$ to $\omega_c + \omega_2$. The bandwidth required for transmission of an amplitude modulated wave is equal to twice the highest frequency component of the modulating signal.

Figure 18.16 shows also the modulated waveform for two modulation factors, one exceeding 100%. In normal modulation systems, it is not desirable for the modulation factor to exceed 100%, this leading to distortion when the carrier is demodulated to recover the information signal.

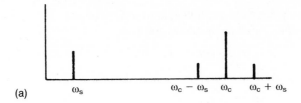

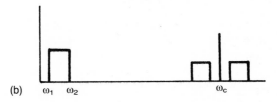

Fig. 18.17 Frequency spectrum for amplitude modulated signal.

It is instructive to determine the signal power contained in each of the components of the modulated signal. If the signal voltage was applied across a resistor R, then the power dissipated due to the carrier term would be given by

$$\text{carrier power} = \left[\frac{V_c}{\sqrt{2}}\right]^2 \frac{1}{R}$$

$$= \frac{V_c^2}{2R} \tag{18.14}$$

Due to each side frequency the power dissipated would be given by

$$\text{side frequency power} = \left[\frac{mV_c}{2\sqrt{2}}\right]^2 \frac{1}{R}$$

$$= \frac{m^2 V_c^2}{8R} \tag{18.15}$$

At 100% modulation, i.e. with $m = 1$, the power contained in each side frequency is equal to $\frac{1}{8}V_c^2/R$ and is one quarter of the carrier power. Because information is contained only in the side frequencies, the transmission of the carrier signal is a power waste. Systems are in use in which the carrier signal is suppressed at the transmitter, resulting in a power saving. Figure 18.16 also shows a suppressed carrier modulated wave. This technique has been extended to single sideband systems in which one sideband is also suppressed, this being possible because the total information content of the wave is contained equally in either sideband. This technique also gives a saving in the required frequency bandwidth. In demodulating such a system, however, the signal intelligence can only be recovered from a single sideband radiation if the exact carrier frequency, suppressed at the transmitter, is known at the receiver.

In practice, amplitude modulation is achieved by applying the signal and the

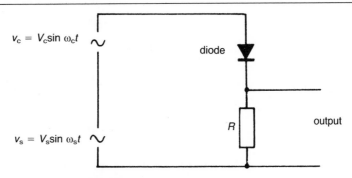

Fig. 18.18 Amplitude modulation.

unmodulated carrier to a non-linear device. Figure 18.18 shows a very simple modulation system in which the two signals are applied to a non-linear circuit made up of a diode and a resistor R in series. The modulated output signal voltage is taken from across the resistor R, whose value is sufficiently low to be negligible in comparison with that of the diode.

A general non-linear characteristic is given by

$$i = a + bv + cv^2 + \ldots \tag{18.16}$$

If the diode characteristic is represented by the first three terms, then the current may be written

$$i = a + b(V_c \sin \omega_c t + V_s \sin \omega_s t) + c(V_c \sin \omega_c t + V_s \sin \omega_s t)^2 \tag{18.17}$$

The first two terms consist of a d.c. component and components at the input frequencies. The third term may be rewritten as

$$c(V_c \sin \omega_c t + V_s \sin\omega_s t)^2 = cV_c^2 \sin^2 \omega_c t + cV_s^2 \sin^2 \omega_s t \\ + 2cV_cV_s \sin \omega_c t \sin \omega_s t$$

The $\sin^2$ terms give further d.c. components and also components at twice the input frequency, because of the expansion

$$\sin^2 \theta = \tfrac{1}{2}(1 - \cos 2\theta)$$

The term $2cV_cV_s \sin \omega_s t \sin \omega_c t$ represents the modulated wave, expanding to give the side frequencies

$$cV_cV_s \cos(\omega_c - \omega_s)t - cV_cV_s \cos(\omega_c + \omega_s)t \tag{18.18}$$

All the component frequencies will exist in the output signal which can be taken as a voltage signal from across the resistor R. The carrier and side frequencies would be selected by a bandpass filter which would be tuned to the carrier frequency and which would have a bandwidth equal to at least twice the highest modulating frequency. A basic modulator of this sort is not very efficient.

Figure 18.19 shows a practical suppressed carrier modulator circuit. This is the double balanced ring modulator consisting of four diodes, which, when driven by a sufficiently large carrier signal, conduct alternately in pairs. The

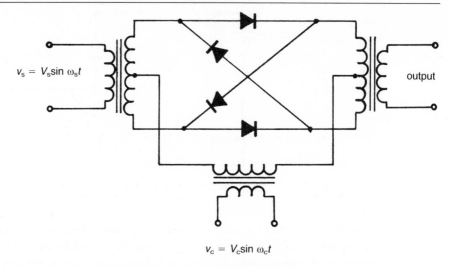

$v_s = V_s \sin \omega_s t$

output

$v_c = V_c \sin \omega_c t$

Fig. 18.19 Suppressed carrier modulation circuit.

input signal is transmitted unchanged during one half of the carrier signal period, and inverted for the remaining half. The figure also shows the output signal waveform. Because of the symmetry of the circuit, currents at the carrier frequency in the two halves of the output transformer primary winding are equal and opposite, and the carrier output is completely suppressed.

Demodulation, or detection of amplitude modulated signals is commonly accomplished using a diode in the circuit of Fig. 18.20. The resistor R forms

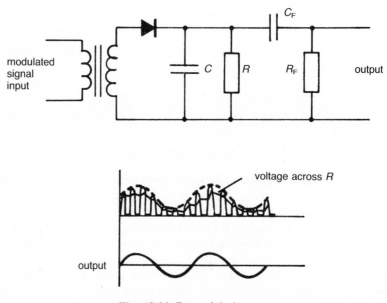

modulated signal input

C_F

C R R_F output

voltage across R

output

Fig. 18.20 Demodulation.

the diode load and is shunted by a capacitor C, the time constant of the combination being long when compared with the period of the carrier signal, but small when compared with the period of the highest frequency modulating signal. The diode acts as a half wave rectifier, removing one half of the modulated wave. The capacitor C charges during the conducting half cycles almost to the peak value of the applied signal voltage, but, because of the relatively long time constant, discharges only slightly between successive peaks. Because the time constant is relatively short in comparison with the modulating signal period, the output voltage follows normal variations in the peak input voltage. The output voltage consists therefore of a varying d.c. voltage; the d.c. component can be removed by the capacitor–resistor high pass filter $C_F R_F$ also shown in the figure. The waveforms across the two resistors are drawn in the figure, together with the half sinewave at carrier frequency for comparison. It should be borne in mind when considering these waveforms that the carrier frequency will, in all practical systems, be of much higher frequency than it is possible to represent in such a drawing.

18.5.2   Frequency modulation

Amplitude modulated systems suffer from the disadvantage that much of the noise and extraneous interference which exists in all communication links appears as random variations in the signal amplitude. It is difficult, therefore, to differentiate between the wanted and the unwanted signal variations.

The unmodulated carrier signal may be written

$$v_c = V_c \sin(\omega_c t + \phi)$$

There are two further possible methods of modulating this carrier signal, while maintaining its amplitude constant; these methods both come under the general name of *angle modulation*.

In a *phase modulation system*, the relative phase of the carrier signal, ϕ, is altered in accordance with the amplitude of the modulating signal. In a *frequency modulation system*, the angular frequency of the carrier signal, ω_c, is altered in accordance with the amplitude of the modulating signal. The two methods are very closely related and give frequency spectra which are very similar. Frequency modulation systems are more commonly used, and consideration will be restricted to this system.

Figure 18.21 shows a constant amplitude carrier signal which is frequency modulated by a sinusoidal modulating signal. At A, C and E the amplitude of the modulating signal is instantaneously zero; at these instants the frequency of the carrier is at its normal unmodulated value. At B, the positive peak value of the modulating signal, the carrier frequency has been increased to a maximum. At D, the negative peak value of the modulating signal, the carrier frequency has been decreased by the same amount to a minimum. The rate at which the carrier frequency is varied is proportional to the frequency of the modulating signal. The magnitude of the carrier frequency variation is known as the *carrier deviation*. In any system, a maximum carrier deviation is specified

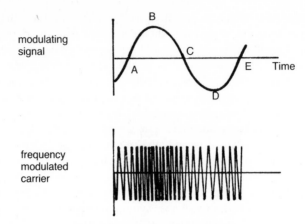

Fig. 18.21 Frequency modulation.

which is reached with the maximum allowed amplitude of the modulating signal. In the BBC VHF transmissions, using frequency modulated carriers in the range 88 to 97.6 MHz, the maximum carrier deviation is ±75 kHz.

As with amplitude modulation, frequency modulation of a carrier wave produces side frequencies. The equation to a frequency modulated wave may be written

$$v = V_c \sin(\omega_c t + m_f \sin \omega_s t) \tag{18.19}$$

where m_f is the *modulation index* and is defined as

$$m_f = \frac{\text{frequency deviation}}{\text{modulating frequency}} \tag{18.20}$$

Equation (18.19) can be expanded and yields a centre carrier frequency and an infinite set of side frequencies, each pair separated by an amount equal to the modulating frequency. Although the sidebands extend to infinity, in practice the magnitudes decrease reasonably quickly. It is, therefore, possible to specify the bandwidth necessary for a frequency modulated transmission, a bandwidth of about 200 kHz being required for the VHF broadcast transmissions. This bandwidth requirement is very high in comparison with that required for an equivalent amplitude modulated transmission. With correct design and adequate signal strength, however, frequency modulation has a major advantage over amplitude modulation in noisy signal channels.

18.5.3 *Multi channel transmission systems*

Many applications occur where simultaneous transmission is required for several data signals. Typical examples are a commercial telephone system where many telephone conversations are being held at the same time over trunk telephone links. In an instrumentation system, the temperature of a particular machine may be measured by thermocouples embedded in the machine at various strategic points, all of which require continuous monitoring.

For such applications there are two basic methods whereby multi channel working may be arranged. One, termed *frequency division multiplex*, involves the positioning of the separate channels side by side in the frequency spectrum. A good example of this technique is given by the use of single sideband modulation to transmit telephone signals. The telephone signals, each with a bandwidth of 300 Hz to 3.4 kHz, are modulated on to subcarriers, separated in the frequency spectrum at 4 kHz intervals. Using single sideband techniques, the sidebands produced by the telephone signal fit neatly into the frequency space between adjacent subcarrier signals. The modulated signals are then combined together to give one complex signal which is itself modulated on to a final carrier signal for transmission over a radio or cable link. At the receiving end, the signal is demodulated to produce the complex subcarrier system, which is split into separate subcarrier signals by a system of bandpass filters, each of which is tuned to one of the subcarrier frequencies and with sufficient bandwidth to pass only the subcarrier and its sidebands. After separation into the separate subcarrier channels, each is demodulated to produce the original telephone signal. In this manner, many hundreds of telephone circuits can be simultaneously established over a single transmission-link.

The other basic multichannel system is termed *time division multiplex* and involves the positioning of the separate channels end to end in time. The system is inherently a sampling system and can only give transmission of sampled values of each signal variable. The principles of sampling, which have already been discussed in Section 18.2, must be applied in determining an adequate sampling frequency. Each channel is connected in turn to the single data transmission link for a period t; after all channels have been so connected, taking a period

$$T = nt$$

where n is the number of channels, the sequence is restarted. It is necessary for some checking signal to be included to allow the synchronization of channels at the transmitter and the receiver. Each channel, therefore, has available to it a very short pulse of carrier signal. The pulse characteristics must be varied in some manner proportional to the signal amplitude at the sampling instant.

18.5.4   *Pulse modulation*

Various methods are in common use whereby the characteristics of a pulse are varied according to a modulating signal.

(1) *Pulse amplitude modulation* (PAM). The amplitude of the transmitted pulse is varied in accordance with the magnitude of the sample of the modulating signal.

(2) *Pulse position modulation* (PPM). With a zero modulating signal the pulses are transmitted in a regular uniform sequence. With applied modulation,

the position of the pulse is varied about its unmodulated position, with the deviation from the zero position being proportional to the magnitude of the modulating signal.

(3) *Pulse duration modulation* (PDM). The modulation is accomplished by fixing either the leading or trailing edges of the pulse and varying the position of the other edge. In some systems the centre of the pulse is fixed and both leading and trailing edges are shifted.

(4) *Pulse code modulation* (PCM). With this type of modulation the sample amplitude is encoded, and the code transmitted as a succession of pulses. The use of this type of code was discussed in Section 14.2.

Problems

1. A three bit digital to analogue converter has a voltage output of +5 V for an input of 0100. What will be the output for an input of 0010?

2. For the ladder resistor network shown in Fig. 18.7, show that the voltage output is $V_0/2 + V_1/4 + V_2/8 + V_3/16$, where V_0, V_1, V_2 and V_3 are the voltages applied when the respective switches S_0, S_1, S_2 and S_3 are closed.

3. A three bit digital to analogue converter, of the form shown in Fig. 18.6, has $R = 2\,k\Omega$ and a reference voltage of +10 V. The operational amplifier has a feedback resistance of $5\,k\Omega$. What is the output voltage for a binary input of 101?

4. A ramp type analogue to digital converter has an eight bit counter and a clock frequency of 1 MHz. What is the maximum conversion time?

5. What is the characteristic impedance of a loss-free line having an inductance of $0.06\,mH/km$ and a capacitance of $0.02\,\mu F/km$?

6. A transducer of output resistance $5\,\Omega$ is to be connected to an amplifier of input resistance $4\,k\Omega$ by a cable of characteristic impedance $600\,\Omega$. What will be the required turns ratios of the matching transformers?

7. What is the modulation factor and frequency spectrum produced when a carrier wave of frequency 10 MHz and peak value 10 V is amplitude modulated by a wave of frequency 6 kHz and amplitude 4 V?

8. A signal voltage of $v_s = 1 \sin 100t$ and a carrier voltage of $v_c = 2 \sin 1000t$ are applied in series to a device whose characteristic is given by

$$i = 10 + 1.5v + 0.06v^2 \, mA$$

Determine the magnitudes of the various frequency components of the device current.

9. For the modulator of Problem 8, determine the resulting modulation factor.

10. An amplitude modulated signal is transmitted with a total power of 5 kW and a modulation factor of 60%. What is the power transmitted by the carrier and each of the side bands?

11. A carrier wave $10\sin 2\pi 10^6 t$ V is amplitude modulated by a signal of $4\sin 2\pi 10^3 t$ V. What is (a) the modulation factor, (b) the sideband frequencies, (c) the fraction of the power transmitted by the carrier, (d) the fraction of the power transmitted in the sidebands?

12. What are the advantages frequency modulation has over amplitude modulation?

13. What is the modulation index for a frequency modulated wave which has a modulating signal of frequency 1 kHz and a peak frequency deviation of 5 kHz?

Chapter 19
Electrical Machines

19.1 Motors and generators

In 1820 Oersted noticed that a compass needle was deflected when placed near to an electric current. A year later, Faraday invented the first direct current motor and ten years later the first generator. During the following years many different types of motors were invented but because the electrical power source was limited to batteries, they could not compare economically with the steam engine. The early direct current generators employed rotating permanent magnets (Pixii's generator 1832), and it was not until 1856 that Siemens designed a machine in which the armature was positioned between the poles of permanent magnets. A major step forward in the field of electrical machines was the replacement of permanent magnets by electromagnets in 1866, for which Siemens was again responsible. In 1883 Crompton, in England, and Edison, in America, introduced heavy duty direct current generators. Ferranti, meanwhile, was working on the design of alternating generators and transformers, and in 1887 two-phase and three-phase induction motors were patented by Tesla. Thus the late 19th century saw the birth of the electrical machine as we know it today.

The function of an electrical machine is to convert mechanical energy into electrical energy (generator) or electrical energy into mechanical energy (motor). The method of driving a generator varies depending on the environmental conditions. The major part of all the electrical energy generated in this country is produced by alternators (a.c. generators) driven by high speed steam turbines, the steam being produced by coal, gas or oil fired boilers or by nuclear reactors. In North America, however, the major proportion of the electrical energy is produced by alternators driven by water wheels (hydro electrical generation). Whatever type of prime mover is used, the important characteristic of an electrical generator is its regulation curve, i.e. the relation between output voltage and output current. Electric motors are used in a wide variety of industrial drives (machine tools, pumps, fans, hoists, transportation, etc.). It is usually the mechanical features of a particular application which determine the choice of electric motor to be used and therefore the torque–speed characteristic of the machine is very important.

The characteristics of various types of machines will now be developed and, for purely historical reasons, the direct current machines will be treated first.

19.2 The d.c. generator

All conventional electrical machines consist of a stationary member called the stator separated by an air gap from a rotating member called the rotor. In direct current machines the stator usually consists of salient poles with coils wound round them so as to produce a magnetic field. The rotor is familiarly called the armature and consists of a series of coils located in slots around its periphery and connected to a commutator.

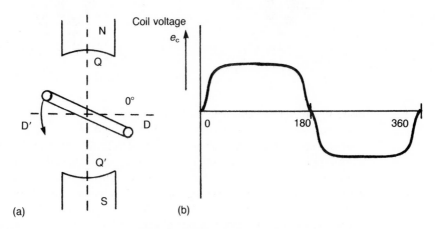

Fig. 19.1 Generation of an e.m.f.

Consider the simplest form of d.c. generator shown in Fig. 19.1(a), in which a single coil is rotated in a two pole field. By Faraday's law the voltage generated in the coil is equal to the rate of change of the flux linking the coil. When the coil lies in plane DD' there is maximum flux linking the coil but minimum rate of change of flux linkage, whereas along plane QQ' there is zero flux linking the coil but the rate of change of flux linkage is a maximum. The variation of the voltage generated in the coil as it moves through 360° is shown in Fig. 19.1(b). It is seen that the coil voltage is alternately positive and negative, i.e. it is an alternating voltage. To convert this alternating voltage into a direct voltage a commutator is used, as shown in Fig. 19.2. The commutator consists of brass segments separated by insulating mica strips. In the simple single coil system shown in Fig. 19.2(a) there are two brass segments and in the two coil system of Fig. 19.2(b) there are four segments.

In a typical machine there may be upwards of 36 coils requiring a 72 or more segment commutator. The commutator is solidly connected to the armature and rotated with it. External connection to the armature is made through stationary carbon brushes which make sliding contact with the commutator as it rotates. Considering Figs 19.1(a) and 19.2(a), as the coil rotates through 180° from the plane DD', coil side A is under a north pole and is connected through its commutator segment to the upper brush, whilst coil side A' is under a south pole and is connected through its commutator segment to the

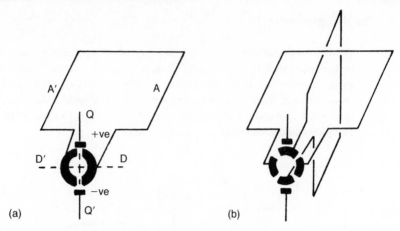

Fig. 19.2 Commutator.

lower brush. As the coil rotates through a further 180° coil side A′ is under a north pole and connected to the upper brush whilst coil side A is under a south pole and connected to the lower brush. It is clear then that the coil side under a north pole is always connected to the upper brush whilst the coil side under a south pole is always connected to the lower brush. The voltage waveform across the brushes is as shown in Fig. 19.3(a), which is not alternating positive and negative as before, i.e. the commutator has converted the alternating coil voltage into a direct brush voltage. Figure 19.3(b) shows that for the two coil system the d.c. armature output voltage is constant.

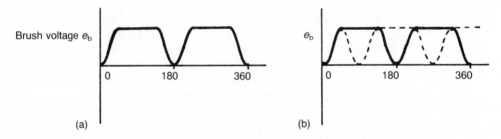

Fig. 19.3 Voltage waveforms.

19.2.1 Voltage and torque equations

Induced armature voltage
Let ϕ be the flux per pole and N be the speed in rps. Consider the single turn armature coil shown in Fig. 19.1(a). When the coil moves through 180° from the plane DD′, the voltage induced in the coil (by Faraday's law) is

$$E_{\text{coil}} = \frac{\text{flux per pole}}{\text{time for } \frac{1}{2} \text{ revolution}} = \frac{\phi}{1/(N \times 2)}$$

$$= 2N\phi \tag{19.1}$$

For a machine is which there are Z_s armature conductors connected in series ($Z_s/2$ turns) and $2p$ magnetic poles, the total induced armature voltage is

$$E = 2N\phi\frac{Z_s}{2}2p$$

$$= 2N\phi Z_s p \tag{19.2}$$

Torque on the armature

The force on a current carrying conductor (Section 1.4.4) is

$$F = BlI$$

The torque on one armature conductor is

$$T_{cond} = FR = B_{av}lI_aR$$

where R is the radius of the armature, I_a is the current flowing in the armature conductors, l is the axial length of the armature and B_{av} is the average flux density under a pole and is given by

$$B_{av} = \frac{\phi}{2\pi Rl/2p}$$

Therefore the torque per conductor is

$$T_{cond} = \frac{2p\phi}{2\pi Rl}lI_aR$$

$$= \frac{1}{\pi}p\phi I_a \tag{19.3}$$

For Z_s armature conductors connected in series the total torque on the armature is

$$T = \frac{1}{\pi}p\phi I_aZ_s \tag{19.4}$$

Terminal voltage

If the terminal voltage is V volts and the armature resistance is R_a, then for a generator

$$V = E - I_aR_a \tag{19.5}$$

and for a motor

$$V = E + I_aR_a \tag{19.6}$$

19.2.2   Methods of excitation

The schematic representation of a d.c. machine is as shown in Fig. 19.4, and the method of excitation depends simply on the interconnection of the field and the armature.

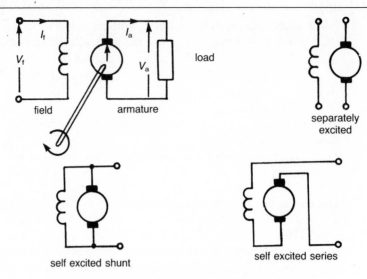

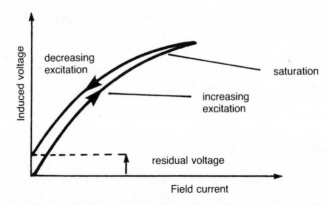

Fig. 19.4 Excitation methods.

19.2.3 Separately excited generator

Open-circuit or no-load characteristic (magnetization curve)

Consider the separately excited generator shown in Fig. 19.4 to be driven at constant rated speed. If the field current (and hence the magnetic field) is increased in steps and the terminal voltage is measured at each step, then a plot of terminal voltage versus field current will yield a curve as shown in Fig. 19.5. Note that because the armature is open-circuited the terminal voltage V_a is equal to the induced voltage E. Assuming that the magnetic circuit of the machine was originally completely demagnetized and that the flux is initially directly proportional to the field current then according to equation

$$E = 2p\phi Z_s N$$

Fig. 19.5 Magnetization curve.

for constant speed

$$E \propto \phi \propto I_f$$

At higher values of field current the iron begins to saturate and the proportionality between the flux and the field current no longer exists, hence the curve no longer approaches a straight line. Because of magnetic hysteresis, the plot of induced voltage versus field current for decreasing excitation is slightly greater than the increasing excitation curve. The voltage at zero excitation is termed the *residual voltage*, without which self excitation would be impossible.

Load characteristic

Let the machine shown in Fig. 19.6 be driven at a constant speed and be supplied with a constant field current I_f. If the load resistance is now varied, the plot of terminal voltage versus armature current obtained is shown in Fig. 19.7. The drop in terminal voltage as the load current increases is due to (a) the armature (or internal) resistance voltage drop and (b) the armature reaction. When current flows in the armature an armature field is established which causes a nett reduction in the field produced by the exciting winding, and as the induced voltage is proportional to the flux, an attendant reduction in terminal voltage is experienced. This effect is termed *armature reaction*. From

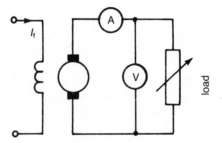

Fig. 19.6 Generator on load.

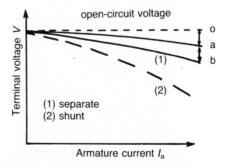

Fig. 19.7 Load curve, oa = armature resistance volt drop, ab = armature reaction volt drop.

the point of view of operating stability, the slightly drooping load characteristic of a separately excited generator is ideal for most applications.

19.2.4 Shunt excited generator

Voltage build-up on no-load

Consider the self excited shunt generator of Fig. 19.4 to be driven at constant speed. If the field is disconnected from the armature the voltage generated across the armature brushes is very small and due entirely to the residual magnetism in the iron. When the field is connected this small residual voltage causes a current to flow in the field winding. If this current is in such a direction as to produce a magnetic flux in the same sense as the residual flux, then the total flux produced by the field winding will gradually build up. The final terminal voltage depends on the total resistance of the field winding and on the magnetization curve of the machine. The magnetization curve of a shunt machine is obtained by connecting it as a separately excited generator. Therefore for the successful operation of a shunt generator there must be residual voltage, the field must be connected the right way round and the field resistance must be less than the critical value.

Load characteristic

The drop in voltage with increased current is more marked in a shunt generator than in a separately excited generator. The increased drop is due to the fact that as the terminal voltage drops then field current also drops. The voltage reduction with increase in load current is the exact reverse of the voltage build-up discussed in the previous section. The load curve for a shunt generator is shown in comparison with a separately excited generator in Fig. 19.8.

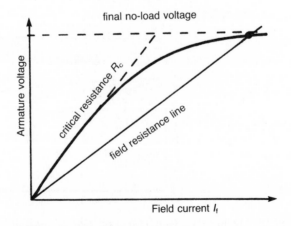

Fig. 19.8 Load curve.

19.2.5 *Series excited generator*

The open-circuit or magnetization curve can only be obtained by connecting the machine as a separately excited generator.

Load characteristic

In the series generator shown in Fig. 19.4 the armature and field are connected in series, therefore the armature current determines the flux. Initially, therefore, the voltage increases as the armature (or load) current increases.

$$E \propto \phi \propto I_a$$

However, at large values of load current the combined armature resistance and reaction effects cause the terminal voltage to decrease, as shown in Fig. 19.9.

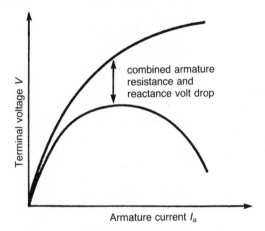

combined armature resistance and reactance volt drop

Fig. 19.9 Load curve.

19.2.6 *Compound generator*

Compound generators are produced by combining the effects of shunt and series excitation. Normally a small series field is arranged to assist (cumulative compounding) the main shunt field, the actual shape of the load characteristic depending on the number of series turns, see Fig. 19.10. If the series field is arranged to oppose the main shunt field (differentially compounded) a fast drooping characteristic is obtained.

19.3 D.C. motors

D.C. motors and d.c. generators are just two versions of the basic d.c. machine, such a machine being essentially a reversable energy converter.

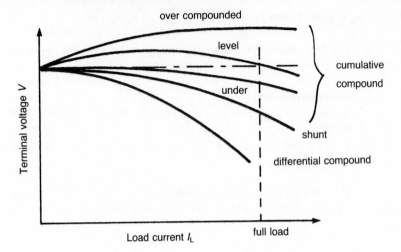

Fig. 19.10 Load curves.

19.3.1 Shunt motor

Typical characteristics of a shunt motor can be obtained by considering the three relevant equations derived in Section 19.2.1, i.e.

$$E = V - I_a R_a$$

$$N \propto \frac{E}{\phi}$$

$$T \propto I_a \phi$$

It can be seen from Fig. 19.11 that the field current I_f will be constant under normal operating conditions. However, when current flows in the armature, the armature reaction effect will weaken the main field, thus tending to increase the speed. Also as I_a increases the induced voltage E will decrease due to the armature resistance volt drop, tending to decrease the speed. The torque

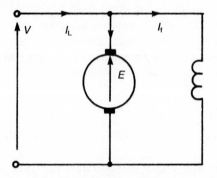

Fig. 19.11 Shunt motor.

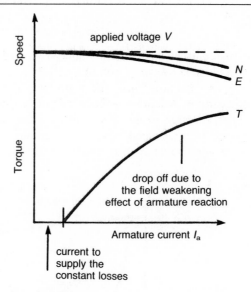

Fig. 19.12 Torque–armature current graph.

increases fairly linearly with the armature current until the armature reaction causes a weakening of the field. Plots of speed and torque versus armature current are shown in Fig. 19.12, and the torque–speed curve derived from them is shown in Fig. 19.13.

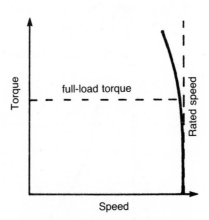

Fig. 19.13 Torque–speed graph.

The torque–speed curve of the shunt motor shows that it could be used for any drive requiring a fairly constant speed from no-load to full-load torque, e.g. machine tools, pumps, compressors, etc.

19.3.2 Series motor

As the load current increases the induced voltage E will decrease due to the armature and field resistance volt drops. Because the field winding is connected in series with the armature, the main flux is directly proportional to the armature current. Armature reaction will, of course, weaken the main field but a fair approximation to the speed versus armature current curve for a series motor would be a rectangular hyperbola, as shown in Fig. 19.14. The torque in this case is approximately proportional to the square of the armature current,

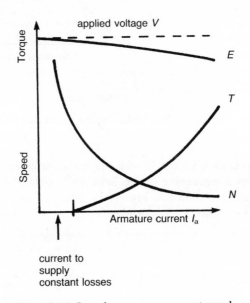

Fig. 19.14 Speed–armature current graph.

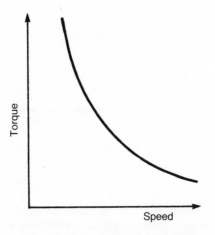

Fig. 19.15 Torque–speed graph.

thus the torque versus armature current curve is approximately parabolic, Fig. 19.14. Figure 19.15 shows the derived torque–speed curve.

19.3.3 Compound motors

Compound motors can be produced by combining the shunt and series windings giving marginal changes to the characteristics shown above.

19.3.4 Starting a d.c. motor

At standstill the induced voltage of a d.c. motor is zero, therefore the applied voltage is equal to the armature voltage drop, I_aR_a. In typical machines the armature resistance is small and hence the armature current at standstill, with rated voltage applied, would be excessive. To limit the starting current, external resistance is connected in series with the armature. As the machine builds up speed, the induced e.m.f. increases and the external resistance is reduced until, at rated speed, all the external resistance has been disconnected.

19.3.5 Stopping or braking a d.c. motor

A d.c. motor will stop if it is disconnected from the supply. The time it takes to reach standstill will depend on its inertia and its friction and windage losses. If fast braking is required then when the motor is disconnected from the supply it is immediately reconnected to a resistor. The inertial energy generated is then quickly dissipated in the resistance.

19.3.6 Speed control of d.c. motors

The speed of a d.c. motor is governed by the applied voltage and the flux.

$$N \propto \frac{E}{\phi} \propto \frac{V - I_aR_a}{\phi} \tag{19.7}$$

It is clear, therefore, that the speed can be varied by varying either the applied voltage or the flux.

Variation of excitation
The main flux of a shunt motor can be weakened by using either a variable rheostat or a potential divider in the field circuit. For a series motor a diverter is used in parallel with the field, see Fig. 19.16. In each of the above methods the field can only be weakened so that only increases in speed above the rated speed can be obtained. It should also be noted that too much field weakening will produce a loss in the torque ($T \propto I_a\phi$).

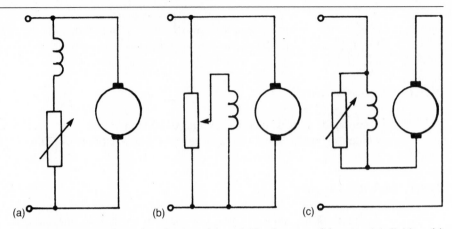

Fig. 19.16 Variation of excitation, (a) variable rheostat, (b) potential divider, (c) diverter.

Variation of armature voltage

The speed can be increased from standstill to rated speed by increasing the armature voltage from zero to the rated value. The main difficulty is in obtaining the variable d.c. voltage.

Potential divider

The potential divider (Fig. 19.17) must be rated at the same current level as the motor, consequently this method would only be used for small d.c. motors.

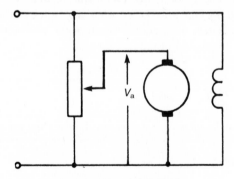

Fig. 19.17 Potential divider.

Ward Leonard drive

In this case the variable d.c. voltage is obtained from a d.c. generator (variable voltage generator, VVG) driven by an induction motor. The field of the VVG is supplied from a centre tapped potential divider, see Fig. 19.18. When the wiper arm of the field potential divider of the VVG is moved from 0 to A, the armature voltage of the main motor is increased from zero and the main motor will increase in speed. If the wiper is moved from A through 0 to B the motor slows down to standstill and then speeds up in the reverse direction. The advantages of the Ward Leonard drive are smooth and complete speed control,

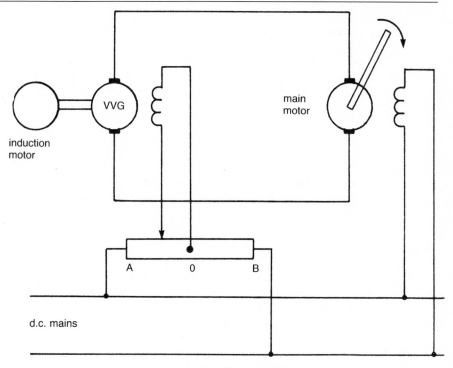

Fig. 19.18 Ward Leonard drive.

operation in the forward and reverse directions, and fast regenerative braking. The disadvantage is the large capital outlay on the induction motor and
variable voltage generator.

Chopper control
Figure 19.19 shows a thyristor circuit in series with the armature of a d.c.
motor. If the thyristor circuit is triggered so that it operates like an on–off
switch (Fig. 19.19(b)), the waveform appearing at the armature terminals is as
shown in Fig. 19.20. The mark/space ratio (time on to time off) can be easily

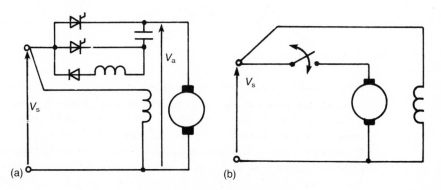

Fig. 19.19 Thyristor control.

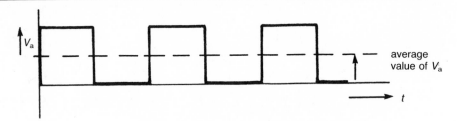

Fig. 19.20 Waveform at armature terminals.

changed and so the average armature voltage can be varied from zero (full off) to the full rated value (full on). The main advantage of this method is its comparative cheapness.

Example A d.c. series motor runs at 1000 rpm when the voltage applied is 200 V and the current taken is 20 A. The armature and field resistances are $R_a = 0.5\,\Omega$ and $R_f = 0.2\,\Omega$. Find the speed for a total current of 20 A at 200 V when a $0.2\,\Omega$ resistor is connected in parallel with the field winding. The flux for a field current of 10 A is 70% of that for 20 A.

e.m.f. of motor taking 20 A at 1000 rpm = $200 - 20(0.5 + 0.2)$

$$= 186\,\text{V}$$

e.m.f. of motor taking 20 A at N rpm with $0.2\,\Omega$ diverter

$$\text{resistor} = 200 - 20(0.5 + 0.1)$$

$$= 188\,\text{V}$$

Note: With the diverter resistor connected the current through the field is reduced to 10 A. From Equation (19.1) for induced e.m.f.,

e.m.f. $\propto$ speed $\times$ flux per pole

therefore $\dfrac{188}{186} = \dfrac{N}{1000} \times \dfrac{0.7}{1}$

$$N = 1444\,\text{rpm}$$

19.4 Three-phase rotating fields

If three coils, physically displaced in space by 120°, are supplied with three-phase currents then a constant magnetic field is produced which rotates at a speed related to the frequency of the currents. Fig. 19.21 shows the arrangement of the coils and the time phase relation of the currents.

Consider the position and magnitude of the resultant flux produced by the three coils at the instants in time shown, using the positive direction of the fluxes as shown in Fig. 19.22. It can be seen that the magnitude of the resulting flux is 3/2 times the magnitude of the phase flux and it rotates at the frequency

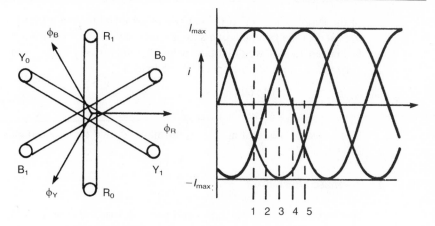

Fig. 19.21 Three-phase rotating fields.

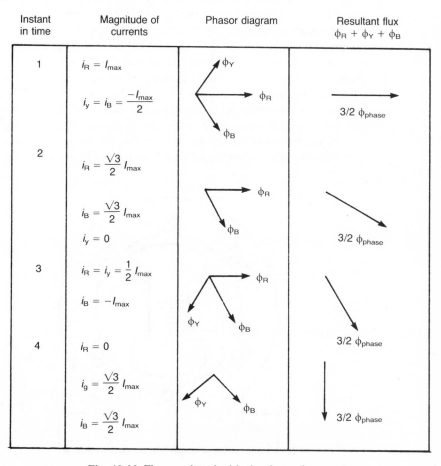

Instant in time	Magnitude of currents	Phasor diagram	Resultant flux $\phi_R + \phi_Y + \phi_B$
1	$i_R = I_{max}$ $i_y = i_B = \dfrac{-I_{max}}{2}$		$3/2\ \phi_{phase}$
2	$i_R = \dfrac{\sqrt{3}}{2} I_{max}$ $i_B = \dfrac{\sqrt{3}}{2} I_{max}$ $i_y = 0$		$3/2\ \phi_{phase}$
3	$i_R = i_y = \dfrac{1}{2} I_{max}$ $i_B = -I_{max}$		$3/2\ \phi_{phase}$
4	$i_R = 0$ $i_g = \dfrac{\sqrt{3}}{2} I_{max}$ $i_B = \dfrac{\sqrt{3}}{2} I_{max}$		$3/2\ \phi_{phase}$

Fig. 19.22 Flux produced with the three-phase system.

of the supply currents. If the coils were arranged to give a four-pole system as opposed to the two-pole one shown, then the speed of the rotating field would be halved (i.e. it would rotate through 180° for one cycle of the alternating current). In general, the speed of the field is

$$N \text{ rpm} = \frac{f \times 60}{\text{pole pairs}} \tag{19.8}$$

The speed of the rotating field is normally termed the synchronous speed.

19.5   Three-phase alternators

A simple form of three-phase, two-pole alternator is shown in Fig. 19.23. On the stator there are three coils spaced 120° apart. The rotor is a salient pole type and is supplied via slip rings with direct current, so producing a uniform magnetic field. If the rotor is now driven by a prime mover voltages will be produced in the coils, and these voltages e_R, e_Y and e_B will be displaced in time phase by 120°. The magnitude of the generated voltages is dependent on the flux produced by the rotor, the number of turns on the coils and the speed of rotation of the rotor. The rotor speed also determines the frequency of the generated voltages. The no-load voltage versus field current and the no-load voltage versus load current characteristics of an alternator are very similar to those of a d.c. separately excited generator. For constant speed operation the terminal voltage will drop due to armature resistance and armature reaction (the term armature being used in this case to denote the stator winding).

It should be noted that as the load on the alternator is increased then the speed of the prime mover would drop, causing the frequency of the generated

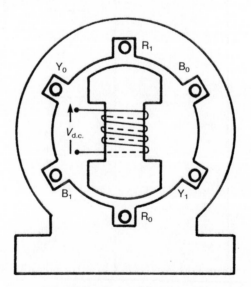

Fig. 19.23 Three-phase alternator.

voltage to fall. In many applications a small change in frequency could be tolerated but if a constant frequency is required some form of governor system must be used on the prime mover to maintain a constant speed on all loads. When alternators are required to operate in parallel, as in the National Grid, the prime movers are always speed controlled and the output voltage is always automatically regulated at the rated value.

In this country most of the electric power generated is produced by alternators driven by steam turbines. The alternators are usually two-pole machines driven at 3000 rpm so as to produce the rated frequency of 50 Hz. In countries where the natural water resources allow, much of the electric power is produced by alternators driven by water wheels (hydroelectric generation). These alternators are normally slow speed machines with large pole numbers. For example, to produce 60 Hz (standard frequency in America) a 36 pole machine would run at 200 rpm.

19.6 Synchronous motors

Synchronous motors are so called because they operate at only one speed, i.e. the speed of the rotating field. They are of exactly the same construction as the alternator discussed in Section 19.5 and in fact the term synchronous machine can be used to describe any machine which has its armature connected to the three-phase mains and its field supplied from a d.c. source. Now if a bar magnet is placed in a strong magnetic field it will always try to align itself with the field (compass needle in the Earth's field). The operation of the synchronous machine is similar in this respect, the constant uniform flux produced by the field (effectively a bar magnet) aligns or *synchronizes* with the rotating flux produced by the armature.

When mechanical load is applied to the shaft the uniform field produced by

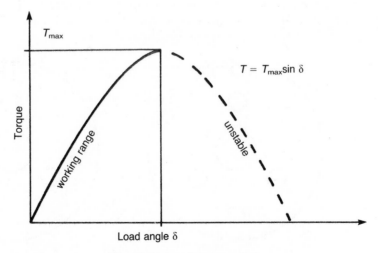

Fig. 19.24 Load characteristic.

the rotor is pulled out of direct alignment with the rotating field produced by the stator, the angle of misalignment being termed the *load angle*. The load characteristic of the synchronous motor is, therefore, the variation of torque with load angle, as shown in Fig. 19.24. The machine possesses no starting torque and it is a necessary requirement with a synchronous motor that its rotor can be run up to synchronous speed by some external means.

The reluctance motor, which is a special form of synchronous machine, has found many applications in recent years. These are listed in Table 19.1 at the end of the chapter.

19.7   Induction motors

The two basic types of commercial three-phase induction motors are the squirrel cage and the slip ring. The stators of both these machines are exactly the same as the alternator discussed earlier, i.e. a conventional three-phase winding, in order to produce a rotating field. In the squirrel cage motor the rotor core is laminated and copper (or aluminium) bars are driven through the slots. These bars are brazed on to solid copper (or aluminium) end rings, producing a completely short-circuited set of conductors. The slip-ring machine has a laminated rotor housing a conventional three-phase winding, similar to the stator, which is connected to three slip rings located on the shaft. The rotor lamination silhouette is shown in Fig. 19.25.

Figure 19.26 shows three stator coils physically displaced by 120°. If these coils are supplied with three-phase currents, a constant rotating field is produced. Consider a single coil AB placed on the rotor. At standstill the rotating field will induce a voltage in the rotor coil because there is a rate of change of flux linking the coil AB. If the coil AB is now short-circuited, the induced e.m.f. will cause a current to flow in the direction shown. There will thus be a force produced on the current carrying conductors A and B (force = BIl), and the rotor will begin to rotate in the direction of this force. The rotor speed will increase until the torque produced by the machine is equal to the mechanical

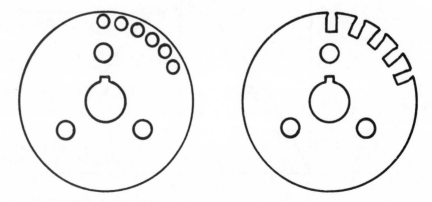

Fig. 19.25 Rotor laminations.

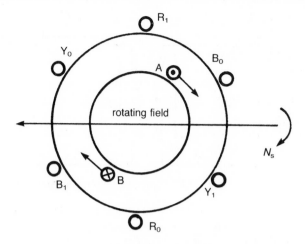

Fig. 19.26 The rotating field.

load torque. The induction motor will never reach synchronous speed because at this speed there would be no relative movement between the rotor and the rotating field, therefore no e.m.f. would be induced in the rotor coils and consequently no torque produced. The ratio of the difference in speed of the rotating field and the rotor to the speed of the rotating field is termed the *slip*, s, where

$$s = \frac{N_s - N}{N_s} \tag{19.9}$$

19.7.1 Transformer similarities

The rotor of the induction motor receives power by induction, as it is not normally connected to the mains. The induction motor can therefore be likened to a transformer and, in fact, at standstill with the rotor not rotating it behaves exactly like a three-phase transformer. However, because of the air gap which exists between the stator and rotor, they are not as closely coupled magnetically as the primary and secondary winding of a transformer. The magnetizing current of an induction motor is therefore larger than that of a transformer.

19.7.2 Development of the equivalent circuit

Let E_2 and f_1 be the standstill rotor induced voltage and frequency. When the rotor is rotating the induced voltage is reduced to sE_2 and the frequency to sf_1, and the rotor equivalent circuit may represented by Fig. 19.27, where one phase only is shown.

The magnitude of the rotor current is

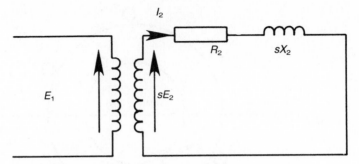

Fig. 19.27 Equivalent circuit for the rotor circuit.

$$I_2 = \frac{sE_2}{\sqrt{[R_2^2 + (sX_2)^2]}}$$

Dividing through by s, the rotor current may be written as

$$I_2 = \frac{E_2}{\sqrt{\left[\left(\dfrac{R_2}{s}\right)^2 + X_2^2\right]}}$$
(19.10)

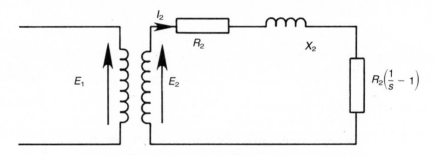

Fig. 19.28 Equivalent circuit for the rotor circuit.

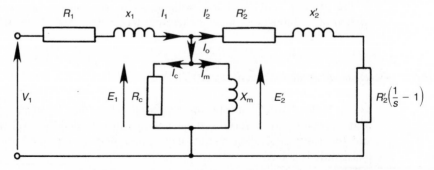

Fig. 19.29 Equivalent circuit referred to the stator winding.

and a new rotor equivalent circuit may be drawn in terms of the standstill values plus an extra resistance of $R_2[(1/s) - 1]$ (Fig. 19.28). Using similar techniques to those explained in Section 12.2, the equivalent circuit of the induction motor referred to the stator is shown in Fig. 19.29.

The parameters of this equivalent circuit can be measured in a similar way to those of the transformer equivalent circuit, by using (1) light running test (no-load test), (2) locked rotor test (short-circuit test).

Light running test

The machine is operated at rated voltage with no external load connected to the shaft. Its running speed will be nearly synchronous. The stator voltage, current and power are recorded. Assuming negligible voltage drop in the stator resistance and leakage reactance.

$$R_c = \frac{V_{oc}}{I_c} \qquad \text{and} \qquad X_m = \frac{V_{oc}}{I_m}$$

$$I_c = I_{oc}\cos \theta_{oc} \qquad \text{and} \qquad I_m = I_{oc}\sin \theta_{oc}$$

Where θ_{oc} is the open circuit power factor angle, and

$$\cos \theta_{oc} = \frac{P_{oc}}{V_{oc}I_{oc}} \qquad \text{hence } I_c = \frac{P_{oc}}{V_{oc}}$$

and therefore

$$R_c = \frac{V_{oc}^2}{P_{oc}}$$

and

$$X_m = \frac{V_{oc}}{I_{oc}\sin\left(\cos^{-1}\dfrac{P_{oc}}{V_{oc}I_{oc}}\right)} \tag{19.11}$$

Locked rotor test

The rotor is locked at standstill ($s = 1$). Reduced voltage is applied at the stator terminals and the short-circuit stator voltage, current and power are measured. The equivalent machine impedance is given by

$$Z_e = \frac{V_{sc}}{I_{sc}}$$

where

$$Z_e = \sqrt{[(R_1 + R_2')^2 + (x_1 + x_2')^2]}$$

and

$$R_1 + R_2' = \frac{P_{sc}}{I_{sc}^2}$$

thus

$$x_1 + x_2' = \sqrt{[Z_e^2 - (R_1 + R_2')^2]} \tag{19.12}$$

It is not possible by this method to separate x_1 and x_2' and these are usually taken as equal. R_1 and R_2' can be taken in the ratio of their d.c. resistances.

19.7.3   Torque and slip relationships

Consider the power distribution in the equivalent circuit of Fig. 19.29. The power input to the motor is $V_1 I_1 \cos\theta$ per phase. The stator and rotor copper losses are $I_1^2 R_1$ and $I_2'^2 R_2'$, and the iron losses are $I_c^2 R_c$.

The power dissipated in the other resistive term corresponds to the only other power sink, i.e. the mechanical load. The mechanical power output per phase is therefore

$$P_m = I_2'^2 R_2 \left(\frac{1}{s} - 1\right)$$

The power supplied to the rotor, per phase, is

$$P_2 = I_2'^2 \frac{R_2}{s}$$

Thus

$$P_m = P_2 - sP_2 = (1 - s)P_2$$

The torque for an m phase machine is given by

$$T = \frac{mP_m}{2\pi N} = \frac{mP_2(1 - s)}{2\pi N_s(1 - s)} = \frac{mP_2}{2\pi N_s} = \frac{m}{2\pi N_s} I_2'^2 \frac{R_2}{s}$$

$$T = \frac{m}{2\pi N_s} \frac{R_2'}{s} \frac{V_1^2}{\left(R_1 + \dfrac{R_2'}{s}\right)^2 + (x_1 + x_2')^2} \tag{19.13}$$

Maximum torque
Assume the stator applied voltage, the machine impedances and the frequency to be constant, and the slip s to be variable, also write X for $x_1 + x_2'$.

$$T = K \frac{1}{sR_1^2 + 2R_1 R_2' + \dfrac{R_2'^2}{s} + sX^2}$$

where K is a constant

$$\frac{dT}{ds} = 0 \quad \text{when} \quad 0 = -R_1^2 + \left(\frac{R_2'}{s}\right)^2 - X^2$$

from which

$$s = \pm \frac{R_2'}{\sqrt{[R_1 + (x_1 + x_2')^2]}} \text{ for maximum torque} \qquad (19.14)$$

The negative sign is for supersynchronous speeds giving a negative (generating torque).

Substituting this value in the equation for torque, the maximum value for the torque, T_{max}, is given by

$$T_{max} = \frac{m}{2\pi N_s} \times \frac{V_1^2}{2\{\sqrt{[R_1^2 + (x_1 + x_2')^2]} \pm R_1\}} \qquad (19.15)$$

This expression is independent of R_2', which only determines the position at which the maximum torque occurs.

Starting torque
At standstill $s = 1$, therefore the starting torque, T_s, is given by

$$T_s = \frac{m}{2\pi N_s} V_1^2 \frac{R_2'}{(R_1 + R_2')^2 + (x_1 + x_2')^2} \qquad (19.16)$$

If the machine is of the slip-ring type, then the starting torque can be increased by adding external rotor resistance via the slip rings.

Torque–slip characteristic
The approximate shape of the torque–slip curve of an induction motor can be obtained by considering the variation of torque when the slip approaches zero, and when the slip approaches unity.

From the expression for torque, $s \to 0$.

$$T \approx \frac{m}{2\pi N_s} \frac{R_2'}{s} \frac{V_1^2}{\dfrac{R_2'^2}{s}}$$

hence T is proportional to s.

Also, when $s \to 1$

$$T \approx \frac{m}{2\pi N_s} \frac{R_2'^2}{s} \frac{V_1^2}{(R_1 + R_2')^2 + (x_1 + x_2')^2}$$

hence T is proportional to $1/s$. The complete torque–slip curve is shown in Fig. 19.30.

Example A three-phase, four-pole induction motor works at 200 V, 50 Hz, on full load of 7.5 kW, when its speed is 1440 rpm. Determine approximately (a) its speed at half full load on full voltage and (b) its speed with an output of 7.5 kW at 190 V.

(a) Over the working range, i.e. at small values of slip,

$$T = Ks$$

Therefore at half full load the slip is given by

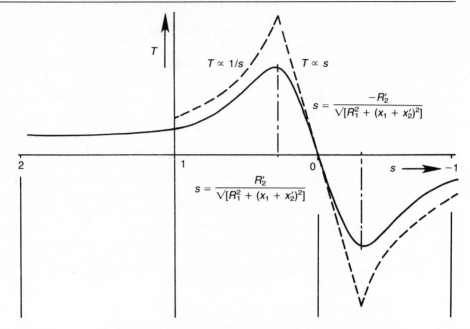

Fig. 19.30 Torque–slip curve.

$$\frac{T_1}{T_2} = \frac{s_1}{s_2}$$

i.e. $s_{\text{half full load}} = 0.02$

(b) From the torque expression derived in Section 19.7.3 for small values of slip

$$T \propto sV_1^2$$

For constant torque, in this case full load torque, the value of slip on reduced voltage is

$$s = 0.04 \times \left(\frac{200}{190}\right)^2 = 0.0443$$

Therefore the speed of the motor is $(1 - 0.0443) \times 1500 = 1434\,\text{rpm}$. (It is assumed that if the power remains constant the torque also remains constant, which is only acceptable for very small speed changes.)

Example A 415 V, six-pole, three-phase, 50 Hz slip-ring induction motor has a star connected stator and rotor, the stator impedance being $(0.6 + j1.5)\Omega$/phase and the equivalent rotor impedance at standstill $(0.5 + j2)\Omega$/phase. The magnetizing branch impedance is $(5 + j35)\Omega$/phase. During starting, the rotor slip rings are connected to a star connected impedance of equivalent value $(1 + j0.2)\Omega$/phase. Determine the starting current and torque. If full load occurs at a slip of 4%, find the normal full load current and power output.

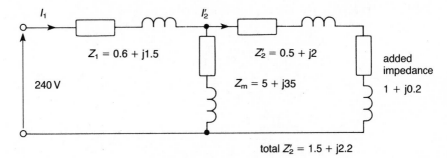

total $Z_2' = 1.5 + j2.2$

Fig. 19.31 Dynamic braking.

From Fig. 19.31 the equivalent circuit for starting, $s = 1$, is

$$I_1 = \frac{V}{Z_1 + \dfrac{Z_2' Z_m}{Z_2' + Z_m}}$$

and

$$I_2' = \frac{Z_m}{Z_2' + Z_m} \times I_1$$

$$I_1 = \frac{V(Z_2' + Z_m)}{Z_1 Z_2' + Z_1 Z_m + Z_2' Z_m}$$

$$I_1 = \frac{240(6.5 + j37.2)}{(-121.4 + j95.57)}$$

Thus the starting current is

$$I_1 = \frac{240 \times 37.8\underline{/80°}}{154\underline{/141.8°}} = 58.7\underline{/-61.8°}\,\text{A}$$

$$I_2' = 58.7\underline{/-61.8°} \times \frac{35.4\underline{/81.9°}}{37.8\underline{/80.1°}}$$

$$I_2' = 55\underline{/-60°}\,\text{A}$$

From Section 19.7.3

$$T = \frac{m}{2\pi N_s} \times (I_2')^2 \frac{R_2'}{s}$$

$$= \frac{3}{2\pi \times \dfrac{1000}{60}} \times 55^2 \times 1.5$$

$$T = 130\,\text{Nm}$$

At a slip of 0.04 and with the added impedance removed, the impedances are

$$Z_1 = 0.6 + j1.5;\ Z_m = 5 + j35;\ Z_2 = 12.5 + j2$$

Therefore the full load current is given by

$$I_1 = \frac{240(17.5 + j37)}{-52.5 + j495.95}$$

$$= \frac{240 \times 40.9 \underline{/64.7°}}{498.75 \underline{/96°}}$$

$$I_1 = 19.7 \underline{/-31.3°} \, \text{A}$$

$$I_2' = I_1 \frac{5 + j35}{17.5 + j37}$$

$$= 19.7 \underline{/-31.3°} \times \frac{35.4 \underline{/81.9°}}{40.9 \underline{/64.7°}}$$

$$I_2' = 17 \underline{/-14.1°} \, \text{A}$$

Therefore the torque produced at full load is

$$T = \frac{3}{2\pi \dfrac{1000}{60}} \times 17^2 \times \frac{0.5}{0.04}$$

$$= 103.5 \, \text{Nm}$$

Power out

$$= T\omega$$

$$= 103.5 \times 2\pi \times \frac{1000}{60} \times (1 - 0.04)$$

$$= 10\,405 \, \text{W}$$

19.7.4   *Starting*

The current drawn from the supply during starting is very large (approximately four or five times the full load current of the motor). For relatively small machines (less than about 15 kW) switching directly on to the supply is permissible. It is usual, however, to obtain the permission of the electricity authorities for the direct on-line starting of larger machines. There are several ways to limit the current during starting, but they all involve auxiliary gear which is often quite costly.

Star-delta starter
Probably the most widely used and cheapest method of starting high powered induction motors is to connect a star-delta starter between the supply and the stator of the machine. With the machine at standstill and the starter in the *start* position, the stator is connected in star. When the machine begins to accelerate the switch is moved to the *run* position which reconnects the stator in delta. By

this star-delta operation the current drawn from the supply is reduced to one third of the current during direct on-line starting.

Auto-transformer

If the stator of the induction motor is fed via an auto-transformer then the current taken from the supply during starting can be considerably reduced from the direct on-line starting current. Unfortunately the starting torque is also considerably reduced. This technique is costly because the auto-transformer has to be of the same rating as the induction motor.

Rotor resistance

With slip-ring machines it is possible to include extra resistance in the rotor circuit. The inclusion of extra rotor resistance not only reduces the starting current but also produces improved starting torque.

19.7.5   Braking

Induction motors may be brought to a standstill very quickly by either plugging or dynamic braking.

Plugging is the term used when the direction of the rotating field is reversed. This is achieved very simply by reversing any two of the three supply leads to the stator. The current taken from the supply during plugging is very large and machines which are to be regularly plugged must be specially rated.

Dynamic braking. In this operation the stator is disconnected from the a.c. supply and reconnected to a d.c. source as shown in Fig. 19.32. The direct current in the stator produces a stationary undirectional field and, as the rotor will tend to align itself with the stator field, it will therefore come to a standstill.

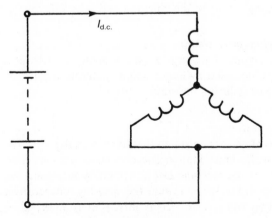

Fig. 19.32 Example.

19.7.6 Speed control of induction motors

The normal running speed of an induction motor is approximately 98% of the synchronous speed of the field at no-load. At full load the speed will have dropped to about 94% of the synchronous speed. It is clear, therefore, that to vary the speed of the induction motor the synchronous speed of the rotating field must be varied. Now the synchronous speed is given by

$$N_s = \frac{\text{frequency} \times 60}{\text{pole pairs}}\text{rpm}$$

therefore changes in the synchronous speed must be brought about by either changing the frequency or changing the number of poles.

Change of frequency
When this type of control was used in the past, the necessary change in frequency was obtained from a specially designed auxiliary machine. With the advent of high power thyristors, static converters have been designed which will produce a variable frequency. At present these variable frequency converters are very expensive (many times the cost of the induction motors they are controlling), but it is hoped that this cost will be reduced as the semiconductor technology increases. Frequency converters do produce a wide range of continuously variable speed control.

Changing the number of poles of the machine
One would expect that once a machine has been built, the number of poles would be fixed. It is possible, however, if the ends of all the stator coils are brought out to a specially designed switch, to change the machine from, say, a four pole to a ten pole. To obtain three different pole numbers and hence three speed ranges requires very complex switching arrangements. It should be noted that the technique of pole changing does not produce an infinitely variable speed, it only gives discrete speed ranges. For some applications, say a two speed fan drive, this is often all that is required, in which case pole changing would be the cheapest and most effective method of speed control.

Marginal speed control
Some applications require only a small variation in the machine speed, which can be obtained by either adding external rotor resistance (slip-ring machines only) or reducing the stator voltage.

Rotor resistance
It has already been stated that increasing the rotor resistance shifts the speed at which the maximum torque occurs. It is clear, therefore, that the operating speed of the machine can be varied by increasing the rotor resistance. This is shown in Fig. 19.33 in that the speed is reduced from N_1 to N_2 and N_3 as the external rotor resistance is increased from R_1 to R_2 and R_3. Although the possible speed variation is small, the cost is relatively low.

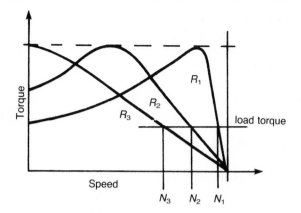

Fig. 19.33 Torque–speed curves for changing rotor resistance.

Reduced stator voltage

If the applied stator voltage is reduced a series of torque–speed curves is obtained as shown in Fig. 19.34. It is seen that for a voltage of V_1 the speed for a particular load torque is N_1, if the voltage is reduced to V_2 the speed is reduced to N_2, etc. The main disadvantage of this method is that as the voltage is reduced so too is the torque, ($T \propto V^2$) and thus this technique is only used for obtaining very small changes in speed.

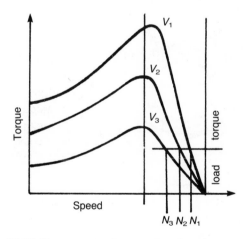

Fig. 19.34 Torque–speed curves for changing stator voltage.

19.8   Single-phase induction motors

The term single-phase induction motor is really a misnomer because the operation of the induction motor depends on the production of a rotating field by the stator winding and this cannot be achieved by a single stator coil alone. A rotating field can be produced by two stator coils displaced in space by 90° and supplied by currents displaced by 90°. (This is the two-phase equivalent of the three-phase case discussed in Section 19.4.) Most single-phase machines attempt to approach the two-phase condition by using some external technique, the most common being the shaded pole motor and the capacitor motor.

19.8.1   The shaded pole motor

The stator consists of a salient pole single-phase winding and the rotor is of the squirrel cage type. Inset in the pole face is a copper ring, as shown in Fig. 19.35. When the exciting coil is supplied with alternating current the flux produced induces a current in the shading ring. The angle θ between the magnetic axes is never more than about 60°, and the phase difference between the currents in the exciting winding and the shading ring is very small, therefore the rotating field produced is far from the optimum two-phase condition. The performance of the machine is consequently poor. The efficiency is low due to the continuous losses in the shading rings and the power factor is poor because the ampere-turns for both fluxes are drawn from the single-phase supply. The main advantage of the shaded pole machine is its simplicity and cheapness of manufacture.

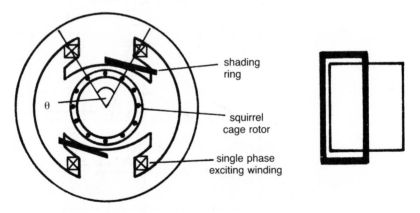

Fig. 19.35 Shaded pole motor.

19.8.2   The capacitor motor

In the capacitor motor shown in Fig. 19.36 the stator has two windings displaced by 90°. A capacitor is connected in series with one winding so that the currents in the two windings have a large phase displacement, thereby closely approaching the optimum two-phase condition. The performance of this type of machine closely approaches that of the three-phase induction motor. Typical test curves are shown in Fig. 19.37.

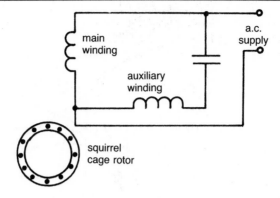

Fig. 19.36 Capacitor motor.

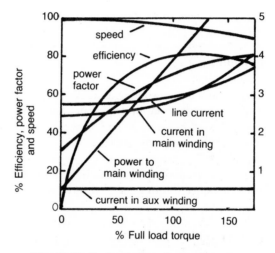

Fig. 19.37 Performance of capacitor motor.

19.9 Universal motor

The universal series motor is usually built in the smaller power sizes (up to about 0.1 kW) and is used for mainly domestic applications. The construction is that of a normal d.c. series motor with a totally laminated magnetic circuit so that it can operate on either a.c. or d.c. supplies (hence the term *universal*). Typical performance curves are shown in Fig. 19.38.

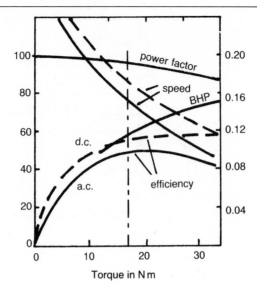

Fig. 19.38 Performance of universal series motor.

19.10 Stepping motors

Stepping motors are now being used extensively in control systems largely as a result of the development of solid state devices which will rapidly and reliably switch the excitation currents. The most common form of stepping motor is the hybrid electromagnetic type which utilizes

(1) The alignment torque produced by a normal doubly excited motor plus.
(2) The reluctance torque produced by a soft iron salient pole rotor.

Firstly consider the basic two-phase permanent magnet device shown in Fig. 19.39. If the stator is not excited, the permanent magnet rotor will select the path of least magnetic reluctance and align with the nearest pair of salient poles.

Let the stator poles now be excited from a d.c. source in the sequence YY' position 1; XX' position 2; reverse currents $Y'Y$ position 3; $X'X$ position 4. As the stator excitation is switched from one set of poles to the other, the rotor realigns itself and moves clockwise through 90° steps to complete one revolution. The response of the rotor as the excitation is changed from position 1 to position 2 is as shown in Fig. 19.40.

If it is assumed that the stator produces a uniformly distributed sinusoidal field, the torque–displacement characteristic is sinusoidal, i.e.

$$T = T_{max}\sin\theta$$

The maximum steady-state or static torque is therefore produced when the magnetic axes of the rotor and stator are in quadrature. Consider the case when the excitation is YY' or position 1 in Fig. 19.39, and the rotor is loaded

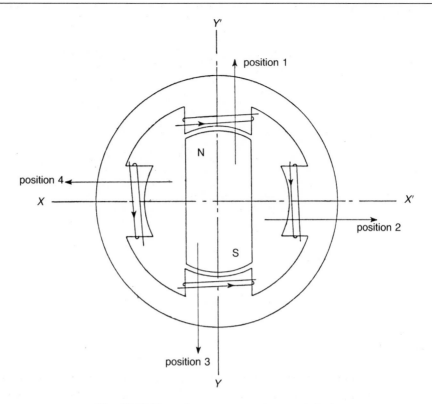

Fig. 19.39 Two-phase permanent magnet device.

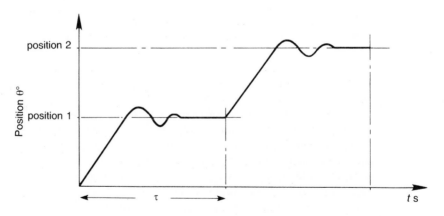

Fig. 19.40 Rotor response.

such that it is producing the maximum torque, i.e. aligned on the $X'X$ axis in position 4. If the step position is then changed from 1 to 2, the torque produced changes from T_{max} to zero. This is undesirable, and under dynamic conditions it is essential to time the switching rate such that the displacement angle θ does not exceed 45° as shown in Fig. 19.41.

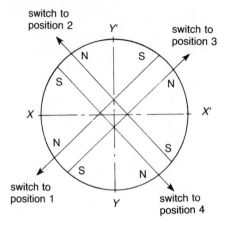

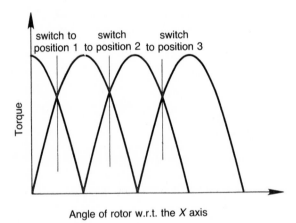

Fig. 19.41 Torque–displacement characteristic.

Consider again the simple arrangement shown in Fig. 19.39, but in this case let the rotor be made from soft iron and therefore not permanently magnetized to a high degree. Again the rotor will align with the pair of stator poles which are magnetized. If the rotor is turned away from the aligned position, a torque is produced tending to realign. As the torque developed is due entirely to there being a position of minimum reluctance, it is called the *reluctance torque*. This torque is zero in the aligned position and maximum midway between the poles, therefore, considering the excitation to be sinusoidally distributed as before,

$$T_{\text{reluctance}} = T_{\text{max}}\sin 2\theta$$

The composite torque–displacement characteristic of the hybrid stepping motor is shown in Fig. 19.42.

To provide the excitation pulses for the coils, a standard transistor bridge circuit can be used as shown in Fig. 19.43. For positive excitation to the *Y* axis

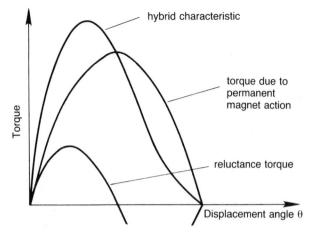

Fig. 19.42 Torque–displacement characteristic.

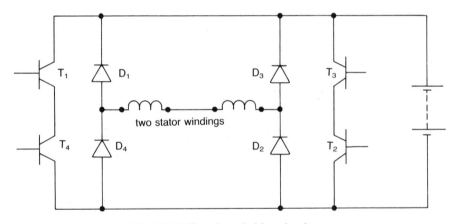

Fig. 19.43 Transistor bridge circuit.

stator coils, transistors T_1 and T_2 are switched on (giving position 1, say). For negative excitation, transistors T_3 and T_4 are switched on (giving position 3). A similar circuit would be used for the X axis stator coils. The diodes provide a freewheeling path to dissipate inductive stored energy when the transistors are turned off.

19.11 Motors and applications

Table 19.1 shows a classification of motors in terms of their speeds and typical applications.

Table 19.1 Classification of motors by speed.

Speed	Motor	Application
Constant	(a) Synchronous (b) Reluctance and hysteresis	Useful for power factor correction Small power applications, clocks, turntables, recently rolling mill tables, conveyors
Almost constant	(a) Induction (b) D.C. shunt	Used when a small speed change can be tolerated – fans, blowers, machine tools, pumps, compressors, crushers
Varying speed	(a) Series (b) Universal, repulsion	Cranes, hoists, traction
Marginal speed control	(a) Wound rotor induction (b) D.C. compound with field control	Cranes, hoists
Variable speed	(a) D.C. motor with armature voltage control (b) Induction motor with variable frequency supply	Used for any drive requiring a wide speed variation, expensive
Multispeed	(a) Cascaded induction motors (b) Pole changing induction motors	Used when two discrete speeds required; two-speed fans
	Single-phase induction motors, fractional horsepower universal motors	All domestic applications, cleaners, washers, tumble dryers, central heating pumps

Problems

1. The open-circuit characteristics of a d.c. shunt generator running at 1000 rpm is given below:

E volts	0	40	80	120	160	200	240
I_f amps	0	0.28	0.56	0.86	1.34	2	3.2

   What is the open circuit voltage of the machine with a field resistance of 85 Ω? Estimate the critical field resistance at 1000 rpm. What is the approximate critical speed with a field resistance of 100 Ω?

2. The d.c. shunt generator of Problem 1 is to maintain a constant terminal voltage of 200 V. At full load the speed drops 10%; and the armature voltage drop is 10 V. Calculate the required change in field resistance from no-load to full-load.

3. If the shunt machine of Problem 1 has an armature resistance of 0.25 Ω, neglecting armature reaction, find the total field resistance required for the machine to run at 1000 rpm as a shunt motor from 200 V d.c. mains if the armature current is 40 A.

4. A four-pole alternator develops 240 V at 50 Hz when the field current is 2.0 A and it is delivering no current. If it is operating on the unsaturated part of the characteristic, what will be the e.m.f. when the speed is 1200 rpm and the field current 3.6 A?

5. A series d.c. motor has an armature plus field resistance of $1.5\,\Omega$ and, when running at 1200 rpm, takes a current of 15 A from a 240 V supply. What will be the speed when a $3.5\,\Omega$ resistor is placed in series with the motor, the load resistance being altered so that the same current is taken?

6. A series motor with negligible resistance and operating on the unsaturated part of the open-circuit characteristic when driving a load takes 40 A at 440 V. If the load torque varies as the square of the speed, find the resistance necessary to reduce the speed by 25%.

7. A d.c. shunt motor has an armature resistance of $0.5\,\Omega$. When connected to a 240 V supply and running at 1500 rpm the armature current is 20 A. What resistance should be placed in series with the armature to reduce the speed to 1000 rpm?

8. A d.c. shunt motor has a Ward Leonard speed control system. When the armature current is 100 A and the armature voltage 440 V, the speed is 1000 rpm. The armature resistance is $0.5\,\Omega$. What is the armature voltage required to reduce the speed to 400 rpm if the load torque varies as the square of the speed.

9. A d.c. series motor has an armature resistance of $0.5\,\Omega$ and a field resistance of $0.2\,\Omega$. It runs at 1000 rpm when the voltage is 500 V and the current 20 A. What resistance should be connected in series to reduce the speed to 800 rpm?

10. A three-phase induction motor has four poles and is connected to the 50 Hz supply. What is (a) the synchronous speed and (b) the rotor speed when the slip is 0.05?

11. A four-pole, 50 Hz, three-phase induction motor on full load has a speed of 1450 rpm. What is the slip?

12. A three-phase induction motor is supplied with a power of 30 kW. If stator losses are 1.5 kW, what is the total mechanical power developed when the slip is 0.05?

13. A three-phase induction motor has a star-connected rotor winding of resistance $0.1\,\Omega$ per phase and a leakage reactance at standstill of $0.9\,\Omega$. If the impedance of the stator winding is negligible, what will be the slip at which maximum torque occurs?

14. A four-pole, 50 Hz, 440 V, three-phase induction motor has a rotor resistance of $0.05\,\Omega$/phase and a standstill reactance of $0.5\,\Omega$/phase. The standstill rotor e.m.f. is 120 V. What is the maximum torque and the slip at which it occurs?

15. A diesel engine drives a six-pole, three-phase alternator at 1200 rpm. The alternator is electrically connected to a four-pole, three-phase induction motor running at 4% slip. What is the speed of the induction motor?

Chapter 20

Electrical Measurements

20.1 Measurement of current and voltage

The basic instruments considered in this section can be used for the measurement of either current or voltage, depending on the connection of external resistance. The moving iron, moving coil and dynamometer instruments are basically current measuring instruments. The current range of such instruments can be changed by connecting shunt resistors in parallel with the instrument so that some of the current is diverted and only a specific fraction passes through the instrument. Since $V = IR$, then a measurement of the current I through a resistance R enables the potential difference V to be ascertained. The resistance of the meter is augmented by a resistor connected in series, this resistor being termed a multiplier, in order to obtain a reasonable voltage range for an instrument.

Example A meter has a resistance of $1\,\Omega$ and gives a full-scale deflection with $100\,\text{mA}$. How can this meter be adapted for the measurement of (a) currents up to $100\,\text{A}$, (b) voltages up to $100\,\text{V}$?

(a) Figure 20.1(a) shows how a shunt can be used to divert some of the current round the meter. Full-scale deflection requires that the current through the meter must be only $100\,\text{mA}$. Therefore the current through the shunt must be $(100 - 0.100) = 99.9\,\text{A}$. Since the potential difference across the meter must equal that across the shunt,

$$0.100 \times 1 = 99.9R$$

Hence the shunt resistance must be $0.001\,\Omega$.

(b) Figure 20.1(b) shows how a multiplier is used to increase the voltage range of the meter. For full-scale deflection the current through the meter is $100\,\text{mA}$. Thus for a potential difference of $100\,\text{V}$ across the arrangement we must have

$$100 = 0.100(R + 1)$$

Hence the multiplier has a resistance of $999\,\Omega$.

20.1.1 Moving iron instrument

There are two types of moving iron instruments, namely the attraction type and the repulsion type. In the attraction type instrument the current is passed

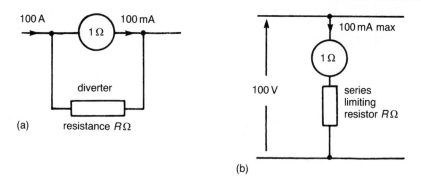

Fig. 20.1 Example.

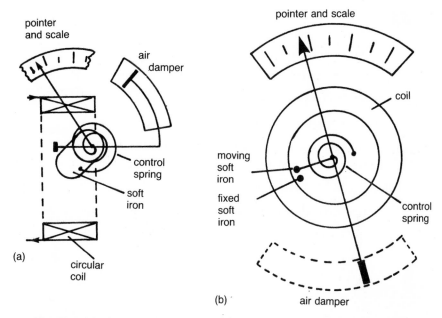

Fig. 20.2 Moving iron instruments, (a) attraction type, (b) repulsion type.

through a coil and produces a magnetic field inside the coil which is proportional to the current. This attracts a piece of soft iron into the coil, causing a deflection of a pointer. Figure 20.2(a) shows the basic form. In the repulsion type instrument, when the current flows in a coil then the resulting magnetic field magnetizes two soft iron rods which are inside the coil; Fig. 20.2(b) shows the type of arrangement. The two rods are magnetized in the same direction and thus there is a repulsive force between them. Since one of the rods is fixed and the other free to move, the latter moves and causes a deflection of the pointer.

To obtain an expression for the deflecting torque of either type of moving iron instrument, consider the energy balance equation when there is a small current change δi. The pointer moves through a small angle $\delta\theta$ and there is a small change in inductance δL due to the movement of iron. In Chapter 1, Faraday's law was written in the form (equation (1.32))

$$e = L\frac{di}{dt}$$

when there was just a current change. When there is both a current and an inductance change then it is written as

$$e = L\frac{d(Li)}{dt} = i\frac{dL}{dt} + L\frac{di}{dt}$$

The energy supplied to the coil as a result of this change in current is $ei\,\delta t$ and thus is given by

$$\text{energy supplied} = ei\,\delta t = i^2\,\delta L + iL\,\delta i \tag{20.1}$$

The energy stored in the coil changes from $\frac{1}{2}Li^2$ to $\frac{1}{2}(L + \delta L)(i + \delta i)^2$ (see Equation (1.47)).

$$\text{energy change in coil} = \tfrac{1}{2}Li^2 - \tfrac{1}{2}(L + \delta L)(i + \delta i)^2$$

Neglecting small terms, this approximates to

$$\text{energy change in coil} = iL\,\delta i + \tfrac{1}{2}i^2\,\delta L \tag{20.2}$$

Work is done by the pointer moving through an angle $\delta\theta$ as a result of the deflecting torque T.

$$\text{Work done by pointer} = T\,\delta\theta \tag{20.3}$$

Therefore the energy balance equation can be written as

energy supplied = energy change in coil + work done by pointer

$$i^2\,\delta L + iL\,\delta i \;=\; iL\,\delta i + \tfrac{1}{2}i^2\,\delta L + T\,\delta\theta$$

Hence we can write

$$T = \tfrac{1}{2}i^2\frac{dL}{d\theta} \tag{20.4}$$

For both the attraction and repulsion types of instrument the deflecting torque is proportional to the square of the current through the coil.

The pointer moves against hair-springs which give a torque proportional to the angle through which it is deflected. The equilibrium position of the pointer thus occurs at an angle at which the deflecting torque is equal in size to the restoring torque provided by the springs. Thus the pointer deflection is proportional to the square of the current. It is possible to affect the square law effect by a design of the rate at which the inductance changes with the angular position.

Because the moving iron meter gives a deflecting torque proportional to the square of the current, the torque is not affected by the direction of the current. Therefore the moving iron instrument can be used for both direct and alternating currents. Other advantages of the instrument are that it is robust and relatively cheap. Disadvantages are that it is affected by stray magnetic fields and liable to hysteresis error, i.e. the instrument reads higher with decreasing than increasing currents, is subject to temperature errors as a result of the resistance change of the coil with temperature and because of the inductance of the coil the readings of the instrument can be affected by changes in frequency.

20.1.2 *Permanent magnet, moving coil instrument*

The permanent magnet, moving coil instrument consists of a light weight coil which is free to rotate in the air gap between a permanent magnet and a soft iron cylinder, as illustrated in Fig. 20.3. When a current I flows through the coil, then the force on each turn which will result in rotation is $F = BIL$ (Equation (1.29)), where L is the length of the coil in the air gap and B the flux density in that air gap. For N turns the force is $NBIL$. If R is the radius of the coil from the central pivot then the torque T causing deflection of the coil is

$$T = NBIL \times 2R \tag{20.5}$$

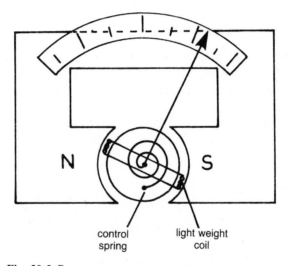

Fig. 20.3 Permanent magnet, moving coil instrument.

The coil rotates against hair-springs which give a restoring torque proportional to the angle through which the coil has been rotated. Thus, at equilibrium when the restoring torque is equal in size to the deflecting torque, the pointer deflection is proportional to the current.

A consequence of this is that if the current is alternating the torque is oscillatory. The instrument can thus be only used for direct currents. The meter can however be used with a bridge rectifier circuit, as in Fig. 20.4. On the assumption of sinusoidal current wave shapes, the scales of such instruments are usually marked in terms of 1.11 times the current actually measured

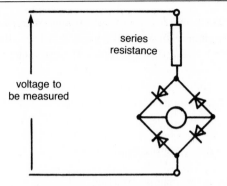

Fig. 20.4 Rectifier instrument.

in order to give the rms value. One of the main disadvantages of rectifier instruments is that they give erroneous readings with non-sinusoidal waveforms. The bridge circuit will include a series resistance when used as a voltmeter. This serves two purposes, one as a multiplier resistance and the other as a way of swamping the non-linear effects of the rectifier resistance.

20.1.3 Dynamometer instruments

The dynamometer instrument is like a permanent magnet moving coil instrument, but with the permanent magnet replaced by two fixed coils. Figure 20.5 shows the basic form of the instrument. The flux density produced in the fixed coils is proportional to the current through the coils. Thus the torque, which is proportional to the flux density multiplied by the current through the moving coil (see Equation (20.5)), is now proportional to the product of the current

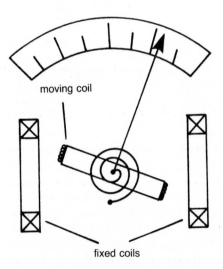

Fig. 20.5 Dynamometer.

through the fixed coils and that through the moving coil. If the fixed coils and the moving coil are in series, then the torque is proportional to the square of the current. It can thus be used with both alternating and direct currents.

20.1.4 *Digital instruments*

The digital voltmeter can be considered to be basically an analogue to digital converter connected to a counter. They can be classified according to the type of converter used. Thus some are sampling meters in that they provide digital values corresponding to the voltage at some instant of time, while others are integrating meters in that they average the voltage value over a fixed measurement time.

Digital meters provide a numerical readout which eliminates interpolation and parallax errors. Displays are generally between $3\frac{1}{2}$ and $8\frac{1}{2}$ digits, the half being because the most significant digit can only take the value 0 or 1.

20.2 Measurement of energy and power

The main instrument used for the measurement of power in electrical circuits, whether d.c., single phase a.c. or three phase, is the watt meter. The conventional form of this instrument is a dynamometer, but electronic versions are now becoming available. The main instrument used for the measurement of energy is the induction instrument that appears in domestic dwellings as the method used by the Electricity Companies to determine the amount of electrical energy used in the home and for which the customer has to pay.

20.2.1 *Wattmeter*

The dynamometer instrument referred to in Section 20.1.3 has a pair of fixed coils and a moving coil with the pointer deflection being determined by the product of the currents in the fixed and moving coils. For use as a wattmeter, the fixed coils can be connected in series with the load and the moving coil in parallel. The fixed coils thus monitor the current through the load and the moving coil the potential difference across it.

The wattmeter can be connected into a circuit in two ways. Figure 20.6 shows one way. The moving coil gives a measure of the potential difference across the fixed coils and the load. The wattmeter thus reads high by the power due to the potential drop across the fixed coils. This method of connection is generally used for low current/high voltage loads. Alternatively the wattmeter can be connected with the moving coil directly across the load. This results in the wattmeter reading high because of the power due to the fixed coils responding to the current through both the load and the movable coil. This method tends to be used for high current/low voltage loads.

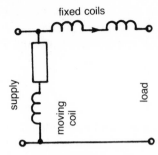

Fig. 20.6 Wattmeter.

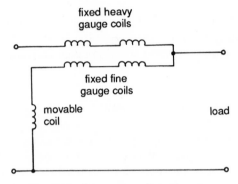

Fig. 20.7 Compensated wattmeter.

These errors can be overcome by using a compensated wattmeter (Fig. 20.7). Such a meter has fixed coils with each having two windings with the same number of turns, one winding using heavy gauge wire and the other fine gauge wire. The winding with the heavy gauge wire is used to monitor the load current while the fine gauge winding is connected in series with the movable coil and so in parallel with the load. The current through this fine gauge winding is, however, in the opposite direction to the load current through the heavy gauge winding and thus cancels out a proportion of the magnetic flux due to the voltage coil current. Consequently the wattmeter indicates the correct power.

20.2.2 *Energy meter*

The watt–hour meter (Fig. 20.8) works on exactly the same principle as the shaded pole induction motor discussed in Chapter 19. Alternating currents in the voltage and current coils produce alternating magnetic fields which result in a rotating magnetic field at the aluminium disc. The magnetic field produced by the voltage coil is 90° out of phase with that produced by the current coil. This is because the voltage coil magnetic circuit is capped by a copper sheet in

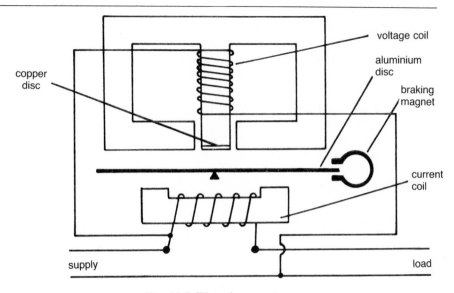

Fig. 20.8 Watt–hour meter.

which the voltage coil produced magnetic fields induce currents out of phase with those in the voltage coil. Hence the magnetic field in the aluminium disc below the copper disc is 90° out of phase with that produced by the current coil which is not capped. The rotating field induces currents in the aluminium disc, which is free to rotate. The interaction of the rotating field and the field produced by the induced currents in the disc produces a torque which causes the disc to rotate.

The average torque acting on the disc is proportional to the power supplied to the coils, i.e.

$$T = kVI \cos \phi$$

where $\cos \phi$ is the power factor and k is a constant. A retarding torque is produced by the braking magnet. The rotating disc is a conductor moving in the magnetic field produced by the permanent magnet and thus currents are induced in it. The induced currents are proportional to the speed with which this disc moves through the magnetic field and thus the retarding torque is proportional to the speed of rotation of the disc N, i.e.

$$T_r = k_r N$$

At equilibrium, when the driving torque is equal in size to the retarding torque, the disc rotates with a constant speed. Then

$$k_r N = kVI \cos \phi$$

The total number of revolutions over a time t is Nt and thus the number of revolutions is proportional to $VIt \cos \phi$, i.e. the total energy supplied during that time. The number of revolutions of the disc can be counted using gears connected to the shaft and a mechanical counter.

20.3   Measurement of resistance

The method used to determine resistance will depend on the accuracy required and also the size of the resistance, different methods being required for low and high values. Simple, low accuracy methods are the ammeter–voltmeter method and the ohmmeter. The Megger is a simple, low accuracy method for very high resistance values and is primarily used for the measurement of insulation resistance. The Wheatstone bridge gives greater accuracy, special versions of the bridge being available for very low or very high resistance values.

20.3.1   Ammeter–voltmeter method

This method involves using an ammeter to measure the current through a resistor and a voltmeter to determine the potential difference across it. The accuracy is determined by the accuracy of the meters and the loading effect of the voltmeter. The ammeter measures the current through both the load and the voltmeter. Thus the voltmeter needs to have a high resistance compared with that being measured if the current taken by it is to be small compared with that through the resistance. Only then will the ammeter give a reasonable measure of the current through the resistance and loading errors be small.

20.3.2   Ohmmeter

Figure 20.9 shows the two forms of the ohmmeter circuit. With the series type ohmmeter, a battery is connected in series with a moving coil meter and a zero adjustment resistor. When the terminals of this arrangement are short-circuited the zero adjustment resistor R_z is adjusted until the meter gives a full-scale reading. This reading then corresponds to zero resistance. Then when a resistor R is connected between the terminals this current is reduced, the value of the resulting current being a measure of the resistance. The current I indicated by the meter is thus

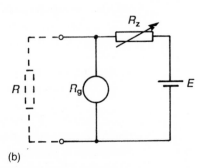

(b)

Fig. 20.9 Ohmmeter, (a) series type, (b) shunt type.

$$I = \frac{E}{R + R_g + R_z}$$ (20.6)

where R_g is the resistance of the meter and E the e.m.f. of the battery. The relationship is non-linear with the scale points very close together for low resistances. Thus this form of ohmmeter is not very useful for the measurement of low resistances.

With the shunt type ohmmeter, the moving coil meter is in parallel with a battery and the zero adjustment resistor and the unknown resistor shunts the meter. With open-circuit between the ohmmeter terminals, i.e. R is infinity, the zero adjustment resistor is adjusted to give a full-scale deflection. When the resistor R is connected as a shunt to the meter then the current through the meter is reduced, the meter reading being a measure of the value of R. The current I taken from the battery is given by

$$E = I\left(R_z + \frac{RR_g}{R + R_g}\right)$$

The current through the galvanometer I_g is I minus the current I_R through the shunt R. But $I_g R_g = I_R R$ and so

$$I_g = I - \frac{R_g I_g}{R}$$

Hence

$$I_g = \frac{RI}{R + R_g} = \frac{ER}{R_z R_g + R R_z + R_g}$$ (20.7)

This type of ohmmeter is particular useful for low resistances.

20.3.3 *Megger*

The Megger is used for the measurement of resistances, such as insulation resistances, which are generally in excess of $1\,M\Omega$. The instrument movement is like that of a permanent magnet moving coil meter, but has two coils (Fig. 20.10). One of the coils, termed the deflecting coil, is connected in series with a fixed resistance and the resistance under test. The other coil, called the control coil, is in series with a fixed resistance and the voltage source. The voltage source is a hand driven generator which is capable of producing a voltage up to $2500\,V$. If the deflecting circuit is open-circuit then the current through the control coil results in the control coil settling down at right angles to the magnetic field of the permanent magnet. The pointer then indicates infinite resistance connected in the deflecting coil circuit. When a resistance is included in the deflecting coil circuit then a current flows in the deflecting coil. The pointer then takes up a position determined by the balance between the torques on the deflecting and control coils. This position is a measure of the resistance.

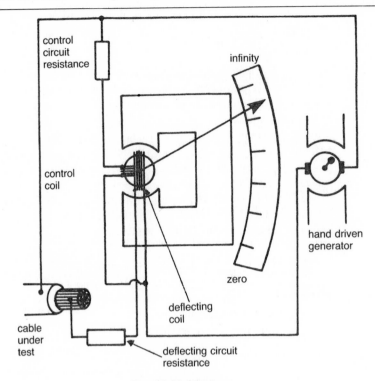

Fig. 20.10 Megger.

20.3.4   *Wheatstone bridge*

Figure 20.11 shows the Wheatstone bridge circuit. When the bridge is balanced, i.e. there is zero current indicated by the galvanometer, then there must be no potential difference between b and d. Therefore the potential drop across P must equal the potential drop across Q. Hence

$$i_1P = i_2Q$$

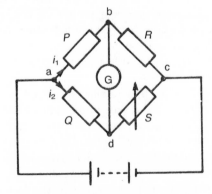

Fig. 20.11 Wheatstone bridge.

Similarly

$$i_1R = i_2S$$

Thus, dividing the above equations,

$$\frac{P}{R} = \frac{Q}{S} \tag{20.8}$$

Thus if R is the unknown resistance, S would be a variable standard resistance, and P and Q are *ratio arms* which may be varied, generally in multiples of 10.

Example   One phase of a three-phase cable is found to be short-circuited to the lead sheath. Suggest a method of locating the fault.

A Wheatstone bridge can be set up in the way shown in Fig. 20.12. The yellow and blue phases are connected together at the remote end. Let the fault occur at a distance x m from the near end and let the cable resistance be R Ω/m. At balance

$$i_1P = i_2Q$$

$$i_1(2L - x)R = i_2xR$$

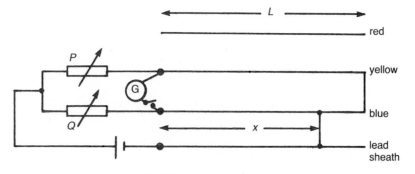

Fig. 20.12 Example.

Thus, dividing the two equations

$$\frac{2L - x}{P} = \frac{x}{Q}$$

Hence

$$x = 2L\frac{Q}{P + Q}\ \text{m}$$

20.4 Measurement of inductance and capacitance

An inductor has both inductance and resistance, with generally negligible capacitance, and can be considered to be a pure inductance in series with a pure resistance. A capacitor has both capacitance and resistance, with negligible inductance, and can be considered as either a pure capacitance in parallel with a pure resistance or a pure capacitance in series with a pure resistance. The inductance plus resistance or the capacitance plus resistance can be measured by the use of a.c. bridges.

There are many forms of a.c. bridges and the two that are given in the following sections are just examples. In general, the a.c. bridge is just like the Wheatstone bridge, but with impedances instead of resistances.

20.4.1 The Maxwell bridge

This bridge (Fig. 20.13) is used for the determination of the inductance and series resistance of an inductor. P and Q are resistive ratio arms, R_1 and L_1 the unknown inductance and resistance, R_2 and L_2 a variable resistance and inductance. When the bridge is balanced, i.e. no current in the detector, then the voltage drop across P must equal that across Q in both magnitude and phase, i.e. the phasors for the two voltages must be equal. Thus

$$\mathbf{I}_1 P = \mathbf{I}_2 Q$$

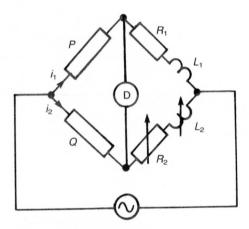

Fig. 20.13 Maxwell bridge.

Similarly

$$\mathbf{I}_1 Z_1 = \mathbf{I}_2 Z_2$$

But Z_1 is the impedance of R_1 in series with $j\omega L_1$ and is thus $R_1 + j\omega L_1$. The impedance Z_2 is that of R_4 in series with $j\omega L_2$ and is $R_2 + j\omega L_2$. Thus, dividing the above two equations and substituting these values for Z_1 and Z_2,

$$\frac{P}{R_1 + j\omega L_1} = \frac{Q}{R_2 + j\omega L_2}$$

$$PR_2 + j\omega L_2 P = QR_1 + j\omega L_1 Q$$

For balance, the real parts must balance and the imaginary parts must balance. Thus

$$R_2 = \frac{Q}{P}R_1 \qquad\qquad (20.9)$$

$$L_2 = \frac{Q}{P}L_1 \qquad\qquad (20.10)$$

20.4.2 *The Wien bridge*

This bridge (Fig. 20.14) is used for the measurement of a capacitance with its resistance in parallel. R_1 and C_1 are the unknown resistance and capacitance, R_2 and C_2 are the variable standard resistance and capacitance, P and Q are the resistive ratio arms. At balance

$$\mathbf{I}_1 P = \mathbf{I}_2 Q$$

$$\mathbf{I}_1 Z_1 = \mathbf{I}_2 Z_2$$

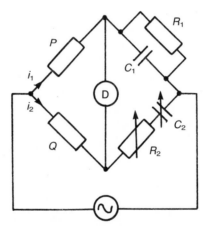

Fig. 20.14 Wien bridge.

Z_1 is the impedance of R_1 in parallel with $1/j\omega C_1$ and is thus $R_1/(1 + j\omega C_1 R_1)$. The impedance Z_2 is R_2 in series with $1/j\omega C_2$ and is thus $R_2 + 1/j\omega C_2$. Hence, dividing the above equations and substituting for Z_1 and Z_2,

$$\frac{P(1 + j\omega C_1 R_1)}{R_1} = \frac{j\omega C_2 Q}{1 + j\omega R_2 C_2}$$

$$1 - \omega^2 R_1 R_2 C_1 C_2 + j\omega R_1 C_1 + j\omega R_2 C_2 = j\omega C_1 \frac{Q}{P} R_1$$

Equating real parts and equating imaginary parts gives for the balance conditions

$$\omega^2 R_1 R_2 C_1 C_2 = 1$$

$$R_1 = \frac{P R_2 C_2}{Q C_2 - P C_1}$$

From which

$$R_1 = \frac{P}{Q} \frac{1 + \omega^2 R_2^2 C_2^2}{\omega^2 R_2 C_2^2} \tag{20.11}$$

$$C_1 = \frac{Q}{P} \frac{C_2}{1 + \omega^2 R_2^2 C_2^2} \tag{20.12}$$

One of the important applications of the Wien bridge is as a frequency selective network in RC oscillators. In which case the bridge would be tuned (or balanced) at frequencies depending on the values of R_1 and C_1.

Example   A single core cable has been broken and there is no connection between the core and the sheath. Show how a Wien bridge may be used to locate the fault.

Let the conductor to sheath capacitance be C μF/m and the resistance be R Ω/m. If the bridge is connected as shown in Fig. 20.15, using standard ratio arms P and Q, a standard variable capacitor C_s and a standard variable resistor R_s, then using Equations (20.11) and (20.12)

$$Rx = \frac{P}{Q} \frac{1 + \omega^2 R_s^2 C_s^2}{\omega^2 R_s C_s^2}$$

$$Cx = \frac{Q}{P} \frac{C_s}{1 + \omega^2 R_s^2 C_s^2}$$

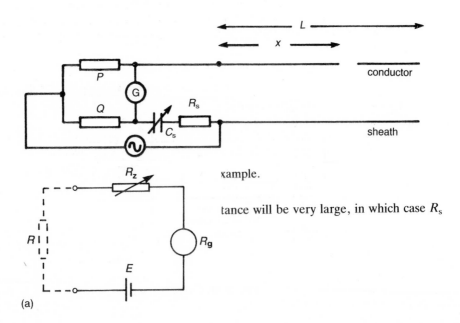

xample.

tance will be very large, in which case R_s

(a)

20.5   Cathode ray oscilloscope

The cathode ray oscilloscope is one of the most commonly used instruments in electronics. Figure 20.16 shows the basic features of the cathode ray tube. Electrons are produced by electrically heating the cathode. The number of these electrons which form the electron beam which travels down the tube, i.e. the brilliance of the spot on the screen, is determined by a potential applied to an electrode, the modulator, immediately in front of the cathode. The electrons are accelerated down the tube by the potential difference between the cathode and the anode, with an electron lens being used to focus the beam so that when it reaches the phosphor coated screen it forms a small luminous spot. The focus is adjusted by changing the potential of the electrodes. The beam can be deflected in the vertical (Y) direction by a potential difference applied between the Y deflection plates, while a potential difference between the X deflection plates causes it to deflect in the horizontal (X) direction.

The vertical deflection circuitry supplying the signal to the Y deflection plates consists of an input selector, a switched attenuator, an amplifier and a delay line. The input selector passes the input signal to the attenuator and enables a d.c. input signal to be passed directly to the attenuator, with an a.c. input signal being passed through a capacitor which blocks off any d.c. components. The switched attenuator is used to change the magnitude of the signal fed to the amplifier and so give different deflection sensitivities. The amplifier is used to give a constant gain. Because there is some delay in the start of the horizontal sweep of the electron beam, it is necessary to delay the vertical system input to the cathode ray tube. The input impedance is typically about 1 MΩ shunted by a capacitance of about 10 to 100 pF.

The X deflection plates can be supplied by an external signal, via a system similar to that for the Y deflection plates or by an internally generated signal. This signal has a sawtooth waveform and sweeps the luminous spot on the screen from left to right at a constant velocity with a very rapid return, i.e. fly

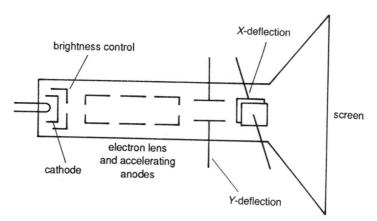

Fig. 20.16 Cathode ray tube.

back. This return is too fast to leave a trace on the screen. The constant velocity movement from left to right means that the distance moved in the X direction is proportional to the time elapsed and so gives a horizontal time axis, i.e. a time base.

For an input signal to give rise to a steady trace on the screen it is necessary to synchronize the time base and the input signal by means of a trigger circuit. This determines when the beam starts to sweep across the screen and can be adjusted so that it responds to a particular voltage level and also to whether the voltage is increasing or decreasing. The time base sweep across the screen can thus be adjusted so that it always starts at the same point on the input signal. The result is that successive scans of the input signal are super-imposed.

20.5.1   Oscilloscope probes

Signals can be fed to oscilloscope inputs by means of just a length of wire, however this has the disadvantages that it may pick up interference and also that it results in the oscilloscope load, which can be considered to be about 1 MΩ shunted by 10 to 100 pF, being connected across the circuit supplying the signal. This loading can upset the normal function of the circuit. Coaxial cable can protect against the pick-up of interference signals, but does not avoid the loading problem. Indeed, the extra capacitance of the coaxial cable will increase the capacitive loading. To overcome these problems oscilloscope probes are used.

These generally consist of three elements: a probe head, the interconnecting cable and the cable termination. The probe head is the point of contact with the signal generating circuit and contains the signal sensing circuit and may be passive, only containing such circuit elements as resistors and capacitors, or active and containing such elements as amplifiers with field-effect transistors. A coaxial cable is used as the transmission line to transmit the signal from the probe head to the termination circuit. This is not always included and the cable may connect directly with the oscilloscope input. The function of the termination circuit is to terminate the connecting cable in the characteristic impedance of the cable and so avoid signals reflecting along the cable and causing signal distortion.

A variety of probes exist. A common one is the passive voltage probe. Such a probe attenuates signals, attenuation ratios of 1-to-1, 10-to-1 and 100-to-1 being common. Active probes tend to have a small amplifier using field effect transistors built into the probe head. This enables the probe and oscilloscope to present a high impedance to the input signal source, typically about 10 MΩ shunted by 0.5 pF, and makes them very useful where long cable runs are necessary. Current probes allow the measurement of alternating current in a conductor without the need to insert any component into the circuit for which the current is being measured, the probe being clamped round the conductor.

20.5.2 Measurements with oscilloscopes

The oscilloscope can be used to display waveforms and measure voltage, current, time and frequency. For the measurement of voltage the signal is connected to the Y input and the peak-to-peak distance of the waveform on the screen determined in terms of the divisions on the screen scale. This can then be translated into a voltage using the calibrated attenuator setting (volts/div) and the probe attenuation.

$$\text{Voltage} = \text{scale divisions} \times \text{volts/div} \times \text{probe attenuation} \qquad (20.13)$$

Direct and alternating currents can be determined by measurement of the potential difference across a known value resistance and the application of Ohm's law. Alternatively, a current probe can be used. The time between two points on a waveform can be determined using the calibrated time base of the oscilloscope.

The frequency of a signal can be determined if it is fed into the vertical input and the time determined between two points one cycle apart on the displayed waveform. A more accurate method, however, is to use *Lissajous figures*, with the unknown frequency being compared with an accurately known frequency. The unknown frequency is fed into the vertical input of the oscilloscope and the known frequency is fed into the horizontal input. The known frequency is then adjusted to give a stationary display on the screen, e.g. as in Fig. 20.17(a). When the display is stationary there is a constant ratio between the frequencies of the two inputs. The frequency ratio is

$$\frac{\text{vertical input frequency}}{\text{horizontal input frequency}} = \frac{\text{no of horizontal loops}}{\text{no of vertical loops}} \qquad (20.14)$$

The number of horizontal loops is the number that could touch a suitably placed horizontal line, the number of vertical loops being the number that could touch a suitably placed vertical line.

Lissajous figures can also be used to determine the phase difference between two signals of the same frequency, the general pattern being an ellipse (Fig.

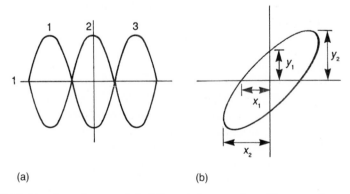

(a)   (b)

Fig. 20.17 Lissajous figures, (a) different frequencies, (b) same frequency but different phases.

20.17(b)). The phase angle can be found from the ratio of the ellipse dimensions (Fig. 11.23), with

$$\sin \phi = \frac{y_1}{y_2} = \frac{x_1}{x_2} \qquad (20.15)$$

Problems

1. A moving coil instrument gives a full-scale deflection with 15 mA. The coil resistance is 5 Ω. What is the value of (a) the shunt to enable it to be used as a 2 A meter, (b) the multiplier to enable it to be used as a 100 V voltmeter?

2. A moving coil instrument gives a full-scale deflection with 20 mA. The coil resistance is 1 Ω. What is the value of the shunt required for the instrument to measure 1000 A?

3. The coil of a moving iron voltmeter has a resistance of 400 Ω and an inductance of 1 H. The series resistor is 2000 Ω. The meter reads 240 V when connected across 240 V d.c. What will it read when connected across 240 V a.c.?

4. A moving iron voltmeter has a total resistance of 1 kΩ and a coil with an inductance of 800 mH. With direct voltage the full-scale reading is 50 V. What will be the percentage error in the full-scale reading when the instrument is used with a 50 V, 200 Hz input?

5. A rectifier type ammeter has been calibrated to read the rms value of a sinusoidal waveform. By what factor will the readings be in error if the waveform has a square waveform?

6. A sinusoidal voltage of 240 V is connected across a series circuit containing a 10 Ω resistor, a silicon diode and a moving coil ammeter. Calculate the reading on the meter.

7. A sinusoidal voltage, maximum value 100 V, is connected across a series circuit containing a resistor in series with a diode and an ammeter. The circuit has a resistance in one direction of 10 Ω and in the reverse direction of 50 Ω. What would be the current indicated by the meter if it was (a) a moving iron meter, (b) a permanent magnet moving coil meter, (c) a permanent magnet moving coil meter with a full-wave rectifier?

8. A dynamometer wattmeter with its voltage coil connected across the load side of the instrument reads 250 W. If the voltage across the load is 240 V and the resistance of the voltage coil is 2400 Ω, what power is being taken by the load?

9. A series type ohmmeter has a moving coil meter of resistance 50 Ω and gives a full-scale deflection with 1.0 mA. The ohmmeter battery has an e.m.f. of 2.0 V and negligible internal resistance. What will be the required value for the zero adjustment resistance and the shunt required for the meter if the mid-scale reading of the ohmmeter is to be 1000 Ω?

10. A shunt type ohmmeter, circuit diagram as in Fig. 20.9(b), has a meter with a resistance of 1 Ω and a full-scale reading of 10 mA. If the battery has an e.m.f.

of 1.5 V and negligible internal resistance, what will be the required value of the zero adjustment resistance and the mid-scale reading of the ohmmeter?

11. A Wheatstone bridge is supplied by a battery of e.m.f. 5 V and internal resistance $1\,\Omega$. At balance $Q = 150\,\Omega$, $S = 10\,\Omega$, $R = 20\,\Omega$. What is the value of P and the current drawn from the battery?

12. A Wien bridge is used to measure the capacitance C_1 and the loss resistance R_1 of an imperfect capacitor. The other arms of the bridge are a standard $0.01\ \mu F$ perfect capacitor in series with a standard $100\,\Omega$ resistor and two perfect resistors, each of $1000\,\Omega$. The bridge balanced at an angular frequency of $1000\ \mathrm{rad/s}$. Calculate the values of C_1 and R_1.

13. An a.c. bridge has arm ab as a capacitor C_1 in series with a variable resistance R_1, bc a resistor R_2, cd has an unknown inductor coil with inductance L and resistance R and da a resistor R_3. Show that at balance

$$L = \frac{R_2 R_3 C_1}{1 + \omega^2 R_1^2 C_1^2}$$

$$R = \frac{\omega^2 R_1 R_2 R_3 C_1^2}{1 + \omega^2 R_1^2 C_1^2}$$

14. An a.c. bridge has at balance arm ab a resistance of $2500\,\Omega$ in series with a $4000\ pF$ capacitance, bc a resistance of $1000\,\Omega$, cd an inductor with inductance L and series resistance R and da a resistance of $2500\,\Omega$. If the angular frequency of the bridge supply is $10\ \mathrm{krad/s}$, what are the values of L and R?

15. The Owen a.c. bridge has arm ab a capacitor C_1, bc a resistor R_3, cd an inductor with inductance L and series resistance R and da a variable capacitor C_2 with a variable resistance R_2. Show that the conditions for balance are

$$L = R_2 R_3 C_1$$

$$R = \frac{R_3 C_1}{C_2}$$

Answers to Problems

Chapter 1

1. 7560 J
2. 50 mA, 5.0 V
3. 200 V, 2000 Ω
4. 4.2 A
5. (a) 12 Ω, (b) 1.1 Ω
6. (a) 0.6 W, (b) 0.13 W
7. 1.1 Ω
8. 7.27 m
9. 0.036 V/m
10. 10 kV/m
11. (a) 50 A, (b) 500 A/m
12. 30 mA
13. 16 mC
14. 4 V
15. 0.288 N
16. 4 T
17. 0.98 A; B horizontal right-to-left, I right angles and into paper
18. 0.6 V
19. 5 mH
20. 2 V
21. 0.4 H, 0.36 H, 1.52 H
22. 0.46 mWb
23. 1.7 Wb
24. 3.4 A
25. 2.64×10^{-4} Wb
26. 39.8 mA
27. (a) 0.42 A, (b) 3.60 A
28. 0.44 A, 2.03 A
29. 48 μC
30. 18 pF, 41 pF
31. 7.5 m
32. (a) 327 pF, (b) 3.27 μC/m^2, (c) 100 kV/m
33. (a) 1.6 μF, (b) 10 μF
34. 2.3 μF
35. 2 V, 8 V
36. 4.9 pF
37. 1290 pF
38. $C = [2\pi\varepsilon_0/\ln(b/a)][H + (\varepsilon_r - 1)h]$

39. Metal sheets connected in series, alternate metal sheets connected together
40. As problem
41. 2.88×10^{-4} J
42. 0.050 J, 0.031 J
43. (a) 26.7 V, (b) 2.1 mJ
44. 1.0 mJ

Chapter 2

1. (a) 2.5 A, (b) 2.0 A
2. (a) 2.4 mA, 26.2 mA, 28.6 mA, (b) 6 A, 2 A, 8 A
3. See problem
4. 0.25 A
5. 0.25 A
6. 17 V
7. 0.15 Ω
8. (a) 12 V, 3.0 Ω, (b) 2 V, 0.083 Ω
9. 4.7 A
10. 50 Ω, 50 W
11. 2.0 Ω, 4.5 W

Chapter 3

1. (a) 3.9 mA, (b) 6.1 V
2. (a) 2 ms, (b) 6.3 V
3. 1.4 H
4. 0.68 mA
5. (a) 3.68 mA, (b) 6.32 V, (c) 3.68 V
6. (a) 5 ms, (b) (i) 0.736 mA, (ii) 0.271 mA
7. 30 μF, 0.965 MΩ
8. 28.9 kΩ
9. 6.44 s
10. 10 kΩ
11. A ramp for t followed by constant signal
12. (a) Critically damped, (b) under damped, (c) under damped
13. 316 Ω
14. $\omega_n = 7.07$ krad/s, $f_n = 1.13$ kHz

Chapter 4

1. (a) 796 Ω, (b) 0.0796 Ω
2. (a) 62.8 Ω, (b) 628 kΩ
3. (a) 62.8 Ω, (b) 3.18 A
4. (a) 318 Ω, (b) 0.785 A

5.   (a) $50 + j31.4\,\Omega$, $59.1\underline{/32.1°}\,\Omega$, (b) $50 - j318\,\Omega$, $322\underline{/-81.1°}\,\Omega$
6.   (a) $23.6\underline{/32.1°}\,\Omega$, (b) $2.12\underline{/-32.1°}\,A$
7.   (a) $167\underline{/-72.5°}\,\Omega$, (b) $0.299\underline{/72.5°}\,A$
8.   (a) $26.1\underline{/-40.0°}\,\Omega$, (b) $1.91\underline{/40.0°}\,A$
9.   (a) $0.8 + j0.4\,\Omega$, (b) $0.89\underline{/26.6°}\,\Omega$
10. $5.59\underline{/26.6°}\,V$
11. $0.28\underline{/26.6°}\,A$
12. $0.368\underline{/57.2°}\,A$
13. $1.12\underline{/63.4°}\,A$
14. $1.14\underline{/-26.0°}\,A$
15. $R_2(1 + j\omega CR_1)/[1 + j\omega C(R_1 + R_2)]$
16. (a) $13.1\underline{/70.1°}\,A$, (b) $10.4\underline{/44.1°}\,A$
17. $39.5\underline{/-1.5°}\,mA$
18. $VZ_2/(Z_1Z_2 + Z_2Z_L + Z_1Z_L)$
19. $57.2\,Hz$, $167\,\Omega$
20. (a) $20\,W$, (b) $10\,W$, (c) 0
21. (a) 0, (b) 0, (c) $R/\sqrt(\omega^2L^2 + R^2)$
22. (a) $25.8\,\mu F$, (b) 0.478 leading
23. $405\,\mu F$
24. $24.9\,\mu F$
25. (a) $225\,Hz$, (b) $10\,\Omega$
26. $\sqrt[(1/LC) - (R^2/L^2)]$, L/RC
27. (a) $712\,Hz$, (b) 22.4, (c) $31.8\,Hz$
28. (a) $20.3\,\mu F$, (b) $2.0\,A$, (c) 31.4

Chapter 5

1. $28.75\underline{/-23.1°}\,A$
2. $33\,\Omega$
3. $99\,\Omega$
4. $1.5\,A$
5. (a) $9.42\,A$, (b) 0.8 lagging, (c) $5.12\,kW$
6. $11.6\,kW$, $34.7\,kW$
7. $I_A = 38.7\underline{/108.3°}\,A$, $I_B = 46.5\underline{/-45°}\,A$, $I_C = 21.2\underline{/190.9°}\,A$
8. $9.2\,kW$
9. $31\,kW$, $42\,kW$

Chapter 6

1. $0.22\,\Omega m$
2. $0.13\,\Omega m$
3. $4.6 \times 10^{-3}\,m^2\,V^{-1}\,s^{-1}$
4. $-1.1 \times 10^{-7}\,A/m^2$
5. $12.3\,mA$

6.   15.4 mA, 18.0 mA
7.   (a) 8.3 Ω, 1.3 Ω, (b) 3.1 Ω, 0.28 Ω
8.   30 mA
9. $a + bI$, $a + 2bI$
10.   121 mA
11.   18.3 pF
12.   10 mA
13.   Series input impedance 1200 Ω, output current source $400I_b$
14.   10.7 kΩ
15.   8.28 V
16.   0.05 mA, 16.2 V
17.   250 Ω, 95 µA, 210, 250

Chapter 7

1.   0.203 V
2.   82.6
3.   2.0 V
4.   750
5.   8256
6.   (a) 100, 40 dB; (b) 200, 46 dB; (c) 46 dB
7.   (a) 8.33 mV, (b) 62.5 mV, (c) 0.741 V, (d) 74.1, 37.4 dB
8.   1:1.73, (a) 8.33 mV, (b) 72.14 mV, (c) 0.855 V, (d) 85.5, 36.8 dB
9.   (a) 39.8 Hz, 121 Hz, (b) 0.0265 µF
10.   49.75, 201 Hz
11.   49.69
12.   0.0483, 600
13.   200
14.   0.0933
15.   2475, 0.0096

Chapter 8

1.   (a) 1 kΩ, (b) 50, (c) 200, (d) 10 000
2.   89.3
3.   (a) 0.040 mA, (b) 1661
4. 9.75×10^{-5} W
5.   (a) 649 Ω, (b) 116
6.   43.5, 91
7.   (a) 45.5, (b) 228, (c) 1.04 mW
8.   240, 129 Hz
9.   1.719 V
10.   3.08 µF

Chapter 9

1. $R_1 = 2\,k\Omega$, $R_2 = 200\,k\Omega$
2. 2.9 V
3. $-(8 + 5 \sin 1000t)$ V
4. As problem
5. Input element $10\,k\Omega$, feedback element $10\,\mu$F
6. 2 cos 1000t V
7. $R_2/R_1 = 19$, e.g. $R_1 = 10\,k\Omega$, $R_2 = 190\,k\Omega$.
8. 2 sin 500t
9. 25, 80 dB
10. 69.5 dB

Chapter 10

1. 47.7 Hz, 10 800
2. 712 Hz, 100
3. 711.8 kHz
4. 19.5 kHz
5. 162 mH to 259 μH
6. 15.4 kΩ, 446 kΩ
7. 6.3 kΩ; $R_2/R_1 = 3$, e.g. $R_2 = 150\,k\Omega$, $R_1 = 50\,k\Omega$
8. 1.59 kHz, 3

Chapter 11

1. (a) 15.75 V, (b) 24.75 V, (c) 49.5 V
2. (a) 49.5 V, (b) 99 V, (c) 49.5 V, (d) 49.5 V
3. (a) 1 : 2.90, (b) 942 V
4. (a) 1 : 1.45 + 1.45, (b) 942 V
5. 1 : 0.81
6. 25 μF
7. 0.80 V
8. 2.16 A
9. 70.7 V, 8 V
10. (a) 14.5 mA, (b) No, power dissipation in the diode with no load is 138 mW.

Chapter 12

1. See text and Fig. 12.6
2. 2.78 + j18.4 Ω
3. 165 V, 0.20
4. 0.969
5. 0.62 kW, 0.984

6. $R_c = 2.18 \, \text{M}\Omega$, $X_m = 165 \, \text{k}\Omega$, $R_1 = R'_2 = 20 \, \Omega$, $X_{11} = X_{12} = 45.83 \, \Omega$
7. (i) (a) 2.73%, (b) 233.45 V, (ii) (a) 4.36%, (b) 229.54 V, (iii) (a) 0%, (b) 240 V
8. (a) 0.96, (b) 0.948, $\frac{3}{4}$ full load
9. 0.965
10. 1.29 T
11. 1080, 45
12. 13 200/2200, 132 kVA; 13 200/11 000, 660 kVA

Chapter 13

1. 87 mA
2. −15 dB
3. (a) −5 dB, (b) 10 mV
4. 44.7 Ω
5. 447 Ω
6. Node analysis can be used
7. −3.7 dB
8. 159 Hz, −3.0 dB
9. 5.0 kHz, 632 Ω
10. 22.1 nF, 15.9 mH
11. $1/\pi\sqrt{(LC)}$, $\sqrt{(L/C)}$
12. 159 Hz
13. As problem

Chapter 14

1. 10111110100_2, 11000010_2, 110001.111_2
2. 3226_{10}, 214_{10}, 37_{10}
3. (a) 101000_2, (b) 11100_2, (c) 571_8
4. (a) 10001_2, (b) 252_8
5. 15D
6. (a) $ABCD = F$, (b) $A + C + C + D = F$
7. $\bar{A}B + AC = F$

Inputs			Output
A	B	C	F
0	0	0	0
0	0	1	0
0	1	0	1
0	1	1	1
1	0	0	0
1	0	1	1
1	1	0	0
1	1	1	1

8. As problem
9. $\bar{A}B$, $C + \bar{D}$, $(\bar{A} + \bar{B})(A + B)$, $\bar{A} + \bar{B}C$
10. HIT $= BC + AC + AB$, MISS $= \overline{\text{HIT}}$
11. $BC + AC + AB = F$ or alternatively $C(B + A) + AB = F$
12. $ID + \bar{I}L = F$
13. $ABCD = \bar{F}$
14. (a) Two AND, one NOR; (b) $ABC = F$, one AND
15. (a) $\bar{A}\bar{C} + AC + F$
    (b) $AC + ACD = \bar{F}$
    (c) $BC + \bar{B}\bar{C} + AC = F$, or $BC + \bar{B}\bar{C} + AB = \bar{F}$
    (d) $AD + AC = F$
16. $F = A + B + C$, two NOR gates in series; $F = ABC$, three NOR gates as inputs to a NOR gate; $F = \bar{A}\bar{B}\bar{C}$, three NOR gates as inputs to a NOR gate leading into another NOR gate; $F = \bar{A}$, one input to a single NOR gate.
17. See Fig. A.1
18. See Fig. A.2

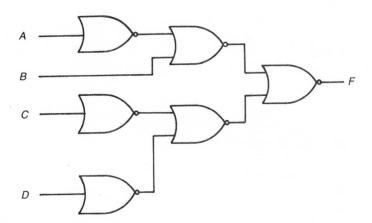

Fig. A.1 Chapter 14, Problem 17.

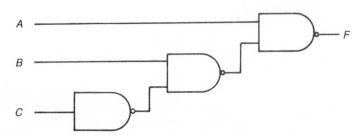

Fig. A.2 Chapter 14, Problem 18.

Chapter 15

1. See Fig. 15.5 and 15.4.
2. As Fig. 15.8 with an extra stage
3. 10
4. See Fig. 15.10
5. See Section 15.2
6. 128
7. $J_A = B + \bar{C}, \bar{J}_B = A, J_C = AB, K_A = 1, K_B = A + C, K_C = B, A$ is the least significant bit
8. As Fig. 15.11 with parallel out lines
9. 7.29 kHz

Chapter 16

1. 26, 91, 0F, 10D6, A0F4
2. (a) 2, (b) relative, (c) 9B, (d) 4, (e) clear memory location, (f) causes content of accumulator A to be changed to one's complement.
3. 0020 96 50   LDAA
   0022 90 60   SUBA
   0024 A7 E1 55 STAA
   0027 3E   WAI
4. 0020 96 50 LDAA
   0022 9B 51 ADDA
   0024 97 60 STAA
   0026 3F   SWI
5. 0020 96 60 LDAA
   0022 D6 80 LDAB
   0024 97 80 STAA
   0026 D7 60 STAB
   0028 3F   SWI
6. 0020 96 50 LDAA
   0022 48   ASLA
   0023 48   ASLA
   0024 9B 50 ADDA
   0026 97 60 STAA
   0028 3F   SWI
7.

	DATA1 EQU	$0050	
	DATA2 EQU	$0060	
	DIFF EQU	$0080	
	ORG	$0020	
0020 96 50	LDA A DATA1	Get minuend	
0022 90 60	SUB A DATA2	Subtract subtrahend	
0024 97 80	STA A DIFF	Store diff. in $0080	
0026 3F	SWI	Program end	

8.
```
                    AUGEND EQU    $0060
                    ADDEND EQU    $0050
          SUM       EQU    $0080
                    ORG    $0020
0020 96 60          LDA  A AUGEND Get first number in A
0022 9B 50          ADD  A ADDEND Add second number
0024 97 82          STA  A SUM       Save the sum
0026 3F             SWI              Program end
```

9.
```
                 HEXNUM EQU    $0050
                 DECNUM EQU    $0060
                    ORG    $0020
0020 96 50          LDA  A HEXNUM Load accumulator A
0022 81 0A          CMP    15$0A       Compare with $0A
0024 2B 02          BMI    SAVE        Branch if minus
0026 8B 06          ADD    15$06        Add $06
0028 97 60 SAVE     STA    DECNUM Store in $0060
002A 3F             SWI              Program end
```

10.
```
                    DRA EQU $8008
                    CRA EQU $8009
                    DRB EQU $800A
                    CRB EQU $800B
                       ORG $0100
0100  4F            CLR  A
0101  B7 80 09      STA  A     CRA   Access to DDRA
0104  B7 80 0B      STA  A     CRB   Access to DDRB
0107  B7 80 0A      STA  A     CRB   Port B all inputs
010A  43            COM  A
010B  B7 80 08      STA  A     DRA  Port A all outputs
010E  86 04         LDA  A     15$04
0110  B7 80 09      STA  A     CRA  Select I/O register A
0113  B7 80 0B      STA  A     CRB  Select I/O register B
0116  B7 80 0A LOOP LDA  A     DRB  Read port B
0119  B7 80 08      STA  A     DRA  Display
011C  20 F8         BRA        LOOP Repeat if required
```

11. CRA XX11X1X0, CRB XX101111 (X = don't care)

Chapter 17

1. As problem
2. As problem
3. As problem
4. As problem
5. 2.475 kHz
6. 15.8%
7. (a) 23.5 mV/°C, (b) 1.0 mV/°C

8.   As problem
9.   (a) 22.1 pF, (b) 38.7 pF, (c) 55.3 pF
10.   −0.89%
11.   +0.79%
12.   As problem
13. $L_1 = L_0/[1 + k(d + x)]$, $L_2 = L_0/[1 + k(d - x)]$
14.   As problem
15.   10.5 mV

Chapter 18

1.   2.5 V
2.   As problem
3.   −31.25 V
4.   255 µs
5.   54.8 Ω
6.   2.58, 11.0
7.   0.4, 10 ± 0.006 MHz
8.   d.c. 11.5 mA; 100 rad/s, 1.5 mA peak; 200 rad/s, 0.03 mA peak; 900 rad/s, 0.12 mA peak; 1000 rad/s, 3 mA peak; 1100 rad/s, 0.12 mA peak; 2000 rad/s, 0.12 mA peak
9.   8%
10.   4.24 kW, 0.38 kW
11.   (a) 40%, (b) 1.001 MHz, 0.999 MHz, (c) 0.926, (d) 0.037
12.   See Section 18.5.2
13.   5

Chapter 19

1.   225 V, 142 Ω, 698 rpm
2.   100 Ω to 70 Ω
3.   111 Ω
4.   346 V
5.   910 rpm
6.   6.42 Ω
7.   3.8 Ω
8.   184 V
9.   4.8 Ω
10.   (a) 1500 rpm, (b) 1425 rpm
11.   0.033
12.   27.1 kW
13.   0.111
14.   276 Nm, 0.1
15.   1728 rpm

Chapter 20

1. (a) $0.0378\,\Omega$, (b) $662\,\Omega$
2. $0.00002\,\Omega$
3. $238\,V$
4. 42%
5. 0.91
6. $10.8\,A$
7. (a) Average of $i^2 = 5.1\,A$, (b) average $i = 2.6\,A$, (c) average $i = 3.8\,A$
8. $226\,W$
9. $950\,\Omega$, $50\,\Omega$
10. $149\,\Omega$, $0.993\,\Omega$
11. $300\,\Omega$, $137\,mA$
12. $0.01\,\mu F$, $1\,M\Omega$
13. As problem
14. $3.96\,mH$, $4.00\,\Omega$
15. As problem

Index